CAMBRIDGE LIBRARY COLLECTION

Books of enduring scholarly value

Botany and Horticulture

Until the nineteenth century, the investigation of natural phenomena, plants and animals was considered either the preserve of elite scholars or a pastime for the leisured upper classes. As increasing academic rigour and systematisation was brought to the study of 'natural history', its subdisciplines were adopted into university curricula, and learned societies (such as the Royal Horticultural Society, founded in 1804) were established to support research in these areas. A related development was strong enthusiasm for exotic garden plants, which resulted in plant collecting expeditions to every corner of the globe, sometimes with tragic consequences. This series includes accounts of some of those expeditions, detailed reference works on the flora of different regions, and practical advice for amateur and professional gardeners.

Notes of a Botanist on the Amazon and Andes

Having previously embarked on a collecting expedition to the Pyrenees, backed by Sir William Hooker and George Bentham, the botanist Richard Spruce (1817–93) travelled in 1849 to South America, where he carried out unprecedented exploration among the diverse flora across the northern part of the continent. After his death, Spruce's writings on fifteen fruitful years of discovery were edited as a labour of love by fellow naturalist Alfred Russel Wallace (1823–1913), whom Spruce had met in Santarém. This two-volume work, first published in 1908, includes many of the author's own illustrations. Showing the determination to reach plants in almost inaccessible areas, Spruce collected hundreds of species, many with medicinal properties, notably the quinine-yielding cinchona tree, as well as the datura and coca plants. Volume 1 contains Wallace's biographical introduction and a list of Spruce's published works. The narrative includes discussion of Pará, Santarém, and the Negro and Orinoco rivers.

Cambridge University Press has long been a pioneer in the reissuing of out-of-print titles from its own backlist, producing digital reprints of books that are still sought after by scholars and students but could not be reprinted economically using traditional technology. The Cambridge Library Collection extends this activity to a wider range of books which are still of importance to researchers and professionals, either for the source material they contain, or as landmarks in the history of their academic discipline.

Drawing from the world-renowned collections in the Cambridge University Library and other partner libraries, and guided by the advice of experts in each subject area, Cambridge University Press is using state-of-the-art scanning machines in its own Printing House to capture the content of each book selected for inclusion. The files are processed to give a consistently clear, crisp image, and the books finished to the high quality standard for which the Press is recognised around the world. The latest print-on-demand technology ensures that the books will remain available indefinitely, and that orders for single or multiple copies can quickly be supplied.

The Cambridge Library Collection brings back to life books of enduring scholarly value (including out-of-copyright works originally issued by other publishers) across a wide range of disciplines in the humanities and social sciences and in science and technology.

Notes of a Botanist on the Amazon and Andes

Being Records of Travel on the Amazon and Its Tributaries, the Trombetas, Rio Negro, Uaupés, Casiquiari, Pacimoni, Huallaga and Pastasa

VOLUME 1

RICHARD SPRUCE
EDITED BY
ALFRED RUSSEL WALLACE

CAMBRIDGE
UNIVERSITY PRESS

University Printing House, Cambridge, CB2 8BS, United Kingdom

Published in the United States of America by Cambridge University Press, New York

Cambridge University Press is part of the University of Cambridge.
It furthers the University's mission by disseminating knowledge in the pursuit of
education, learning and research at the highest international levels of excellence.

www.cambridge.org
Information on this title: www.cambridge.org/9781108069205

This edition first published 1908
This digitally printed version 2014

ISBN 978-1-108-06920-5 Paperback

The original edition of this book contains a number of colour plates,
which have been reproduced in black and white. Colour versions of these
images can be found online at www.cambridge.org/9781108069205

Selected botanical reference works available in the CAMBRIDGE LIBRARY COLLECTION

al-Shirazi, Noureddeen Mohammed Abdullah (compiler), translated by Francis Gladwin: *Ulfáz Udwiyeh, or the Materia Medica* (1793) [ISBN 9781108056090]

Arber, Agnes: *Herbals: Their Origin and Evolution* (1938) [ISBN 9781108016711]

Arber, Agnes: *Monocotyledons* (1925) [ISBN 9781108013208]

Arber, Agnes: *The Gramineae* (1934) [ISBN 9781108017312]

Arber, Agnes: *Water Plants* (1920) [ISBN 9781108017329]

Bower, F.O.: *The Ferns (Filicales)* (3 vols., 1923–8) [ISBN 9781108013192]

Candolle, Augustin Pyramus de, and Sprengel, Kurt: *Elements of the Philosophy of Plants* (1821) [ISBN 9781108037464]

Cheeseman, Thomas Frederick: *Manual of the New Zealand Flora* (2 vols., 1906) [ISBN 9781108037525]

Cockayne, Leonard: *The Vegetation of New Zealand* (1928) [ISBN 9781108032384]

Cunningham, Robert O.: *Notes on the Natural History of the Strait of Magellan and West Coast of Patagonia* (1871) [ISBN 9781108041850]

Gwynne-Vaughan, Helen: *Fungi* (1922) [ISBN 9781108013215]

Henslow, John Stevens: *A Catalogue of British Plants Arranged According to the Natural System* (1829) [ISBN 9781108061728]

Henslow, John Stevens: *A Dictionary of Botanical Terms* (1856) [ISBN 9781108001311]

Henslow, John Stevens: *Flora of Suffolk* (1860) [ISBN 9781108055673]

Henslow, John Stevens: *The Principles of Descriptive and Physiological Botany* (1835) [ISBN 9781108001861]

Hogg, Robert: *The British Pomology* (1851) [ISBN 9781108039444]

Hooker, Joseph Dalton, and Thomson, Thomas: *Flora Indica* (1855) [ISBN 9781108037495]

Hooker, Joseph Dalton: *Handbook of the New Zealand Flora* (2 vols., 1864–7) [ISBN 9781108030410]

Hooker, William Jackson: *Icones Plantarum* (10 vols., 1837–54) [ISBN 9781108039314]

Hooker, William Jackson: *Kew Gardens* (1858) [ISBN 9781108065450]

Jussieu, Adrien de, edited by J.H. Wilson: *The Elements of Botany* (1849) [ISBN 9781108037310]

Lindley, John: *Flora Medica* (1838) [ISBN 9781108038454]

Müller, Ferdinand von, edited by William Woolls: *Plants of New South Wales* (1885) [ISBN 9781108021050]

Oliver, Daniel: *First Book of Indian Botany* (1869) [ISBN 9781108055628]

Pearson, H.H.W., edited by A.C. Seward: *Gnetales* (1929) [ISBN 9781108013987]

Perring, Franklyn Hugh et al.: *A Flora of Cambridgeshire* (1964) [ISBN 9781108002400]

Sachs, Julius, edited and translated by Alfred Bennett, assisted by W.T. Thiselton Dyer: *A Text-Book of Botany* (1875) [ISBN 9781108038324]

Seward, A.C.: *Fossil Plants* (4 vols., 1898–1919) [ISBN 9781108015998]

Tansley, A.G.: *Types of British Vegetation* (1911) [ISBN 9781108045063]

Traill, Catherine Parr Strickland, illustrated by Agnes FitzGibbon Chamberlin: *Studies of Plant Life in Canada* (1885) [ISBN 9781108033756]

Tristram, Henry Baker: *The Fauna and Flora of Palestine* (1884) [ISBN 9781108042048]

Vogel, Theodore, edited by William Jackson Hooker: *Niger Flora* (1849) [ISBN 9781108030380]

West, G.S.: *Algae* (1916) [ISBN 9781108013222]

Woods, Joseph: *The Tourist's Flora* (1850) [ISBN 9781108062466]

For a complete list of titles in the Cambridge Library Collection please visit:
www.cambridge.org/features/CambridgeLibraryCollection/books.htm

NOTES OF A BOTANIST

ON THE

AMAZON AND ANDES

MACMILLAN AND CO. LIMITED
LONDON · BOMBAY · CALCUTTA
MELBOURNE

THE MACMILLAN COMPANY
NEW YORK · BOSTON · CHICAGO
ATLANTA · SAN FRANCISCO

THE MACMILLAN CO. OF CANADA, LTD.
TORONTO

Walter L. Colls.

Yours very faithfully
Richard Spruce.

NOTES OF A BOTANIST

ON THE

AMAZON & ANDES

BEING RECORDS OF TRAVEL ON THE AMAZON AND ITS TRIBUTARIES, THE TROMBETAS, RIO NEGRO, UAUPÉS, CASIQUIARI, PACIMONI, HUALLAGA, AND PASTASA; AS ALSO TO THE CATARACTS OF THE ORINOCO, ALONG THE EASTERN SIDE OF THE ANDES OF PERU AND ECUADOR, AND THE SHORES OF THE PACIFIC, DURING THE YEARS 1849-1864

BY RICHARD SPRUCE, PH.D.

EDITED AND CONDENSED BY

ALFRED RUSSEL WALLACE, O.M., F.R.S.

WITH A

BIOGRAPHICAL INTRODUCTION

PORTRAIT, SEVENTY-ONE ILLUSTRATIONS

AND

SEVEN MAPS

IN TWO VOLUMES—VOL. I

MACMILLAN AND CO., LIMITED

ST. MARTIN'S STREET, LONDON

1908

To sit on rocks, to roam o'er flood and fell,
To slowly pace the forest's shade and sheen,
Where things that own not man's dominion dwell,
And mortal foot hath ne'er or rarely been;
To climb the trackless mountain all unseen,
With the wild flocks that never need a fold;
Alone o'er crags and foaming falls to lean;
This is not solitude; 'tis but to hold
Converse with Nature's charms, and view her stores unroll'd.

BYRON.

PREFACE

It was Dr. Spruce's intention to leave all his manuscripts and notes to Mr. Daniel Hanbury, as stated in one of his letters to that gentleman; but the unexpected death of his friend, and his own occupations together with his continuous ill-health, led him, apparently, to give up all expectation of his Journals being published. He knew that I was fully occupied with work of my own, and probably did not like to ask me to undertake so great a task; especially as he was quite aware that much of his writings were of a fragmentary nature, and so full of contractions as to be sometimes, in his own words, "hieroglyphic," and that it would be impossible for any one but himself to properly combine and fully utilise them.

Shortly after Spruce's death, I offered to do what I could to put together a narrative of his travels from his Journals and letters, if, on examination of the materials, it seemed possible to do so. His executor, Mr. M. B. Slater, was anxious that I should undertake the duties of a literary executor; but, partly owing to both of us being fully occupied with our own affairs, it was only after a delay of eleven years that I was able to begin the preparation of the present volumes.

The first eight chapters of Spruce's proposed

Notes of a Botanist, etc. (as given on the title-page), had been carefully written out (in a large account-book) during his last years in South America, and were apparently ready for publication after being copied and finally corrected. With considerable condensation, this constitutes the first six chapters of the present work.

I have omitted the first chapter—a mere journal of the voyage from Liverpool to Pará—with the exception of two short introductory paragraphs, and have combined the two following chapters, which deal with the district of Pará. The journals of the voyages to Santarem, to the Trombetas river, and to Manáos, have been condensed by large omissions, and a number of historical and geographical notes of little general interest have also been omitted. With these exceptions, the whole narrative is exactly as Spruce left it; and I have been careful to preserve his frequent north-country or archaic words and expressions (though these have often been "queried" by the printer's reader) in order that his individuality of style may be preserved.

Wherever I have found it necessary to insert connecting phrases or paragraphs, or to make any explanatory interpolations, these are indicated by being enclosed in square brackets, while the omissions are shown by rows of small dots so as not to disfigure the pages, and this rule is followed throughout the entire work.

The remainder of the two volumes is of a very composite nature, and the materials I had to put in order are sufficiently stated in the introductory notes to the various chapters. I may add here that of the whole quantity of material—Journals, Letters, printed or written Articles, and scattered Notes—that I have had to examine, only about one-third have been found suitable for a work of combined general and botanical interest and of moderate bulk.

It has been my endeavour to bring together whatever might be useful to botanists, and also to include all matters of interest to general readers. This task has been to me a labour of love; and I have myself so high an opinion of my friend's work, both literary and scientific, that I venture to think the present volumes will take their place among the most interesting and instructive books of travel of the nineteenth century.

I have to thank Sir Clements Markham and Sir Joseph Hooker for their interest in obtaining a grant of £10 from the Royal Society towards the expense of copying Spruce's letters preserved at Kew and some of the less legible of the Journals. The Pharmaceutical Society has also allowed me to copy such as were suitable among the great mass of letters which Spruce wrote to Mr. Daniel Hanbury; while Messrs. John Teasdale and George Stabler have lent me others of great interest.

In order to render the work as useful as possible to botanists, the generic and specific names of every plant mentioned by Spruce have been carefully indexed, the species alone being in italics; and to avoid errors they have been compared in all doubtful cases with the copious Index in Lindley's *Vegetable Kingdom*, which was nearly contemporary with Spruce's travels.

For the convenience of non-botanical readers, most of the longer passages which are wholly botanical, as well as some others of purely anthropological or historical value, have been printed in smaller type, so that they may be readily skipped by those who are chiefly interested in the actual narrative of Spruce's travels as told by himself.

I have endeavoured to make the Biographical Introduction as complete as possible, within the limits suitable to such a work as the present. I think it will be acceptable to all who knew Spruce either personally or through his writings; while to those who here make his acquaintance for the first time, it will reveal something of the life of a very enthusiastic student of Nature, under difficult conditions, as well as of a refined and attractive personality.

The illustrations are mostly from Spruce's own pencil sketches and drawings. Most of the larger of these were in very delicate outline, but a few were highly finished; and from these, as indicating the type of scenery, the outlines have been shaded by a skilled artist under my directions, so as to produce very lifelike and attractive views in districts quite beyond the sphere of the travelling photographer.

For the photographs of forest-scenery I am indebted to Dr. J. Huber of the Pará Museum, who has kindly sent me the issues of his *Arboretum Amazonicum*, from which I have selected for reproduction such as illustrate plants or scenes referred to by Spruce. The remaining illustrations are from the works of recent travellers on the Orinoco and in the Andes, the use of which has been obtained by the publishers.

The beautiful portrait of Spruce forming the

frontispiece was taken by a friend of Spruce's four years before his death. The photo-plate (made to illustrate Dr. Balfour's obituary notice in the *Annals of Botany*) has been kindly lent for the present work by the Clarendon Press, Oxford.

I also have to thank the Royal Geographical and Linnean Societies for permission to make use of Articles and Maps which were first published in their Journals.

ALFRED R. WALLACE.

Thanks to the human heart by which we live,
Thanks to its tenderness, its joys, and fears,
To me the meanest flower that blows can give
Thoughts that do often lie too deep for tears.

WORDSWORTH.

Oh for a lodge in some vast wilderness,
Some boundless contiguity of shade,
Where rumour of oppression and deceit,
Of unsuccessful or successful war,
Might never reach one more.

COWPER.

CONTENTS

CHAPTER I

PARÁ AND THE EQUATORIAL FORESTS

CHAPTER II

VOYAGE TO SANTAREM AND FIRST RESIDENCE THERE

CHAPTER III

TO OBYDOS AND THE RIVER TROMBETAS

CHAPTER IV

RESIDENCE AT SANTAREM

CHAPTER V

GEOLOGY AND BOTANY OF SANTAREM

CHAPTER VI

FROM SANTAREM TO THE RIO NEGRO

CHAPTER VII

RESIDENCE AT MANÁOS

CHAPTER VIII

VOYAGE UP THE RIO NEGRO TO SAÕ GABRIEL

CHAPTER IX

CATARACTS AND MOUNTAIN-FORESTS OF SAÕ GABRIEL

CHAPTER X

CATARACTS AND UNEXPLORED FORESTS OF THE UAUPÉS RIVER

CHAPTER XI

AT SAN CARLOS DO RIO NEGRO

CHAPTER XII

IN HUMBOLDT'S COUNTRY: VOYAGE UP THE CASIQUIARI, THE CUNUCUNÚMA, AND PACIMONI RIVERS

CHAPTER XIII

TO THE CATARACTS OF THE ORINOCO AND RETURN TO SAN CARLOS

CHAPTER XIV

SAN CARLOS TO MANÁOS (BARRA)

But oh! the free and wild magnificence
  Of Nature in her lavish hours doth steal,
In admiration silent and intense,
  The soul of him who hath a soul to feel.
The river moving on its ceaseless way,
  The verdant reach of meadows fair and green,
  And the blue hills that bound the sylvan scene,—
These speak of grandeur, that defies decay,—
Proclaim the eternal architect on high,
Who stamps on all his works his own eternity.

LONGFELLOW.

ILLUSTRATIONS

MAPS

ERRATA

Vol. I. p. 154, *for* "Enkylista" *read* "Eukylista" (also in Vol. II. pp. 4 and 28).

,, 268, "*Mauritia carinata* of Humboldt" seems to be a mistake for "*Mauritia aculeata*" (=*Mauritia gracilis* Wall.). *See* Spruce's *Equatorial American Palms*, p. 169.

,, 433, *for* "*grandiflora*" *read* "*grandifolia.*"

,, 467, *for* "Schieckia" *read* "Schiekia."

Vol. II. p. 100, *for* "Rhacophilum" *read* "Rhacopilum."

,, 210, *for* "*Nickera*" *read* "*Neckera.*"

BIOGRAPHICAL INTRODUCTION

About fifteen miles to the north-east of York there are three small villages, each being about two miles distant from the others, thus forming a nearly equilateral triangle within which lies the fine park and mansion of Castle Howard. Ganthorpe, the most westerly, was Spruce's birthplace; in Welburn, to the south, he lived for some years before going to South America, and again after his return; while in Coneysthorpe, close to the north-east boundary of Castle Howard park, he passed the last seventeen years of his life.[1]

The district in which these villages lie is somewhat elevated, being from 300 to 400 feet above the sea, broken into hill and dale, with abundance of woods and a few small streams. Being situated on the Middle Oolite beds, while the Upper Oolite and Lias are within a few miles to the north and south of it, there is a considerable variety of soils—clayey sand and calcareous rocks of various degrees of hardness—highly favourable to a varied and interesting vegetation; and it still offers to the visitor a charming example of English rural scenery. It is an ideal home for a botanist and student of nature, and it was here that Richard Spruce acquired that deep love of flowers, and especially of the lowliest plants—the Mosses and Hepaticæ—which was the joy of his early manhood and the consolation of his declining years.

Spruce's father (also named Richard) was the highly-respected schoolmaster at Ganthorpe, and afterwards at Welburn, both schools being partially endowed by the Howard family. Mr. G. Stabler, who was for some time at his school, informs me that he was a very good mathematician, but less advanced in the classics, and that he was a wonderfully fine penman, a characteristic in which his son resembled him, as his remarkably clear and uniform handwriting, even under the most adverse conditions,

[1] Richard Spruce, born Sept. 10, 1817; died Dec. 28, 1893.

sufficiently shows. His mother was one of the Etty family, a relative of the great painter, who was born at York.

Spruce appears to have been educated wholly by his father. He was an only child, but his mother died while he was young, and when he was about fourteen his father married again, and had a family of eight daughters, only two of whom survived their half-brother. This circumstance rendered him unable to do anything for his son but help him to follow his own profession, with which object Spruce took lessons in Latin and Greek from an old schoolmaster named Langdale, who had been educated for the priesthood and whose scholarship was of a high order. His influence may be seen in some of Spruce's letters to Mr. Borrer and Mr. Bentham, when he has occasion to discuss questions of Latin construction, being always able to give reasons or quote authorities in support of his own views.

Although he disclaimed any linguistic ability or love of philology, he evidently had a considerable natural aptitude for languages, since he not only taught himself to read and write French fairly well, but in after years was able to acquire the Portuguese and Spanish languages with great facility, so as to be able to write them grammatically as well as to speak them; and also to acquire some colloquial skill in three different Indian languages—the Lingoa Geral, Barré, and Quichua—which, in one case, was probably the means of saving his life.

He appears to have remained at home, studying and assisting his father, till he was of age, about which time he became tutor in a school at Haxby, four miles north of York, and a year or two later (at the end of 1839) obtained the post of mathematical master at the Collegiate School at York, which he retained till the school itself was given up, at midsummer of 1844. At this time he was quite undecided as to his future, and made some efforts to get another position of the same kind. One such opportunity occurred, with a fairly liberal salary, but he found it would involve residence in the school, with supervision of the boys out of school hours, so as to leave him little or no leisure; and as this was very distasteful to him, besides involving too much mental strain for his very delicate health, he gave up the idea. In fact, during the whole time he had been at York he had had repeated illnesses, especially in the winter. His lungs were affected, and he believed that he should not have lived another year if he had continued at school-work, the confinement and mental worries of which were very prejudicial to his constitution. In the next winter he wrote that he "was wearing a perpetual blister and found much benefit from it." In the following year he had a serious attack of congestion of the brain; and in 1848 he had another illness from gall-stones, causing, he declared, "the most excruciating pain it is possible to conceive," and which left him very weak for a long time. These serious illnesses, together with great liability to severe colds and constantly recurring winter cough, indicate the great delicacy of his organisation, and render more remarkable the amount of labour and privation he afterwards endured.

The breaking up of the York Collegiate School was the turning-point in Spruce's life, resulting in his becoming a botanist and botanical explorer of the first rank. We must therefore go back a few years to relate what is known of his early life as a student of plants.

Mr. G. Stabler, who was also a native of Ganthorpe, tells us that, when quite a child, Spruce "showed much aptitude for learning, and at an early age developed a great love of nature. Amongst his favourite amusements was the making lists of plants, and he had also a great liking for astronomy." In 1834, when sixteen years old, he had drawn up a neatly written list of all the plants he had found around Ganthorpe. It is arranged alphabetically and contains 403 species, the gathering and naming of which must certainly have occupied some years. Three years later he had drawn up a "List of the Flora of the Malton District," the MSS. of which is in the possession of his executor, Mr. Slater, and this contains 485 species of flowering plants. Several of Spruce's localities for the rarer plants are given in Baines's *Flora of Yorkshire*, published in 1840.

By this time it is evident that he had not merely collected plants but had studied them carefully, as shown by the fact that in 1841 he discovered, and identified as a new British plant, the very rare sedge *Carex paradoxa*. He had also now begun the study of mosses, since in the same year he found a moss new to Britain, *Leskea pulvinata*, previously known only from Lapland. Among his early friends or correspondents were Ibbotson, Baines of York, and Slater of Malton, while he himself tells us (in a letter to Mr. Borrer) that Sam Gibson was his first adviser in the study of mosses. This Gibson was a whitesmith or "tinman" at Hebden Bridge, about six miles west of Halifax, and was one of a considerable number of North-country working-men botanists of the early nineteenth century. Spruce probably visited him during his first residence near York, when he would have the necessary leisure during his vacations, since Gibson speaks of him as his "friend" in 1841, and Spruce told Mr. Slater that he had seen Gibson in his workshop with Hooker's *British Flora* on the bench by his side, and that it was in parts so begrimed and blackened as to be almost illegible.

During his first year at the Collegiate School, however, he gave himself to the study of mathematics with so much ardour as for a time to neglect botany; but Mr. Stabler tells us that in one of his summer vacations he found, on Slingsby Moor, a few miles north of his home, "one of the uncinate Hypna in splendid fruit. His love of plants, from which he had been weaned for a short time by his mathematical studies, returned with such force that he vowed on the spot that henceforth the study of plants should be the great object of his life." I think we can fix the date of this incident by the first entry in a little "List of Botanical Excursions," which is: "1841, June 19. Slingsby Moor and Terrington Carr." Similar entries are made during the remainder of his residence in England, his visits to Ireland and the Pyrenees, as well as throughout his South American travels; while the remainder of his life was equally devoted to "the study of plants."

The Phytologist was started in 1841 as a monthly magazine for British Botany especially, and Spruce contributed to it in the first and succeeding years accounts of his botanical excursions and notes on rare plants; and it was probably his critical remarks on Carices, Mosses, and Hepaticæ that led to a correspondence with Dr. Thomas Taylor, one of the joint authors of the *Muscologia Britannica*, with Mr. William Wilson of Warrington, and with Mr. Borrer of Henfield. With all these eminent botanists he soon became intimate, and each in turn invited him to visit them. In the summer vacation of 1842 he stayed three weeks at Dunkerron, near Killarney, with Dr. Taylor, and visited a few other places, but the weather was bad, he had a severe cold, and he spent most of his time in the study of British and exotic mosses in his host's rich herbarium.

Early in September of the same year, Mr. William Borrer, one of the most acute and enthusiastic British botanists, called upon him at York, and Spruce took him to Clifton Ings, on the banks of the Ouse, a locality for *Leskea pulvinata* and other rare mosses. In the following September (1843) Mr. Borrer again visited him, and they went together to Castle Howard to examine some of Spruce's favourite haunts in search of rare plants. From the date of their second meeting a correspondence began which continued at short intervals till within a few months of Spruce's departure on his South American expedition. After Mr. Borrer's death in 1862, a parcel of mosses, together with a packet of Spruce's letters, were given to Mr. W. Mitten, and the latter have thus come into my hands, as Mr. Mitten's executor. The series appears to be complete from August 25, 1843, to August 5, 1848—66 in all. This last letter, like the great majority of the series, is about details in the structure and classification of Mosses and Hepaticæ, but a postscript states that, as he is soon coming to London to superintend the sale by auction of the late Dr. Taylor's herbarium, he hopes to meet Mr. Borrer. The letters are, however, full of interest, and enable me both to give a connected sketch of his occupations after giving up scholastic work, and also, by suitable extracts, to give some idea of his character and opinions.

More than thirty years later, when describing a new genus of Amazonian Hepaticæ in the *Journal of Botany*, and noticing that a British species (*Odontochisma Sphagni*) grows with it though not on a Sphagnum, he gives us in a footnote the following very interesting bit of nature-study combined with archæology, which is so characteristic that I will here quote it nearly entire, especially as it refers to one of his excursions with Mr. Borrer. He writes:—

"On our own moors I have far oftener seen *Odontoschisma Sphagni* growing on *Leucobryum glaucum* than on Sphagna. Now that the steam-plough is fast obliterating the small remnant of moors in the Vale of York, it is worth while recording something about the Leucobryum, as seen on Strensall Moor, five to six miles north of York. There it forms immense rounded hassocks, some of which in my youth were as much as three feet high; and although the ground whereon they grew is now drained and ploughed out, I am told that on another part of the moor there are still left a few hassocks about two feet high. When the late Mr. Wilson first saw them, thirty years ago, he took them at a distance for sheep; as he approached them he changed his mind for haycocks; but when he actually came up and saw what they were he was astonished, and declared he had never seen such gigantic moss-tufts elsewhere. During seven consecutive years that I saw them frequently, I could observe no sensible increase in height. The very slight annual outgrowth of the marginal branches is comparable to the outermost twigs of an old tree, and is almost or quite counterbalanced by the soft, imperfectly elastic mass incessantly decaying and settling down at the base; so that these tufts of Leucobryum may well be almost as secular as our Oaks or Elms; and some of them might even be coming into existence, if not so far back as when the warders of Bootham Bar and Monk Bar (the northern entrances to York) used to hear the wolves howling beneath their feet on the bleak winter nights, at least whilst the 'last wolf' was still prowling in the Forest of Galtres.

"Strensall Moor, Stockton Forest, Langwith Moor, etc., are all relics of the Forest of Galtres, an ancient royal demesne of the Saxon kings, in which roamed the stag, bear, wolf, and wild boar. A perambulation made in the ninth year of Edward II. found it to extend from the walls of York northwards nearly twenty miles, viz. to Isurium (Aldburgh), and eastwards to the river Derwent. Several hamlets had sprung up on it, and a few solitary granges—moated round to protect the inmates from wolves, biped and quadruped. (One of these moated granges was still the only habitation on Langwith Moor in 1842, when I showed Mr. Borrer *Jung. Francisci* in fruit growing close by.) Camden calls it 'Calaterium Nemus, vulgo *The Forest of Galtres* . . . arboribus alicubi opacum, alicubi uliginosa planitie madescens.' In his time it stretched northwards only to Craike Castle and the source of the river Foss: '*Fossa*, amnis piger . . . originem habet ultra Castellum Huttonicum, terminatque fines Calaterii nemoris,' etc. (*Brit.* fol. 1607, p. 588). What remains of it now is only here and there a fragmentary 'uliginosa planities' —still rich in Sphagna, bog Hypna, and numerous other Mosses and Jungermanniæ—to say nothing of nobler plants—and in the drier parts adorned with wide beds of *Cetraria islandica* and *Cenomyce rangiferina*, associated with *Dicranum spurium*, *Bartramia arcuata*, *Racomitrium lanuginosum* (often fertile), and other tall Mosses.

"Tradition reports—but adds no date to the supposed fact—that the last wolf in England was killed on the borders of the Forest of Galtres, at Stittenham, two miles from where I am writing, by one of the Gowers, of which noble family Stittenham was (and still is) an ancient possession. The crest of the Gowers is 'a wolf passant argent,' etc., and over the family vault in the neighbouring church of Sheriff Hutton are suspended the funereal trophies of a Gower, viz. a casque, gauntlets, etc., and a pennon, now faded, but said to have been blazoned with the representation of a combat between a man and a wolf. Whether, however, the badge was assumed from that heroic action, or the tradition was founded on the badge, let the heralds decide.[1]

"I conclude this note by earnestly beseeching our local botanists to lose no time in exploring the moors that still remain untouched by cultivation in the Vale of York and elsewhere. On the wide plain between the Ouse and the foot of the wolds there

[1] Spruce has added in pencil on the margin "1660," as if he had since ascertained the date of this event.

are still left several patches of moor which have never been thoroughly examined for Cryptogamia. On one of these—Barmby Moor—I found the rare *Scalia Hookeri*, Lyell, in fruit on November 5, 1842, and I suppose I and Mr. Curnow are the only living botanists who have gathered it in Britain; but Gottsche finds it near Hamburg, and Lindberg at Helsingfors. In 1856 I gathered a second species, *Scalia andina*, MSS.—thrice the size of its European congener—in the Eastern Andes of Peru."

In his letter to Mr. Borrer of August 25, 1843, Spruce apologises for not having written before about some flowering plants Mr. Borrer had sent him several months earlier, and then adds: "But my attention was then, and continues to be, so entirely engrossed by the Musci and Hepaticæ, that I could not expect to gather anything for you that you would consider at all interesting. As my wish is to study the plants I collect, and not merely to amass an extensive collection, my small amount of leisure obliges me to confine my botanical pursuits within very narrow limits." In the next letter (September 9, 1843), written after Mr. Borrer's second visit, when they gathered mosses together around Castle Howard, he determined one of their gatherings to be *Bryum intermedium*, Brid., a moss which had previously been confounded with other species, but which he had named from the very accurate descriptions in the work on European Mosses by Bruch and Schimper then in course of publication. He here shows his critical faculty and confidence in his own results by adding, "With *B. turbinatum*, to which Hooker and Taylor united it, it has nothing to do!" By this time he had so impressed his friend with his extensive knowledge and the accuracy of his judgment, that Mr. Borrer sent him many of his doubtful Mosses and Hepaticæ to determine, and though Spruce disclaimed being "an authority" (as Mr. Borrer had termed him), he was always ready to give his opinion when he had sufficient materials on which to form one.

In March 1844 he wrote to Mr. Borrer in regard to certain species of Bryum: "Mr. Wilson was formerly of opinion that we should never be able to distinguish *Br. cæspititium* from these species *by the eye*, but I find now not the slightest difficulty in doing this—in fact, we appear to have had no eyes for seeing the Brya till operated upon by Bruch and Schimper!" And at the end of this letter he says: "I shall certainly not 'declare off' from receiving your doubtful mosses (unarranged). I love to combat with difficulties, knowing that the solution of every 'crux' brings me nearer to a finished botanist."

His paper on the Musci and Hepaticæ of Teesdale, the result

of a three weeks' excursion in the preceding summer, showed him to be one of the most lynx-eyed discoverers of rare species, as well as an accurate discriminator of them. In Baines's *Flora of Yorkshire* (1840) only four mosses were recorded from Teesdale, though no doubt many more had been collected. Spruce at once raised the number to 167 mosses and 41 hepaticæ, of which six mosses and one Jungermannia were new to Britain. In April 1845 he published in the London *Journal of Botany* descriptions of twenty-three new British mosses, of which about half were discovered by himself and the remainder by Mr. Borrer and other botanists.

In the same year he published, in the *Phytologist,* his "List of the Musci and Hepaticæ of Yorkshire," in which he recorded no less than 48 mosses new to the English Flora and 33 others new to that of Yorkshire.

By the liberality of Mr. Borrer, and by exchanges with other botanists, he had now obtained specimens of nearly every known British moss; and he had also been in correspondence with Bruch and some other Continental botanists, and had received from them a large number of European species, which were of great value to him for comparison. As it was his practice to make a careful microscopical study of all the species he possessed, and as his whole spare time for the three years 1842-44 was devoted to this work, we can accept his statement to Mr. Stabler, that before he went to the Pyrenees he was so thoroughly familiar with them that he could give from memory the distinctive characters of almost every species.

In the latter part of 1844, when he had to leave the York school, his future was very unsettled. A plant agency in London and the curatorship of some Colonial botanical garden were successively discussed with Mr. Borrer and Sir William Hooker, and rejected as either unsuitable or uncertain of attainment. The latter gentleman then suggested his going as a plant-collector to Spain, that being a rich and comparatively little known part of Europe; but on inquiry the country was found to be in so disturbed a state that travelling would be dangerous, and collections difficult to preserve and to transport safely to England. The matter was decided by the suggestion of the Pyrenees, which Mr. George Bentham had visited a few years before, and where it was thought that a really good collector, such as Spruce had proved himself to be, could easily pay his expenses by the sale of sets of dried plants well preserved and accurately named. This expedition was decided on in December 1844, chiefly, as Spruce told Mr. Borrer, because it would gratify "his irresistible inclination to study his beloved mosses." From some passages in his letters it is evident that Mr. Borrer advanced him a sum of money to be repaid by the first set of his Pyrenean collections. He had intended to leave in April, but at the beginning of that month scarlet fever attacked Welburn, and three of his little half-sisters out of four who were attacked, died of it.

He was able to leave, however, at the end of April, and after spending a few days in the neighbourhood of Bordeaux, he reached Pau early in May, and devoted his whole time and energies till the following March in collecting and studying the beautiful flowers and unexpectedly interesting mosses of the Pyrenees. All his previous inquiries had led him to believe that mosses were few in number and of common species; and the French collections he examined before reaching the mountains showed this to be the case. In a letter to Mr. Borrer dated October 29, after he had been four months in the mountains, he writes: "As the result of my wanderings I have now to show most of the rarest flowers of the Pyrenees, a great many of which have been gathered at heights of from 9000 to 10,000 feet, and even upwards, and cannot, in fact, be obtained without climbing thus high; while my Cryptogamic *récolte* may now fairly be called *immense*. . . . The rotten trunks of trees furnish quite a garden of Jungermanniæ throughout the Pyrenees. . . . The two best stations for Cryptogams I have found to be Cauterets and Bagnères de Luchon; at the former place I stayed three and at the latter above five weeks. I can easily conceive why so few mosses have been gathered in the Pyrenees; for the flowers are so numerous, so varied, and so beautiful, that no person who was not, like myself, quite *entêté* of Bryology would deign to pick up a humble moss! In a guide-book to the environs of Luchon, of some scientific pretensions and containing some two or three chapters on Botany, it is even said 'la famille des Mousses n'existe pas dans les Pyrénées'! Yet of all places in the Pyrenees, the valleys, lakes, and cascades in the vicinity of Luchon are the most prolific in mosses. Above the region of forests mosses become very scarce; the rocks are too exposed to the heat of the Pyrenean sun to permit them to flourish. It is in the immense forests of beech, elm, etc., that I have made my richest harvest." And towards the end of the same letter he writes: "From what I have said you will begin to see that I am satisfied with my expedition. It is true I have traversed many a weary mile and found very little in the way of mosses, but this must always be the case with a first explorer, and on the whole I am well content with my success. Whether or not my collection be considered a rich one by others, I do not think many more mosses remain to be found in the Pyrenees; some there are undoubtedly, for there are many localities I had not time to examine, but it will perhaps be difficult to find a person who will search for them so carefully and patiently as I have done."

In a later letter to Mr. Borrer (dated January 5, 1846), after four pages about mosses and the difficulty and cost of sending home his large boxes full of plants, he adds a postscript, which, as an echo of a now almost forgotten mania, it may be interesting to quote. "*P.S.*—I am afraid I shall find no one but yourself when I return to England who will deign to look at Pyrenean plants—you appear to be all going mad about Railways. I get hold sometimes of a *Times* or *Morning Chronicle*, with Supplement on Supplement of Railway advertisements, and I turn over page after page until I am quite in despair of arriving at any *news*. And when at last I come to something that looks readable, I still find it to consist almost entirely of accounts of Railway meetings, etc. You appear also to have undergone such *strange metamorphoses*! pour exemple, when I read 'the *Railway King* is determined to have a narrow gauge into Cornwall,' it is scarcely credible that his said majesty is no other than my old acquaintance George Hudson, *quondam* Linen-draper of York! Alack! alack! what will this world come to!!"

He returned to England in April 1846, and at once made his long promised visit to Mr. W. Borrer at Henfield, Sussex. They explored together all the best collecting grounds in the district, after which Mr. Borrer took him to Tunbridge Wells and St. Leonard's Forest; and, after a three weeks' delightful excursion, accompanied him to London, where Spruce had to make some arrangements as to the disposal of his Pyrenean collections. He had obtained in considerable quantities between 300 and 400 species of the choicer alpine plants, and these all had to be named, arranged into sets, and sent to the various purchasers in Great Britain and the Continent, a work which fully occupied him for the remainder of the year.

In his favourite groups the Mosses and Hepaticæ he had done for the Pyrenees what he had previously done for Teesdale—shown them to be exceptionally rich in these plants. A list published by Léon Dufour in 1848 contained only 156 mosses and 13 hepaticæ, though of course many more may have been gathered by botanical collectors from other parts of Europe. Spruce at once raised the number to 386 mosses and 92 hepaticæ. My friend Mr. M. B. Slater informs me, from an examination of the latest work on the Mosses of France, that 17 of the species discovered by Spruce were absolutely new to science, and that 73 more had never previously been gathered in the Pyrenees. Of the Hepaticæ he described four species as altogether new, while a still larger proportion than of the mosses were new to the Pyrenees, and of these a considerable number were only known elsewhere in our own islands, which are the richest part of Europe in this group.

After the flowering plants had been distributed, he began his elaborate work—*The Musci and Hepaticæ of the Pyrenees*, which occupied all his spare time during the next two years, and was only published after his departure for South America. It occupies 114 pages of the *Transactions* of the Botanical Society of Edinburgh, and besides giving the names of all the species carefully identified, describes fully all that were new or doubtful, and gives particulars of the local and geographical distribution of each. He had already given a more general account of his whole excursion in two letters to Sir William Hooker, which were published in the *London Journal of Botany* for 1846, under the title *Notes on the Botany of the Pyrenees.* These are very interesting reading for every lover of plants, besides giving an excellent idea of Pyrenean scenery and inhabitants. During this visit to France he made the acquaintance of several botanists, and from them and from Bruch, with whom he had been corresponding for some years, he received such a quantity of mosses that he was able to inform Mr. Borrer in 1846 that his European mosses were nearly complete, and he was thus enabled, by comparison with authentic specimens, to name all the known species in his Pyrenean collections.

His thorough knowledge of the British species and his habits of carefully verifying every point in the descriptions of his predecessors, enabled him to detect errors which had been long overlooked. Mr. Borrer had sent him a copy of Bridel's later description of *Hypnum catenulatum*, on which Spruce remarks: "I find his description to be concocted of Hooker and Taylor's, Schwaegrichen's, and his own in *Muscol. Recent.*, combining the errors of all three! I have before heard of this trick of Bridel's, and also of his drawing up descriptions from figures alone." And in a succeeding letter (October 20, 1846), he writes about some moss which had got mixed up with other species, and after tracing out the sources of the confusion between several eminent botanists, he adds: "Here is a pretty mess—there seems to have been a contest between Schwaegrichen, Bruch, *e.a.*, which of them could most excel in getting the wrong sow by the ear." A few months later he writes that he finds even Bruch and Schimper to be "not infallible!" And again, that Sir W. Hooker had been "exceedingly indignant that I should have presumed to demur against Mr. Wilson's judgment." But it is pleasant to know that he remained on terms of friendship with both for the rest of their lives.

In the first letter to Mr. Borrer from the Pyrenees, Spruce tells him how wonderfully his health had been improved by the continual outdoor work and mountain air. When he first arrived a walk of three miles fatigued him dreadfully, but after two or three months he was able to walk 25 or 30 miles over rough mountain roads without any discomfort, and he also appears to have gone through the winter at Bagnères de Bigorre, always collecting mosses on fine days, without any serious attacks of his usual ailments.

When he had got back to his home in Yorkshire and settled to work at his mosses, he naturally became very anxious as to his future, and more than ever disinclined to go back to teaching or to any kind of work that involved confinement to the house, which he felt sure would be fatal to him in a few years. On June 4, 1846, he writes to Mr. Borrer: "I yearn to be independent, and I hope the next time I go out it will be to settle in some comfortable office; but I must be contented to wait until an opening occurs, and in the meantime what my hand has found to do I will do with all my heart, for my heart is in it."

The correspondence with Mr. Borrer came to an end in 1848. It consists of five letters written during this year, mostly about mosses and private affairs. His father was very ill in the early part, and Spruce had for two months to do his school work. Then, in June, Spruce himself had a severe liver-attack, with gall-stones (already mentioned), from which he did not completely recover till August. In a letter written in July he says: "I have engaged to go up to London in the early part of September, to superintend the sale of Dr. Taylor's herbarium and books, which his son is going to send thither." And in the last letter (dated August 5), which is mainly about the determination of difficult mosses, he says: "When I come up to London to superintend the sale of Dr. Taylor's herbarium I will endeavour to bring with me all the books I still have of yours. Perhaps I may have the pleasure of seeing you then."

From this time letters to his botanical correspondents are wanting, but this is easily explained. When in London in September he must have had ample opportunities of consulting his chief friends, Mr. Borrer and Sir William Hooker, and was no doubt also introduced to Mr. George Bentham, and through their advice and encouragement determined to undertake the botanical exploration of the Amazon valley. It is probable, also, that he heard from some of our entomological friends at the British Museum how successful Bates and myself had already been, and how highly we spoke of the climate and the people, showing that there were no real difficulties in the way of a naturalist collector.

His decision having been taken, his whole time must have been fully occupied till the date of sailing for Pará on June 7, 1849. Letters from Sir W. Hooker in October and November 1848 show that this journey was under discussion, and that by December it was finally decided upon. A letter to Mr. G. Stabler shows that Spruce came to Kew in April 1849, and spent about two months there. During this time Mr. Bentham agreed to receive all his botanical collections, name the already described species, sort them into sets under their several genera, and send them to the various subscribers in Great Britain, as well as in different parts of Europe. He also undertook to describe the more interesting new species and genera, and to collect the subscriptions and keep all accounts, in return for which invaluable services he was to receive the first (complete) set of the plants collected.

Later letters show that only eleven subscribers were obtained at first; but that after the early collections arrived and were reported on by Sir W. Hooker in the *Journal of Botany*, and by so great a botanist as Mr. Bentham, subscribers were at once found for twenty sets, which, a few years later, when the great novelty of the collections and their admirable condition as specimens became more widely known, increased to over thirty.

As will be seen by some of the letters printed in these volumes, Spruce highly appreciated the great service Mr. Bentham rendered him in thus undertaking the laborious duties of a botanical agent; and that he fully expressed his gratitude for it, as well as for the numerous letters Mr. Bentham wrote to him on botanical subjects connected with his work, which were his chief solace and encouragement during his long and often solitary wanderings.

LIFE IN ENGLAND AFTER THE RETURN FROM SOUTH AMERICA

June 1864 *to December* 1893

The opening paragraphs of this biography, together with the first pages of Chapter XXIII. of the present work, sufficiently indicate where this section of Spruce's life was spent; while the first six and the last six chapters constitute a portion of the literary tasks which occupied him during the first four or five years after his return to England. This was the period during which his health was somewhat improved and he could take short walks (to the extent of half a mile or so) amid the rural scenes endeared to him by the memories of early youth. But during the last twenty years of his life he rarely went far outside his small cottage, alternating only from chair to couch, with an occasional walk round the room, or in the very small patch of garden.

What was especially trying to him was, that for months or even for years together, he was unable to sit up at a table to write or to use a microscope, and could never do so for more than a few minutes at a time with intervals of rest on a couch. There seems little doubt that this extreme prostration might have been much alleviated, perhaps even cured, had the precise cause of it been discovered as soon as he arrived home. Yet although, by Mr. Hanbury's advice, he consulted Dr. Leared, the most eminent specialist of that time on diseases of the digestive organs, both he and other physicians who attended Spruce at Hurstpierpoint and in London appear to have entirely misunderstood his case, and paid little attention to his own account of his sufferings and his localisation of their origin.

But, four years after his return to England, Dr. Hartley of Malton found that almost all Spruce's distressing symptoms were due to a stricture of the rectum, which none of his other doctors had discovered or even suspected. He says, in a letter to Mr. Hanbury: "I have always signalised the seat of the pain to my previous physicians, but none of them—not even Dr. Leared—ever thought of passing a bougie into the rectum. They found it so much easier to hide their ignorance under the accusation of hypochondria, and to prescribe brandy-and-water every three hours." Under very simple treatment—enemas and gentle opiates—he so far recovered as to be able to work at the microscope for short periods, and even to walk half a mile in fine weather. But through neglect the disease had become incurable, and the consequent weakness and continuous discomfort lasted during the remainder of his life.

His condition before Dr. Hartley's discovery is shown by the following extract from a letter to Mr. Stabler (in 1867): "I can hardly write in any other way than reclining in my easy-chair with a large book across my knees by way of a table, and consequently I rarely write anything but what is absolutely necessary." And in October 1869: "I have made two attempts to complete my monograph of the South American Plagiochilæ, but the sitting up to the microscope has brought on bleeding of the intestines to such an extent that I fear I must renounce the task altogether, to my deep regret. I have not looked through the microscope for many weeks."

Yet during the succeeding seven years, with only slightly improved health, he did much botanical work. The most important was a paper on the Palms of the Amazon valley and of equatorial South America, for the purpose of which Dr. Hooker sent him all the Herbarium specimens at Kew, those in the Museum being too bulky to render their removal advisable. The result was a paper in the *Journal of the Linnean Society*, occupying 118 closely-printed pages, containing a very interesting account of the geographical distribution of the species, and a new classification of the genera, founded mainly on characters of the spathe, the fruits, and the leaves, as examined by himself during his fourteen years' wanderings. He enumerates and characterises 118 species, of which more than half are fully described as new and for the most part discovered by himself, the characters having been carefully noted from the living plants. Dr. J. B. Balfour, Keeper of the Edinburgh Royal Botanical Gardens, speaks of this essay as a "classical one"; while Sir Joseph Hooker informs me that it is "full of suggestions, some of which have been taken up by later authors."

But his greatest work, and that which has established his reputation among the botanists of the world, is his massive volume of nearly 600 closely-printed pages, on the *Hepaticæ of the Amazon and the Andes of Peru and Ecuador*. This appeared in 1885, as a volume of the *Transactions and Proceedings* of the Botanical Society of Edinburgh. It contains very full descriptions of more than 700 species and varieties, distributed in 43 genera and a large number of new sub-genera, all precisely characterised and defined. Of these 700 species nearly 500 were collected by himself (the number in the first four sets distributed being 493), and of these more than 400 were quite new to the science of botany.

The whole of Spruce's Mosses—a group only second in his estimation to the Hepatics—were placed in the hands of Mr. William Mitten of Hurstpierpoint, for classification, description of new species, and distribution; and were all included in this botanist's great work on South American Mosses, published by the Linnean Society in 1867. In a volume of 632 pages, 1710 species of mosses are described from the whole of the continent of South America. Of these 580 species were collected by Spruce, 254 of them being entirely new. For these figures and those of the Hepaticæ I am indebted to Mr. Matthew B. Sclater (Spruce's sole executor), who has taken the trouble to extract all the necessary items from the two bulky volumes referred to.

Spruce's work on the Hepaticæ brought him a large correspondence from every part of the world, and for the remainder of his life he was sufficiently occupied with this, with the determination of specimens sent him, and with a few special papers, among which were the description of a new hepatic from Killarney, in the *Journal of Botany* in 1887; and one of 18 pages in the *Memoirs* of the Torrey Botanical Club, on a collection made in the Andes of Bolivia.

Turning now to the less exclusively botanical subjects, a few extracts from letters to his more intimate friends will serve to illustrate Spruce's habits and interests during the period of his secluded Yorkshire life.

In June 1869 he wrote to Mr. Stabler with a characteristic joke on his own infirmities: "One day last week a dentist relieved me of four teeth, and I belong now to the genus Gymnostomum; but by the time you come over I hope to have developed a complete double peristome."

In May 1871 he writes: "It has been very 'hard times' with me all this year; nevertheless, I lately plucked up courage to disinter my microscope—after it had lain away out of sight full eighteen months, and I have gone thoroughly over all my South American Plagiochilas, have described all the forms, and have made up my mind as far as possible about the species. The result has been to make me more Darwinian than ever. I feel certain that if we had all the forms now in existence, and that have ever existed, of such genera as Rubus, Asplenium, Bryum, and Plagiochila, we should be unable to *define* a single species—the attempt to do so would only be trying to separate what Nature never put asunder—but we should see distinctly how certain peculiarities had originated and become (temporarily) fixed by inheritance; and we could trace the unbroken pedigree of every form."

About this time Spruce had instructed his former landlord at Ambato, Manuel Santander, how to collect orchids and butterflies for Mr. James Backhouse of York. These collections were not very successful, and when they came to an end he received the following characteristic letter, which, as it gives a few facts about Baños which Spruce himself had omitted to state, and also in its concluding paragraph shows what an impression Spruce had made on these kind-hearted people, I will give here. Other letters equally enthusiastic are given in Chap. XXIII. of the present work.

Extracts from Letter from Manuel Santander, September 1870

"On the 13th we went to the village of Baños to inquire for the guide Juan. . . . We went to see the hot springs, which are truly prodigies of nature, seeing that at only 12 feet from them is a well of the coldest water. The proximity of the steaming springs made us perspire abundantly, and it is impossible to bear one's hand in them.

. . . .

"All your old friends salute you and are well. Don Pedro Mantilla tells you that we have now a coach road to go and eat pears and peaches at Lligna, and I say to you—'Come to your Ambato to lay your bones along with ours.' There is now a coach road from Quito all the way to Riobamba. The coach comes from Quito in one day, and you might now travel without agitating yourself much. Oh if we had you at our side we should be happy!"

When Lindberg, the Swedish botanist, was about to visit him, Spruce writes to Mr. Stabler:

"*July* 1, 1872.—I shall be very glad indeed that you come whilst Lindberg is here, for I am still in such indifferent health that, without your aid and Mr. Slater's, I fear I shall be able to entertain him very poorly indeed."

. . . .

"On the 4th inst. I was agreeably surprised by a visit from three Bryologists, Messrs Slater, Anderson, and Braithwaite. I have also lately had other botanists here, especially Inchbald and Giles Munby—the latter resided fifteen years in North Africa and has written a Flora of Algeria. I knew him in York nearly thirty years ago."

In March 1873 he writes to the same friend: "I have only

just resumed microscopic work again, for in the very cold weather I had to give it up. But I have gone completely through all my South American Hepaticæ, and have selected and classified type-specimens for ulterior analysis. In Lejeunea alone—in its widest sense, that is, including Phragmicoma, etc.—I have no fewer than 460 'forms.' I have also gone over all my old European herbarium and have brushed away the excreta of destructive insects, so that (except to myself) the signs of their ravages are now scarcely apparent."

Again in October 1873: "I hammer away, as well as I can, at Lejeuneas and their relatives. It serves to beguile pain; whether it will ever be completed, time will show."

More than a year later, in December 1874, he writes: "My work is now limited to chewing the cud of partially digested observations made during the past summer"—indicating under what difficulties and painful conditions he continued to labour at the great work (and enjoyment) of his life—the minute and exhaustive study of the Hepaticæ. It was about this time that his long correspondence with Mr. Daniel Hanbury was brought to a close by the lamented death of his friend. The following extracts from some of Spruce's latest letters to him are of general interest:—

Richard Spruce to Daniel Hanbury

"WELBURN, *Feb.* 10, 1873."

[In reply, apparently, to some depreciatory remarks upon his favourite Hepatics, Spruce writes as follows:—]

"The Hepaticæ are by no means a 'little family.' They are so abundant and beautiful in the tropics, and in the Southern Hemisphere generally, that I think no botanist could resist the temptation to gather them. In equatorial plains, one set creeps over the living leaves of bushes and ferns, and clothes them with a delicate tracery of silvery-green, golden, or red-brown; and another set, along with mosses, invests the fallen trunks of old trees. In the Andes they sometimes hang from the branches of trees in masses that you could not embrace with your arms. I have some species with a stem half a yard long, and others so minute that six of them grow and fruit on a single leaflet of an Acrostichum. Then, as to number and variety, I suppose that the working up of my South American Hepaticæ may entail equal labour to that of monographing the world's Rubiaceæ. In the largest genus, Lejeunea, I have not merely thousands of specimens, but thousands of papers covered with specimens; and all these must be analysed under the microscope, without which one cannot accurately discern any of their features.

"I like to look on plants as sentient beings, which live and enjoy their lives—which beautify the earth during life, and after death may adorn my herbarium. When they are beaten to pulp or powder in the apothecary's mortar they lose most of their interest for *me*. It is true that the Hepaticæ have hardly as yet yielded any substance to man capable of stupefying him, or of forcing his stomach to empty its contents, nor are they good for food; but if man cannot torture them to his uses or abuses, they are infinitely useful where God has placed them, as I hope to live to show; and they are, at the least, useful to, and beautiful in, themselves—surely the primary motive for every individual existence."

He then goes on to show that these little plants are not always without sensible properties. Some possess colouring matters, and yield a yellow or brown dye; others give out fragrant odours, and some a pungent taste comparable to that of camphor or pepper. But such species are as yet few in number.

In a previous letter he had described how he had had to make up for lost time, as during his travels he had no leisure to study the plants which he collected in detail. "I have therefore had first to 'fetch up' those who have had the start of me; and now, after working constantly at Hepaticæ, and thinking of little else for eighteen months, I begin to feel I know something about them. I have now worked up all the more difficult genera except one, and the Hon. Mrs. Howard offers to be at the expense of a few illustrative figures, so that if I am spared to complete the task I hope to have done something likely to be permanent. All this study has been carried on, accompanied by quite as much pain and cramping as of yore; but to be occupied on sensible objects dulls the feeling of pain much more than purely mental occupation.

"Since I came to Welburn I have also reduced all my meteorological and hypsometrical observations, and 'done' the native languages and ethnography, besides a few minor matters—all, however, chiefly written in pencil, and often hieroglyphically, and they want putting in order and writing out *au net*."

The following passage from a succeeding letter shows curiously his love for all living things:—"Neither these nor any other Mosses or Hepatics are ever likely to become of much direct importance to man—at least I hope not, for if they should, unfortunately, then the little birds and the beetles would be put to their pins for shelter and bedding." This is the last letter of any general or botanical interest.

Mr. Hanbury died of typhoid fever, March 24, 1875. Spruce's last letter to him in the collection of the Pharm. Soc. is May 26, 1874.

The next letter (to Mr. Stabler, May 9, 1875) gives Spruce's own account of his great loss:—"It has been a time of trouble and suffering. First I had the great grief of losing one of my oldest and best friends, Daniel Hanbury, from what seemed at first only a slight attack of typhoid, but whose fatal progress could not be arrested. I have lost in him a town correspondent who was always ready to execute any little commission—the expense of which he has often generously borne himself—and to look up for me the latest information on any subject. Add to this his uniformly kind and genial disposition, and you will see that such a friend cannot easily be replaced. I enclose two notes for your perusal from his venerable father, who is eighty years old. Then I fell ill myself, with bronchitis accompanied by intermittent fever—every alternate day twelve hours' fever—and was some weeks before I shook it off. Lastly, within these few days there is something the matter with my right eye, which has prevented my using the microscope, and I am fearful I may be going to lose the sight of that eye."

The only records of the last fifteen years of Spruce's life which are available are in the continuous series of letters to his life-long friend Mr. G. Stabler, who, both as schoolmaster, invalid, and botanist, was in complete sympathy with him. This gentleman—now afflicted with complete blindness—has kindly furnished me with copious extracts from these letters, from which I will now give such selected passages as seem of general or personal interest. They are largely occupied with his ever-varying degrees of capacity for study, with the progress of his great work on the Hepaticæ, and succeeding papers on the same group, with often amusing records of his various botanical and other visitors, among whom were several foreign botanists, his valued friend Sir Clements Markham, the late Duke of Argyll, the Duchess of Argyll, Lady Lanerton, and Lady Taunton. With the Duke he had two hours' talk, not only on natural history, for he says: "Besides these subjects, we chatted on many others, from the undulatory theory of light to Spanish and Russian politics; and my guest was just as frank and simple as our valued friend Matthew Slater."

Of one of his more distinguished visitors he writes (Oct. 1878): "I had a visit yesterday from Lord Northbrook, a former Governor-General of India, and I did not hesitate to put him through his catechism on Indian matters; but I cannot here detail his views and opinions: they were *very different* from those of Lords Lytton and Beaconsfield."

In Nov. 1875 he writes: "I have just been writing five long letters for M. André to take to South America. He starts on the 7th, and goes direct to Loja and the Equatorial Andes." M. André is the well-known French botanist and enthusiastic traveller and plant-collector.

The following note is of interest:—"I am sorry to hear of Lindberg's sufferings in the head and eyes. I had the same thing myself for months in South America, and from a similar cause. I had had some smoking-caps made of black satin, lined with red silk. The red dye came out and stained my forehead, but it was long ere I found out that it was really causing the atrocious pains which almost drove me crazy. Perhaps I swore about it in Spanish quite as emphatically as Lindberg does in English."

The following extract from a letter of April 23, 1886, shows how well he could introduce that rather rare thing, an appropriate yet unhackneyed quotation:—"I was very sorry to hear of poor Bishop Hannington's death—botanical bishops are so rare! We once had one, however—Dr. Goodenough, Bishop of Carlisle, whose monograph of British Carices is still a classic. Though so sound a botanist (and divine, I presume), he was a dreary preacher. Once on a time he had had to preach to the peers, and Peter Pindar wrote of him:

> 'Twas well enough that Goodenough before the Lords should preach,
> But sure enough full bad enough for those he'd got to teach."

After Spruce's work on the Hepaticæ was published, he was occupied from 1889 to 1892 in the very tedious but to him interesting task of sorting out his immense collection of South American Hepaticæ into sets of species for distribution, writing labels for names, etc., the whole of which was completed and twenty-five sets sent off before the end of the year. The first four sets contained 493 species each, and the first eleven over 400, while the last five were reduced to about 200 or 300, showing the rarity of many of these delicate little plants, which were often found only once, and then perhaps in minute patches, either mixed with or growing upon other species.

The following extracts from his two last letters (the second written within two months of his death) show that his interest in botany continued to the last.

On October 27, 1892, he wrote to Mr. Stabler: "Last

month I completed my seventy-fifth year, and am become almost a fixture. Only my eyes do not fail me. In the winter of 1889 I had a paralytic attack, accompanied by almost complete incapacity for two entire months. Since then I have only been able to write very little, and I have been occupied principally in revising my collections and in preparing the exsiccata of them. I have a few last words to say on the Hepatics, but I do not know if I shall have the courage to complete them."

And on Oct. 13, 1893, as follows:—"Slater and I have discovered two lady botanists in our own neighbourhood—or rather they have discovered us. Mrs. Tindall's husband is brother of the proprietor of Kirby Misperton, but their home is in the south. Miss Lister, her cousin, is a clever botanical artist. Her home is in Dorset. They are very quiet, unassuming ladies—fine scholars (I envied them their familiarity with German)—and have both a fair knowledge of British flowers and mosses, but are comparatively new to Hepaticæ."

Shortly after writing the above he had a severe attack of influenza, which caused his death on the 28th of December, at the age of seventy-six years and three months.

Richard Spruce's life was spent in continuous labour for science and humanity—as a teacher, an explorer of nature, and more directly by his successful work in the introduction of the valuable red bark into India. Although his labours for this last object, extending over two years, were largely contributory to his permanent loss of health, his friends had the greatest difficulty in obtaining for him, first the small Government pension of £50 a year in 1865, and in 1877, through the long-continued and earnest representations of Mr. (now Sir Clements) Markham, a further pension of £50 from the Indian Government. Having lost the greater part of his savings through the failure of a mercantile house of the highest standing in Guayaquil, his means on his return to England were exceedingly scanty, so that he had to spend the last twenty years of his life in a small cottage sitting-room about 12 feet square, with a bedroom of equally limited proportions. Here he was carefully looked after and nursed by a kind housekeeper and a little girl attendant, who were also his friends and companions; and in this humble dwelling he received visits from his numerous friends, and, amid all his pains and infirmities, was cheerful and contented. He was well acquainted with general literature, including the old travellers and poets—Shakespeare and Chaucer being always among his small collection of books. He was a musician and a chess-player, and was especially fond of a joke, even at his own expense. And, in the words of his life-long friend, Mr. George Stabler, "he was always courteous and gentlemanly in his bearing, and ever affectionate, kind, and sympathising as a friend."

As a friend and admirer for more than forty years, I may be allowed to give here my own estimate of him from a short obituary notice I wrote for *Nature* (February 1, 1894):—

Richard Spruce was tall and dark, with fine features of a somewhat southern type, courteous and dignified in manner, but with a fund of quiet humour which rendered him a most delightful companion. He possessed in a marked degree the faculty of order, which manifested itself in the unvarying neatness of his dress, his beautifully regular handwriting, and the orderly arrangement of all his surroundings. Whether in a native hut on the Rio Negro or in his little cottage in Yorkshire, his writing-materials, his books, his microscope, his dried plants, his stores of food and clothing—all had their proper places, where his hand could be laid upon them in a moment. It was this habit of order, together with his passion for thoroughness in all he undertook, that made him so admirable a collector. He was full of anecdote, and even when suffering from his complicated and painful illnesses, an hour would rarely pass without some humorous remark or pleasant recollection of old times. He was a man who, however depressing were his conditions or surroundings, made the best of his life. He was a Liberal in politics as in religion, a true lover of work and workers of whatever class or country; and nothing more excited his indignant wrath than to hear of the petty but often cruel persecutions to which the labouring classes were (and still are) so often subjected. He was an enthusiastic lover of nature in all its varied manifestations, from the grandeur of the virgin forest or the glories of the sunset on the snowy peaks of the Andes, to the minutest details of the humblest moss or hepatic. In all his words and ways he was a true gentleman, and to possess his personal friendship was a privilege and a pleasure.

He was buried at Terrington (the parish in which he was born, Ganthorpe being only a hamlet) beside his father and mother, in accordance with his own directions to his executor, Mr. Matthew B. Slater of Malton.

It now only remains to give some indication of his scientific labours as judged by his fellow-botanists.

His great characteristic was the thoroughness of his work. As a British botanist he quickly made his mark, and very soon became an authority on our indigenous Flora, especially as regards his favourite groups the Mosses and Hepatics. A little later, when he went to the Pyrenees, he made such beautiful collections of the rarest alpine plants, and so many discoveries among the hitherto little known mosses, as to prove his capacity both as collector and painstaking student of a new flora. I cannot help thinking that it was this thoroughness of Spruce's work in everything that he undertook that so greatly impressed Mr. Bentham (who had himself collected in the Pyrenees and published a catalogue of its plants) as to cause him to undertake the enormous labour and responsibility of acting as his agent in the naming and distribution of his South American plants, a labour which, notwithstanding his other botanical work, including the *Flora of Hongkong* and the *Handbook of the British Flora*, which were being written and published at the same time, he continued to the very last, that is, during the twelve years that Spruce was able to send home collections.

No sooner did his early consignments from the neighbourhood of Pará reach England than the expectations of his friends were fully justified; and Mr. Bentham wrote to him: "The specimens are excellent, and being so well packed, they have arrived in admirable order. . . . It is one of the best tropical collections as to quality of specimens that I have seen." Sir William Hooker wrote to the same effect, and this high quality was maintained throughout his whole expedition, except in those cases where delays or exposure to damp or floods when they had passed out of Spruce's hands caused more or less injury.

Sir Joseph Hooker writes me on this point, as regards some of the later collections: "I can remember the arrival of one consignment to Bentham at Kew, and marvelling at the extraordinary fine condition of the specimens, their completeness for description, and the great fulness and value of the information regarding them inscribed on the tickets."

Professor Daniel Oliver, who assisted Mr. Bentham in the work of distributing the specimens, also writes me on these particulars: "Mr. Spruce's specimens were most carefully collected, dried, and packed, extraordinarily so considering the difficulties of all kinds with which he had to contend; and what was of special value, they were accompanied by beautifully legible labels giving precisely the information as to locality, habitat, habit, etc., required to supplement the dried specimens. I may add, the duties of a trained collector could not have been better done. The collections were specially rich in arborescent species, the obtaining of which must often have been of considerable difficulty."

No praise can be higher than this from two botanists who have for many years had the charge of the largest collections of plants in the world.

His botanical knowledge, his accuracy, and his judgment in the classification and description of the plants which he specially studied, have also been recognised by the most competent judges.

In the very condensed record of the works of eminent botanists and botanical collectors, given in the last volume of the great *Flora Braziliensis*, he is said to have "most accurately examined and published" the Pyrenean Mosses and Hepaticæ; while of the volume on the Hepaticæ of the Amazon and Andes it is said that he "most sagaciously elaborated and described" the whole of the known species.

On Spruce's return to England, the veteran botanist Von Martius invited him to undertake the elaboration of one of the Natural Orders for his great work on the Flora of Brazil, showing that he must already have had the highest opinion of his competence as a botanist. This Spruce was obliged to decline on account of his ill-health; but several letters passed between them on botanical subjects, showing on the part of Martius the highest appreciation and even enthusiastic friendship. In 1866 he writes to him as "My dear Spruce," and concludes with this amusingly pathetic appeal: "Por la misericordia de Dêos, I beg you to exhilarate me by an answer. Your very attached friend and admirer—MARTIUS." In 1867 he signs himself "For ever your affectionate devoted friend"; and in August 1868, very shortly before his death, "Your affectionate and admiring friend."

Of the general results of Spruce's botanical exploration and study in South America, the late Mr. George Bentham, who knew more of his work than any one else, thus wrote: "His researches into the vegetation of the interior of South America have been the most important we have had since the days of Humboldt, not merely for the number of species which he has collected (amounting to upwards of 7000), but also for the number of new generic forms with which he has enriched science; for his investigation into the economic uses of the plants of the countries he visited; for several doubtful questions of origin as to interesting genera and species which his discoveries have cleared up; and for the number and scientific value of his observations made on the spot, attached to the specimens preserved, all which specimens have been transmitted to this country, and complete sets deposited in the Botanical Herbarium at Kew."[1]

Mr. John Miers, a great authority on South American plants, wrote two very long letters to Spruce in 1874, full of botanical details, and accepting many of Spruce's suggestions and his corrections of the statements of other botanists, etc.

In an obituary notice by Dr. Isaac Bayley Balfour, the writer speaks of his essay upon the Palms of the Amazon as being "classical," and that his work upon the Hepaticæ of the Amazon and Andes "is now generally recognised as the most important book upon the group that has appeared in recent years"; and he adds that, "though ostensibly descriptive and systematic, his writings are weighty in the discrimination of characters and in the adjustment of boundaries; but over and above this they have the charm of deserving to be read between the lines, for they abound with interjected suggestions, often most pregnant. For instance, the question of the evolution of the leafage of the Hepaticæ and its relation to that of the higher plants may be raised in a footnote; the water supply and the biological relation-

[1] Mr. Stabler could not tell me where he found this statement by Mr. Bentham given in his "Obituary Notice of Richard Spruce" in *Proceedings* of Botanical Society of Edinburgh (Feb. 1894). Dr. B. Daydon Jackson kindly searched all the Presidential Addresses to the Linnean Society, as well as the articles referring to Spruce in the *Journal of Botany*, without finding it. I then applied to Mr. A. Gepp of the Natural History Museum (who had written an obituary notice of Spruce), and he informs me that his colleague, Mr. James Britten, has found it in a seven-page pamphlet, without author's or printer's name, headed—*Statement of the Results of Mr. Richard Spruce's Travels in the Valley of the Amazon, and in the Andes of Peru and Ecuador.*

Then follows a description, year by year, of Spruce's work, concluding with a "Note by Mr. Bentham, President of the Linnean Society, on Mr. Spruce's Services to Botany"—which, after referring to his work in Great Britain and the Pyrenees, concludes with the passage quoted above.

This pamphlet bears the MS. inscription:

"J. J. Bennett, Esq., with Mr. Clements R. Markham's compliments, June 10, 1864."

As this was only a fortnight after Spruce's return to England, the statement appears to have been one of the documents used by his friend Mr. (now Sir) Clements Markham, for the purpose of obtaining for him the small Civil List Pension of £50, which was granted him the following year, as stated by Mr. Stabler.

These circumstances may account for the fact that Mr. Bentham has published no general estimate of Spruce's work in any of his papers describing portions of his collections, or elsewhere, and renders it more important that the above statement should be preserved here.

ships of the group may be incidents of the description of the finding of a new species, and so forth."

Another botanist, Mr. Antony Gepp of the British Museum, in an article "In Memory of Richard Spruce" in *The Journal of Botany* (February 1894), writes as follows:—"His *Hepaticæ of the Amazon and the Andes* is the most logical and scientific classification of the group that has been evolved, and is based entirely upon broad and constant characters that had previously been overlooked or underrated.

. . . .

"Mr. Spruce delighted to lead his readers on from the immediate subject to kindred matters, illustrating his arguments with copious instances, analogies, and original observations. Thus, after describing the new Irish hepatic *Lejeunea Holtii*, he proceeds to contrast the comparative wealth of Lejeuneæ found at Killarney (13 species) with the three known to occur in the rest of Europe; and so passes on to a general consideration of the phenomena of distribution and the part played by animals as carriers of seeds and spores, quoting an anecdote told him by an Indian of the Rio Negro of the revels held by the beasts of the forest upon a clearing, immediately after it had been deserted by its owners."

And lastly, the distinguished veteran botanist Sir Joseph D. Hooker writes me the following brief but very high appreciation:

"No doubt his (Spruce's) monumental work on the Hepaticæ is his crowning one, and will ever live."

LIST OF BOOKS AND PAPERS PUBLISHED BY RICHARD SPRUCE, PH.D.

The following list of Dr. Spruce's published works has been compiled from the following sources :—(1) A list appended to the Obituary Notice by Mr. G. Stabler in the Transactions of the Botanical Society of Edinburgh. (2) A MS. list kindly drawn out for me by Sir Clements Markham. (3) Lists and copies of books and papers from Spruce's library, furnished by his executor, Mr. Matthew B. Slater. It is, I believe, quite complete.

1. Three Days on the Yorkshire Moors. Phytologist, i. 101-104 (1841).

2. Discovery of *Leskea pulvinata*, Wahl. Phytologist, i. 189 (1842).

3. List of Mosses, etc., collected in Wharfedale, Yorkshire (contains 19 Mosses, 16 Hepaticæ, 8 Lichens)—Note on *Didymodon flexicaulis*—Mosses near Castle Howard. Phytologist, i. 197-198 (1842).

4. *Bryum pyriforme*. Phytologist, i. 429 (1842).

5. On the Folia Accessoria of *Hypnum filicinum*, Linn. Phytologist, i. 459 (1842).

6. A List of Mosses and Hepaticæ collected in Eskdale, Yorkshire. (Contains 91 Mosses and 28 Hepaticæ.) Phytologist, ii. 540-544 (1843).

7. Note on *Carex paradoxa*. Note on *Carex axillaris*. Phytologist, i. 842-843 (1843).

8. On the Branch-bearing Leaves of *Jungermannia juniperina*, Sw. Phytologist, ii. 85-86 (1845).

9. A List of the Musci and Hepaticæ of Yorkshire. Phytologist, ii. 147-157 (1845).

10. On several Mosses new to the British Flora. Hooker's Lond. Journ. of Bot. iv. 345-347, 535 (1845).

11. The Musci and Hepaticæ of Teesdale. Trans. Bot. Soc. Edin. ii. 65-89 (1846).

12. Notes on the Botany of the Pyrenees. Hooker's Lond. Journ. Bot. v. (1846). Two papers. Also in the Ann. Mag. Nat. Hist. iii. and iv.

13. The Musci and Hepaticæ of the Pyrenees. Trans. Bot. Soc. Edin. iii. 103-216 (1850). Also in the Ann. Mag. Nat. Hist. 2nd series, vols. iii. and iv.

14. Extracts of Letters from Richard Spruce, Esq., written during a Botanical Mission on the Amazon. Hooker's Journ. Bot., May 1851, Nov. 1851, Oct. 1852, Nov. 1852, July 1853, Aug. 1853, Feb. 1854, April 1854 (vols. ii. iii. iv. v.).

15. Botanical Objects contributed to the Kew Museum from the Amazon. Hooker's Journ. Bot. vols. v. and vii. (1851-1853).

16. Edible Fruits of the Rio Negro. Hooker's Journ. Bot. v. 180 (1853).

17. Extract of a Letter relating to Vegetable Oils. Hooker's Journ. Bot. vi. 333 (1854).

18. Note on the India-rubber of the Amazon. Hooker's Journ. Bot. vii. 193 (1855); Journ. of Pharmacy, xxviii. 382.

19. On Five New Plants from Eastern Peru. Linn. Soc. Journ. iii. 191-204 (1859).

20. On *Leopoldinia Piassaba.* Linn. Soc. Journ. iv. p. 50 (1860).

21. Notes on a Visit to the Chinchona Forests on the Western Slopes of the Quitonian Andes. Linn. Soc. Journ. iv. 176-192 (1860).

22. On the Mode of Branching of some Amazon Trees. Linn. Soc. Journ. v. p. 14 (1861).

23. Mosses of the Amazon and Andes. Linn. Soc. Journ. v. 45-51 (1861).

24. Report on the Expedition to procure Seeds and Plants of the *Chinchona succirubra* or Red Bark Tree. With a Map of the Bark Regions of Ecuador by Mr. Clements R. Markham. Printed for the India Office (1862), pp. 112. Also in the Chinchona Blue Book, pp. 1-63.

25. On the Mountains of Llanganati in the Eastern Cordillera of the Quitonian Andes. Roy. Geog. Soc. Journ. xxxi. 163 (1862).

26. Notes on the Valleys of Piura and Chira in Northern Peru, and on the Cultivation of Cotton therein (pp. 81). London: Eyre and Spottiswoode (1864).

27. On the River Purus, a Tributary of the Amazon. By Mr. Richard Spruce. (Dated) June 18, 1864. (Signed) Richard Spruce. (13 pages.) (No name of publisher or printer.)

Dr. J. S. Keltie informs me that this paper was published as a note on page 339 of the "Travels of Pedro de Cieza de Leon," edited by (Sir) Clements Markham for the Hakluyt Society in 1864. A separate reprint was taken, and is as bound up by Spruce in a volume of his "Opuscula."

28. Note on the Volcanic Tufa of Latacunga at the Foot of Cotopaxi, and on the Volcanic Mud of the Quitonian Andes. Geol. Soc. Quart. Journ. xxi. 249 (1865); Phil. Mag. xxix. 401.

29. On the Fertilisation of Grasses. American Naturalist, iv. 239.

30. "THE WHITE ISLAND." The English Leader, No. 63. London. (Oct. 27, 1866.) An Apologue on Sabbatarianism, in the style of Swift.

31. Notes on some Insect and other Migrations observed in Equatorial America. Linn. Soc. Journ. Zool. vol. ix. 346-367 (1867).

32. Catalogus Muscorum fere omnium quos in Terris Amazonicis et Andinis per annos 1849-1860 legit Ricardus Spruceus. Londini, 1867. (Nos. 1-1518.)

33. Notes on Papayaceæ. By Joaquim Corre de Mello and Richard Spruce. Linn. Soc. Journ. Bot. vol. x. 1. (1869).

34. Palmæ Amazonicæ, sive enumeratio Palmarum in itinere suo per regiones Americæ Equitoriales lectarum (183 pages). Auctore Ricardo Spruce, Ph.D., F.R.G.S. Linn. Soc. Journ. Bot. vol. xi. (1870).

35. On some Remarkable Narcotics of the Amazon Valley and Orinoco. Geographical Magazine, August 1872.

36. Personal Experiences of Venomous Reptiles and Insects in South America. Geographical Magazine, July 1873.

37. Zum geographischen Verständnis der americanischen Reisepflanzen. Botan. Zeitung, col. 28 (1873).

38. On Anomoclada, a New Genus of Hepaticæ, and on its Allied Genera Odontoschisma and Adelanthus. Journ. of Bot., 1876 (32 pages).

39. Musci Præteriti. Journ. of Bot. (Dec. 1880, No. 216, and Feb. 1881, No. 218).

40. On *Marsupiella Stableri* (n.s.) and some Allied Species of European Hepaticæ. Rev. Bryologique, viii. 89-104 (1881).

41. The Morphology of the Leaf of Fissidens. Journ. of Bot. No. 220 (April 1881).

42. On Cephalozia (a Genus of Hepaticæ), its Sub-genera and some Allied Genera (vi and 99 pages). Malton: Printed for the Author (1882).

43. Hepaticæ Amazonicæ et Andinæ. Trans. Bot. Soc. Edin. xv. 1-590, t. i.-xxii. (1885).

44. Précis d'un voyage d'exploration botanique dans l'Amérique équatoriale, pour servir d'introduction provisoire à son ouvrage sur les Hépatiques de l'Amazon et des Andes. Par Richard Spruce. (20 pages.) (Extrait de la Revue Bryologique, août 1886.)

45. *Lejeunea Holtii*, a new Hepatic from Killarney. Journ. of Bot. vol. xxv. (Feb. 1887).

46. On a new Irish Hepatic (*Radula Holtii*). Journ. of Bot. xxv. 209-211 (1887).

47. Hepaticæ in Provincia Rio Janeiro, a Glazion lectæ. Rev. Bryologique, xv. 33-34 (1888). (List only.)

48. Hepaticæ Paraguayenses, Balanza lectæ. Revue Bryologique, xxv. 34-35 (1888). (List only.)

49. *Lejeunea Rosseltiana*, Mass. Journ. of Bot. xxvii. 337-338 (1889).

50. Hepaticæ Novæ Americanæ, tropicæ et aliæ. Bull. Soc. Bot. de France, xxxvi. cxxxix. ccvi. (1889).

51. Bescherelle et Spruce. Hépatiques nouvelles de Colonies françaises. Bull. de Soc. Bot. de France, xxxvi. clxxvi. clxxxix. Pls. xiii. xvii. (1889). (New species from Guadaloupe, French Guiana, New Caledonia, and Réunion Island.)

52. Hepaticæ Spruceanæ, Amazonicæ et Andinæ, annis 1849-1860 lectæ (Malton, 1892). (Specimens distributed by Spruce, with copious notes.)

53. Hepaticæ Bolivianæ, in Andibus Boliviæ Orientalis annis 1885-6 a cl. H. H. Rusby lectæ. Mem. Torrey Bot. Club, i. 113-114 (1890).

54. Hepaticæ Elliottianæ, insulis Antillanis S[ti] Vincentii et Dominicæ a clar. W. R. Elliott annis 1891-92 lectæ, Ricardo Spruce determinatæ. Linn. Soc. Journ. Botany, vol. xxx. Pls. xx.-xxx., pp. 331-372 (1904).

CHAPTER I

PARÁ AND THE EQUATORIAL FORESTS

(*July* 12 *to October* 10, 1849)

I EMBARKED at Liverpool on the 7th of June 1849, on board the brig *Britannia*, of 217 tons, Edmund Johnson commander, with a crew of twelve men. My fellow-passengers were Mr. Robert King, a young man who had agreed to brave the wilds of the Amazon as my companion and assistant, and Mr. Herbert Wallace, who was going out to join his elder brother, then engaged in the exploration of the Amazon valley, of which he has given so pleasant an account.

When day broke this morning (July 12) the city of Pará lay distinctly before us, in a line of houses of striking appearance stretching along the right bank of the river; the Custom-house, with a rather paltry mole in front of it, standing about midway, with the towers of the church of Mercés peering over its roof; the church and convent of St. Antonio occupying the extreme left (as seen from our ship); and the cathedral nearly on the extreme right. It was 10 o'clock when we came to our anchorage near the mole, and visits from the Custom-house authorities kept us on board until 1 P.M., when we went on shore, and after dining with Mr. Miller, the consignee of the *Britannia*, we waited on Messrs. Archibald and James Campbell, colonists of long standing and extensive possessions at Pará, to whom we had been furnished with letters of introduction from friends in England. By these gentlemen we were most cordially received, and were immediately installed in the house of Mr. James Campbell, the elder of the two brothers.

I remained at Pará only three months, and even of that short period part was spent at Mr. A. Campbell's farms, in the environs. Botany occupied me so completely that my notes on the city are of the scantiest description, and I must refer to the accounts of preceding travellers for more detailed accounts of it and its inhabitants.

The beginning of the dry season is a sort of spring in the Amazon valley. As the rains abate and the rivers subside, the trees begin to flower, first those of the gapó or inundated river-margins, then those of the terra firme or dry land. Some trees flower ere the old leaves fall off, others along with the young leaves. In either case the trees are never denuded of leaves, except in a few cases of extreme rarity, the old leaves hanging on until the young ones are developed, exactly as in evergreens at home. A few months later and it is the height of summer; flowers are scarce, and most trees are ripening their fruits and seeds. Both flowers and fruits of the real forest trees were for a long time "sour grapes" to me. Like Humboldt, I was at first disappointed in not finding agile and willing Indians ever ready to run like cats or monkeys up the trees for me, and in seeing how futile must be the attempt to reach with hooked knives fastened to poles flowers which grew at a height of a hundred or more feet, on trees whose smooth trunks (far too thick to be "swarmed") rose to 50 or 60 feet before putting forth a branch. At length the conviction was forced upon me that the best and sometimes the only way to obtain the flowers or fruits was to cut down the tree; but it was long before I could overcome a feeling of compunction at having to destroy a magnificent tree, perhaps centuries old, merely for the sake of gathering its flowers. By little and little I began to comprehend that in a forest which is practically unlimited—near three millions of square miles clad with trees and little else but trees—where even the very weeds are mostly trees, and where the natives themselves think no more of destroying the noblest trees, when they stand in their way, than we the vilest weeds, a single tree cut down makes no greater a gap, and is no more missed, than when one pulls up a stalk of groundsel or a poppy in an English cornfield. I considered further that my specimens would be stored in the principal public and private museums in the world, and would serve to identify any particular tree with its products, as well as for studying the peculiarities of its structure. In fine, I reconciled myself to the commission of an act whose apparent vandalism was, or seemed to be, counterbalanced by its necessity and utility. In the same way I suppose a zoologist stifles his qualms of conscience at killing a noble bird or quadruped merely for the sake of its skin and bones. I know not whether Alexanders and Napoleons make use of any such process of reasoning to justify to themselves the waste of human life entailed by their victories; but if the bodies of the slain at Arbela or Austerlitz could all have been collected and preserved—stuffed and set up in attitudes of mortal agony—under glass cases in one vast museum, what instructive specimens they would have been of the fruits of war!

I was thus reduced at first to vegetation that was easily accessible, but having never before seen tropical plants in their homes, all were to me new and beautiful, although I knew that most coast plants have a wide distribution in the tropics, so that a very small proportion of them would be of any value in the eyes of botanists at home, many of them having already been gathered elsewhere. In marshy places, and at the muddy mouths of igarapés,[1] there was great store of handsome but rank and corpulent grasses, and of sedge-like plants, especially of those tall Cyperi which form extensive beds in such situations, and look at first exceedingly beautiful with their umbels of polished brown or green-and-gold spikelets, but soon tire from their monotonous abundance. Mangroves cause almost the same impression—everybody admires their fresh and uniform green at first sight, and yet nothing can be more dreary and wearisome than to live near, or sail along, a coast where no trees but mangroves are visible. Mangroves are abundant enough from Pará downwards, especially on islands that are flooded with every tide, but from thence upwards, where the water becomes less and less brackish, they gradually disappear.

[1] Igarapé (Lingoa Geral), from *igara*, a canoe, and *pé*, a way, is the general term for brooks and small rivers.

Low moist flats were often partially covered with the Páo de lacre or Sealing-wax tree (*Vismia guianensis*, Pers.), a bush of 12 to 15 feet high, of the same family as our St. John's worts, and like them having the leaves, flowers, etc., studded with glandular dots. From the wounded stem exudes a thick reddish juice, which being collected and allowed to dry, forms a very good substitute for sealing-wax. Of taller trees in such sites there were several species of Inga, some with large, flat, scimitar-like pods; others with slender, cylindrical, furrowed and twisted pods a yard long, hanging from the branches like rope's-ends or portions of some twining stem (whence their Indian name Ingá-sipó). With them were several Monkey-pods (species of Pithecolobium), nearly related to the Ingas in habit and character, but with the leaves twice (instead of only once) pinnate, and with smaller pods often curled into a ring, or at least with the valves rolling back when ripe so as to simulate a monkey's tail. Over these and other trees climbed Malpighiaceæ, adorned with racemes of yellow or pink flowers with elegantly fringed petals and usually a pair of large glands (or tubercles) at the base of each segment of the calyx; and still more showy Combretaceæ, whereof one species (*Cacoucia coccinea*, Aubl.) was all in a flame with its long spikes of brilliant scarlet flowers.

Waste places, with a drier soil, were often clad with a vigorous but weedy vegetation, the predominant plants being rank prickly Solana, with large woolly leaves and apple-like fruits, and several species of Cassia, gay with golden flowers, which were followed by long pods whose loose seeds kept up a continual rattling as one pushed through the interwoven branches. There also grew sensitive plants in great variety and abundance; plants allied to our mallows; others to our sweet-peas and kidney-beans, and amongst them various species of Centrosema, with large white or purple flowers more or less orbicular in outline. On the ground and over the bushes trailed and twined the milky stems of various Convolvulaceæ (chiefly species of Batatas) and Apocyneæ (Echites), the former with large funnel-like white, purple, or violet flowers, the latter with yellow flowers in the form of a bell or trumpet. There also clambered by its tendrils *Passiflora fœtida*, Cavan., one of the commonest of tropical weeds, and unique in a tribe whose flowers exhale such exquisite odours for its heavy narcotic smell, quite recalling that of the roosting-places of the Urubú or turkey-buzzard, whence its Indian name Urubú-muracajá. Herbs of humbler growth and less roving habits were chiefly Labiates and other kindred plants.

Sometimes in similar sites Peppers of various kinds monopolised the largest share of the soil, many of them (species of Artanthe) rising to shrubs or even trees, and notable for the numerous rib-like veins springing at an acute angle from each side of the midrib of their aromatic (or sometimes fetid) leaves, and for their minute flowers being arranged on tessellated spadices similar to those of many Aroids. Other Peppers (species of Peperomia) looked like minute ferns, as they crept with thread-like stems over the trunks of trees, and put forth their roundish fleshy leaves mottled with green and brown.

In the virgin forest, the few plants whose flowers did not hang beyond my reach were chiefly shrubs and low bushy trees of the orders Clusiaceæ, Melastomaceæ, and Rubiaceæ; but I was partly consoled for the scarcity of accessible flowers by the abundance of ferns and even of mosses. However interesting the latter were in my eyes, I should despair of giving any account of them which would interest the general reader, and I shall content myself with mentioning one feature which was new even to me, namely, how in warm, moist, and shady equatorial forests the very leaves on the trees get covered with beautiful lichens and Hepaticæ. The former show usually a whitish crust, dotted over with the black, red, or yellow shields; but there are some species which, notwithstanding their minuteness, are as perfectly foliaceous as the Parmelias and Stictas that adorn our secular oaks. The epiphyllous Hepaticæ are to the naked eye merely patches, or slender intricate threads, of a white, green, pink, or brown colour, but the lens shows them to have distinct leaves, closely and symmetrically set on to the stem in two ranks, and flowers (or perianths) of various forms, but usually pentagonal or tubular.

. . . . . .

At Caripi

When we had been at Pará a little more than a month we were glad to accept an invitation from Mr. A. Campbell to accompany his family to Caripi, one of his farms, about thirty miles away up the river Pará. We started in Mr. Campbell's galiota on the morning of August 21, with the fag-end of the ebb, which carried us down beyond the islands in front of Pará, and then the flood-tide, aided by several oars, carried us up all the way to Caripi, where we arrived well on in the afternoon. The river Pará is at least ten miles wide there, appearing more like an inland sea or large lake, and the coast of the isle of Marajó is dimly seen on the opposite side, without any intervening islet. The shore is a spacious and gently sloping beach of white sand, which at low water we could traverse in an upward direction for a distance of several miles, without any obstacle except having to ford the igarapés which here and there intersect it. At a little way up, the beach begins to be bounded by low cliffs of a ferruginous, coarse-grained sandstone in horizontal strata, the same as is to be seen near Pará, on the river Guamá and elsewhere, being, in fact, the common building stone; but great was my surprise to see also large detached blocks of a honeycombed rock, with a reddish vitrified surface, quite resembling masses of slag, and plainly of igneous origin. I saw one instance of the contact of the two rocks, where the trap had penetrated the clefts of the sandstone and partially fused it. We shall see that I afterwards came on the same sort of blocks at various points in the Amazon valley.

The estate of Caripi embraced, I believe, many square leagues, but with the exception of a small space kept open near the house for the grazing of cows and goats, and of a few mandiocca clearings away at the back, tenanted chiefly by Indian squatters, the ground was all forest. Caripi was, in fact, merely a place of convalescence for Mr. Campbell's family and friends, its salubrity depending on the dryness of the site and the cool breezes that sweep across the bay, enhanced by the facilities for bathing and the absence of carapanás. . . .

The large and commodious house had been shut up for a few months during the absence of the family, and when the room destined for my reception was reopened there appeared in the middle of the floor a heap of fresh earth near 3 feet high, as if thrown out of some newly-opened grave, but in reality the work of that great excavator and roadmaker, the saúba ant—the navvy of the Amazon valley, of whom we shall see more hereafter. Hordes of bats were disturbed and flew wildly about when the light was admitted; some of them were killed by the negroes, and the rest returned to their roosting-places in the roof. There were amongst them some that looked very formidable, being about two feet across the expanded wings, although I afterwards saw far larger ones on the Upper Rio Negro. Neither these nor the smaller kinds were known to bite; but as undoubted vampires sometimes entered the house at nightfall, it was customary, as a preservative from their attacks, to sleep with a light in the room, and this I afterwards found to be a common practice all through the Amazon. . . .

On the second or third night of our sojourn at Caripi, happening to awake a little after midnight, I saw King lying with his head out of his hammock and nearly touching the ground, while close by his ear sate a sooty imp, which from its size might be a big toad, like Eve's dream-prompter; but the lamp which burnt dimly in a corner of the room gave too little light to allow me to see clearly what it was. I leaped from my hammock, seized my terçado, sprang across the room, and as I pinned the monster to the ground, he opened wide his wings and showed himself to be a young bat of the largest kind. I had scarcely performed this feat when the two parent bats sallied forth from the roof and attacked me; and when I beat them off, they flew round and round the room, attempting to strike me with their wings every time they passed me, and I them with my terçado. By this time King was wide awake, and seeing the odd combat that was going on, but not knowing how it had originated, sat up in his hammock convulsed with laughter, in which I heartily joined.

. . . . . .

On the 24th of August we visited an Indian settlement by an igarapé, about five miles inland from Mr. Campbell's house, in order to see the manufacture of fireproof pottery, and especially the Caraipé tree, in whose bark (mixed with the clay) was said to reside the fire-resisting property. The identification of this tree had been specially recommended to me from England, where, from the similarity of the name, it had been supposed to be a species of Caraipa, a genus of the order Ternströmiaceæ. . . .

One of Mr. Campbell's mulattos accompanied us as guide. Leaving the beaten track, he took us by a short cut through the forest, along a hunter's trail, where my unpractised eyes could scarcely distinguish any semblance of a path. We reached the igarapé, which was not very wide, but as such conveniences as bridges were almost unknown in that region, we should not have been able to get across if our guide had not swum over and brought us a canoe from the other side. A few steps beyond stood the four or five cottages we were in quest of, embosomed in a grove of orange trees and plantains. I surveyed them with interest, for they were the first abodes of the dwellers of the forest I had seen, although there were some of mongrel character (like their inhabitants) in the Nazaré and other suburbs of Pará. They wore an air of neatness and comfort, and made me think of Will Atkins's house on Robinson Crusoe's island. The walls were of palm leaves, closely woven into a sort of matting. The roofs were covered with a sort of shingles, made by tying several of the broad flat fronds of a small palm called Ubím (Geonoma) on to a stick so as to closely overlap each other. A roof of Ubím looks pretty, keeps out the rain well, and lasts a long time. At a short distance was the essential mandiocca plantation, covering several acres. An old Indian pointed out to me eight or nine varieties of that most useful vegetable (the *Manihot utilissima* of botanists), each grown in a plot kept carefully separate from the rest; he professed to distinguish them by the leaves, but I confess I was unable to do so; however, there is no doubt that the roots vary much in shape and colour, some being whitish, others deep yellow; that some kinds ripen sooner than others, and that some suit best for making farinha de agua, others for farinha secca. Farinha de agua is made by macerating the mandiocca roots in water till they are soft enough to be broken up by hand.

Farinha secca is made entirely from the fresh grated roots. The former contains nearly all the starch in combination with the other nutritive constituents; but the latter has parted with most of the starch in the repeated washings and squeezings the pulp undergoes to free it from the poisonous juice. When the main object is to have the tapioca or mandiocca starch separate, the pulp of the grated root is alone employed.

I was then shown the Caraipé pottery, which comprised almost every kind of cooking utensil. It was made of equal parts of a fine clay, found in the beds of igarapés, and of calcined Caraipé bark; but in other places where I have seen the manufacture carried on (and there is no Indian's house in the Amazon valley where it is not familiar) a much smaller proportion of the bark was used. The property which renders the bark available for this purpose is the great quantity of silex contained in it. In the best sorts—such as I afterwards saw on the river Uaupés—the crystals of silex may be observed with a lens even in the fresh bark; and the burnt bark turns out a flinty mass (with a very slight residuum of light ash, which may be blown away), so that for mixing with clay it requires to be reduced to powder with a pestle and mortar. The bark I saw at Caripi is, however, much less siliceous, and when burnt may be broken up with the fingers.

. . . . . .

Having satisfied my curiosity as to the pottery, we started into the wood to see the Caraipé tree, and after much searching found one—a straight slender tree, whose height I estimated at 100 feet; and it was branched only near the summit, so that it was impossible from below to say what the leaves were like. A young Indian offered to procure them for me, and I then witnessed for the first time the Indian mode of climbing any tree not of inordinate thickness. A handkerchief is tied by the two opposite corners, or a bit of rope about 2 feet long by the two ends, or, better still, because everywhere obtainable in the forest, a ring of sipó is made of the same size. The climber, standing at the foot of the tree, puts the toes of each foot into the ring and stretches it to its full extent; then, embracing the tree with his arms—or grasping it with his hands if it be very slender—he draws up his legs as far as he can, and holding the ring tight to the tree with his feet, so as to form a sort of step, he straightens himself out and repeats the process; so that by a series of snail-like movements (I mean as to the attitudes, not the pace), he soon reaches the top of the tree. Many Indians, without any apparatus at all, will walk up a slender smooth tree, monkey fashion, especially if it lean over a little, and in this way I have seen the Tapuyas climb Coco and Assaí palms for the sake of their fruit. . . .

The Indian brought down branches of the Caraipé, but they unfortunately possessed only leaves, no flowers or fruit. Defective as they were, my dried specimens were placed in the hands of Mr. Bentham, and his vast knowledge of what may be called comparative vegetable anatomy enabled him to assign them, nearly with certainty, to the order Chrysobalaneæ, and even to indicate the genus (Licania) to which they probably belonged. I afterwards fell in with several sorts of

Caraipé trees, and was fortunate enough to gather flowers and fruits of some of them, which confirmed Mr. Bentham's opinion of their being species of Licania. The leaves are mostly like those of our apple and pear trees, although the Licaniæ are in reality more nearly related to the plum tribe (Drupaceæ), and the small sub-globose drupes are not unlike very small and prematurely-ripened peaches in their downy skin, usually painted on one side with carmine or purple, but they are very dry and scarcely edible.

— I may add here that, besides what I saw of Caraipé bark on the Amazon proper, I found it applied to the same use on the Upper Rio Negro, the Uaupés, the Casiquiari, and the Orinoco as far downwards as to the cataracts, and that I saw Caraipé ware brought from the Guaviari. In this region it is mostly known by its Barré name of Canída, and utensils of very large size, such as stills and coppers, are made of it. Finally, I saw it in use also along the eastern roots of the Andes of Peru and Ecuador, or in the ancient provinces of Maynas and Canelos, where it is called Apacharáma.

. . . . . .

During our stay at Caripi, Mr. Campbell had a small clearing made not far from the house for planting mandiocca. I examined the trees as they were cut down, and secured flowers of a good many of them. I had also two Mirití palms cut down, for the sake of truncheons of their trunks to send to the Museum of Vegetable Products at Kew. There were two forms, considered distinct species by Von Martius, viz. *M. flexuosa*, which has the fruits nearly globose; and *M. vinifera*, which has oblong fruits. The former seems confined to the submaritime region; but the latter is abundant all through the Amazon valley, and perhaps more so along the eastern roots of the Andes than anywhere else. Neither of them reappears on the western side of the Andes. One of the palms cut down (*M. vinifera*) measured 80 feet to the top of the fronds; the trunk as far as to the base of the fronds being 71½ feet long by near 16 inches diameter. Each of the fan-shaped fronds was 9½ feet across, and its stalk or petiole was a pole 13 feet long and thick in proportion, so that a single leaf or frond was no light load under a hot sun; but the spadix of fruits was a heavy load for two men. These Mauritias formed a large grove at the mouth of an igarapé, and along the adjacent white beach. The two specimens cut down were among the smallest that bore perfect fruit, but some of the others would be at least half as high again, or say 120 feet. Viewed by moonlight, the effect was indescribably grand and striking, reminding me of the lofty pillars and "high embowered roofs" of the cathedrals of my native land.

Visit to Tauaú

On the 4th of September we left Caripi for Tauaú, another of Mr. Campbell's farms, whither he and family had gone before us. Our way lay down the main river to the mouth of the Guajará, then up this river and its tributary the Acará to a little above the junction of the latter with the Mojú.

. . . . . .

Tauaú is said to take its name from abounding

in clay, of which advantage had been taken to establish a pottery, where all the coarser kinds of earthenware were manufactured on a large scale. As the clay was apparently of very good quality, the proprietor had two or three times tried to produce glazed crockery, and with that view had got out skilled workmen from Europe and North America; but either the materials or the workmen were not so good as they were supposed to be, for the project did not succeed. The pottery of Tauaú is, however, famous throughout the Amazon, and I recollect seeing large waterpots, with "Tauaú" stamped on them, even on the Casiquiari in Venezuela, whither they had been taken from Pará probably with wine or cachaça, and would be sent down thither again full of turtle oil or balsam capivi.

The pottery and the clay-pits occupied a low marshy flat which extends down the river for several hundred yards; but in the port, where we landed, the ground rose abruptly from the water-side, and a flight of steps led up to the house, which stood on a terrace some 60 feet above the river. At the back was a considerable extent of open pasture, reclaimed from the forest, rising on one side into positive hills, whereof the highest point might be 130 feet high. By a broad road leading from the house across the campo there were rows of fine young Castanha trees (*Bertholletia excelsa*, H. B. K.), on which grow the well-known Pará or Brazil nuts of commerce, called in their native country castanhas or chestnuts. These trees had been planted by the Jesuits, the founders and former possessors of Tauaú.

The Primeval Forests

At Tauaú I first realised my idea of a primeval forest. There were enormous trees, crowned with magnificent foliage, decked with fantastic parasites, and hung over with lianas, which varied in thickness from slender threads to huge python-like masses, were now round, now flattened, now knotted, and now twisted with the regularity of a cable. Intermixed with the trees, and often equal to them in altitude, grew noble palms; while other and far lovelier species of the same family, their ringed stems sometimes scarcely exceeding a finger's thickness, but bearing plume-like fronds and pendulous bunches of black or red berries, quite like those of their loftier allies, formed, along with shrubs and arbuscles of many types, a bushy undergrowth, not usually very dense or difficult to penetrate. The herbaceous vegetation was almost limited to a few ferns, Selaginellas, sedges, here and there a broad-leaved Scitaminea, and (but very rarely) a pretty grass (Pariana), whose broad leaves set on closely in two ranks quite resemble the pinnate frond of a palm, to which family there is a positive approach in the spikes of large polyandrous flowers. In some places one might walk for a considerable distance without seeing a single herb, or even rarely a fallen leaf, on the bare black ground. It is worthy to be noted that the loftiest forest is generally the easiest to traverse; the lianas and parasites (which may be compared to the rigging and shrouds of a ship, whereof the masts and yards are represented by the trunks and branches of trees) being in great part hung too high up to be much in the way; whereas in woods of recent growth (caapoera), and in the low gapó that sometimes skirts the rivers, they have not yet got hoisted high enough to allow one to pass beneath them, but bar the way with an awful array of entangled, looped, and knotted ropes, which even the sword itself can sometimes with difficulty unloose.

The noblest trees in the forests of Tauaú were the Bertholletiæ, and one specimen was perhaps as large a tree as I have anywhere seen in the Amazon valley. Its nearly cylindrical trunk, not at all dilated at the base, measured 42 feet in circumference, and at 50 feet from the ground it seemed almost fully as thick. It began to branch at about 100 feet, so that its crown rose high above the surrounding trees, but I could not see it distinctly enough to be able to form an idea of the entire height. I suppose the Bertholletiæ and Eriodendra (Silk-cotton trees, in Lingoa Geral, Samaúma) to be the loftiest trees in the Amazon valley; but I unfortunately never saw an entire trunk of any well-grown specimen prostrate; I was, however, assured by the Messrs. Campbell and others that trees of these genera had been measured and found full 200 feet long. In the forest at the back of Pará I measured a fallen, leafless tree (genus and species unknown) which was 157 feet long, and when the top was entire it might have been 10 or even 20 feet longer. On the Rio Negro I have cut down and measured so many trees, including some of the very largest, that I possess data for deducing very accurately the average and extreme heights of the forests in various parts of that region; but truth compels me to admit that I have never anywhere measured a loftier tree than that at Pará. In height, then, the forest trees of the Amazon must yield to the pines of North America and even to the gum trees of Australia.

Whilst on this head, I may say a word about the height of palms. Humboldt having seen, at two or three points of his South American journey, the crowns of palms standing so completely above the surrounding forest as to give (to use his own words) the idea of *a forest above a forest*, that has been rashly assumed by some writers as a universal characteristic of South American palms. A traveller approaching by sea the cities of Guayaquil, Panama, and many others within the tropics, will see groves of Coco palms towering far above the bushy spreading Mangoes and Guavas (Ingas) that nestle at their base; but the latter are by no means forest trees, nor is the Coco a forest palm. Let him, however, leave the coast and penetrate the virgin forest beyond, and he will see that the loftiest palms do not usually exceed the exogenous trees of average height; and that, except on the river-banks, they are often quite hidden from view until closely approached. From the bald granite hills of the Rio Negro and Orinoco, and from some of the Lower Andes, I have looked over perfect oceans of forest, and am able to assert that very rarely do palms domineer over all other trees; so rarely, indeed, that I believe I have only noted it twice, and then on a very limited area, during the whole course of my travels. On the contrary, the foliage of a grove of gregarious palms, such as the Piassaba and the great Caraná, is usually depressed below the top of the surrounding forest.[1]

Buttresses

A brief sketch of the most marked types of vegetation observed at Tauaú, Caripi, and Pará may perhaps be found interesting, and will serve as a standard of comparison in treating of the aspects of nature in other regions. To begin with the forest trees. Almost the first thing that strikes the observer is the enormous dilatation at the base of many of the trunks, in the shape of broad, flat, subtending buttresses, more or less triangular in outline, and rarely exceeding 6 inches in thickness, set around each trunk to the number of from four to ten. These buttresses are really exserted roots, or, as the Indians correctly call them, sapopemas (*sápo*, a root; *péma*, flat); and among European trees the lime perhaps shows them most distinctly, but on a vastly smaller scale than in many Amazon trees, where they often

[1] In faithfully recording my own experience, I have no thought of impugning the testimony of other, and no doubt equally conscientious, observers. Humboldt and Bonpland assure us that they saw Wax palms (*Ceroxylon andicola*) 180 feet high in the cool forests of the Andes of New Grenada, and therefore, no doubt, surpassing every other tree in their neighbourhood. Dampier, in his graphic account of Campeachy, says: "As the [Silk] Cotton is the biggest tree in the woods, so the Cabbage tree [or palm] is the tallest; the body is not very big, but very high and strait. I have measured one in the Bay of Campeachy 120 feet long as it lay on the ground, and there are some much higher. . . . Those trees appear very pleasant, and they beautify the whole wood, spreading their green branches above all other trees" (*Travels*, i. p. 165). Here he plainly speaks of the appearance of the forest from the sea, and his testimony does not contradict my own; for I concede that the low forest, such as usually grows at the swampy head of bays, and along inundated river-margins, is overtopped by Cocos, Mauritias, and other maritime and riparial palms.

extend on the ground to 15 feet from the base of the trunk, and the same distance up it—indeed, I have occasionally seen a sapopema stretch upwards to a height of 50 feet before it fairly ran out. A slight roof of palm leaves being made to rest on the hypotenuses of two adjacent sapopemas, the intermediate space has often served me as a temporary hut. An idea of their size may also be formed from the fact that I have seen a table-top, in a single piece, 8 feet long by 4 feet wide, cut out of a sapopema, and in the Andes of Maynas I once saw a circular tray of the same material very nearly 6 feet in diameter. Sometimes they fork once or oftener before plunging into the ground, and sometimes they are free beneath towards the centre, so as to present a combination of arch and buttress. Not infrequently they are fantastically twisted, and the outer edge may be either straight or bulged outwards; but in all cases their woody fibre is in a state of extreme tension, so that on striking an axe or cutlass suddenly into them they give out a sound like the breaking of a harp string. On examining attentively trees which have sapopemas notably developed, it will be found that they have no central or tap root at all, nor do the lateral roots dip deep under the soil. It is clear, indeed, that the roots of lofty trees which do not take deep hold of the ground must either run a long way on or near the surface, of which we have an example in the spruce fir (*Abies excelsa*), or else must extend vertically as well as horizontally, so as to perform the office of both buttresses and stays, as in the sapopemas of which we are treating. When I afterwards explored the great granite region of the Rio Negro and Orinoco, and saw there tall trees growing on perfectly bare rock, or where the earthy covering was only a few inches thick, so that the roots were necessarily either wholly or in great part above the surface, I understood how sapopemas might have originated; for those peculiarities or seeming anomalies of structure, which we (to hide our ignorance) are too often contented to call "freaks of nature," have no doubt arisen, in the first instance, from the adaptation of organisation to the accidents of existence, and have been continued through descendants of the original stock even when no longer exposed to the influence of such accidents.

A few sorts of trees, including some palms, are supported on exserted or superterraneous roots, which differ only in that particular from ordinary subterraneous roots, that is to say, they are round or cylindrical, and not flattened and dilated vertically like the sapopemas. In England, an old willow or other tree standing by a river, whose floods have washed away nearly all the earth from its roots, may give an idea of this form; which, however, is constant in many Amazon trees whose roots have never been exposed to denudation by the action of water, whatever may have been the case with the prototypes of those trees. These examples led me to conjecture, at first, that the sapopema form itself might have taken its rise from denudation, in the remote ancestors of the existing types of trees; or at least that sapopemas were at first a sort of scaffolding to raise the *crown* of the root above the reach of inundations; and I am still willing to believe that to this cause their origin may be partly traced. But that it has not been the sole, or even the principal, cause is plain from the fact that of the trees flourishing at the present epoch in inundated grounds, where their trunks are under water for months to the height of 10, 20, or more feet, very many have no sapopemas at all. As suggested above, a rocky matrix, bare or thinly covered with earth, may have been the main origin of sapopemas, for it is in such sites that the most numerous and perfect examples of them exist at this day; and if we suppose it combined with inundations and denudations, I think we may thereby explain most of the modifications of exserted and dilated roots.

Sapopemas exist on trees of many genera and families, but they seem to attain their greatest size in Bombaceæ, Leguminosæ, Lecythideæ, Moraceæ, and Artocarpeæ. There is, however, one family, Lauraceæ, consisting almost entirely of forest trees, yielding to no others in their noble aspect and the usefulness of their products, in which I have never seen more than a rudimentary development of sapopemas; their roots, in fact, penetrate deeper than most others, and wherever laurels predominate it is a sure indication of a good depth of soil. There are instances in a single family of some species of trees having large sapopemas and others none at all; as in Lecythideæ, where the gigantic Bertholletia buries its roots almost entirely, and the species of Lecythis, some of which are trees of vast size, have the roots raised high out of the ground.

Aerial Roots

In Moraceæ, especially in the parasitic (or, properly speaking, epiphytal) fig trees, we have another type of sapopemas, whose origin is plain enough. The excrement of a bird, containing seeds of figs on which it has fed, falls on the fork of a tree, or even on the bare trunk or branches, to which it adheres; there a seed germinates, and as its stem grows upwards, its root, in the form of a broad plate—soon enlarging into a sheath, if the mother tree be slender—pushes downwards, diverging a little from the vertical on all sides, and dividing into a number of forks, seeks the ground. If the height be great, the forking is repeated several times, giving the appearance of so many pairs of marauding legs descending from the upper part of a habitation, to which they had gained access one does not at first see how, and feeling for the ground with their toes. Having reached the ground, they plunge therein, increase rapidly in breadth, by the addition of matter to their outer edge, but scarcely at all in thickness, so as to form plank-like buttresses, and the parasite having thus gained an independent footing, straddles over the too often lifeless trunk of the friend whom he has crushed to death in his embrace, when his support is no longer needed. In both the eastern and western roots of the Andes, the trees which have the largest sapopemas are mostly figs. In the plain of Guayaquil, figs are the giants of the forest; and it is notable that when they grow upon any exogenous tree, they soon squeeze it to death; but if on a palm, the latter resists the pressure, and seems mostly to live out the natural term of its existence. There is a noble Attalea, in particular, which is often seen growing (as it were) out of a gigantic fig tree, but in reality the fig has grown on the palm. Parasitical trees of whatever family (Moraceæ, Clusiaceæ, etc.) are in Spanish America expressively called Mata-palos or Tree-killers. On the Amazon I never heard any collective name for them. Only a few figs grow precisely in the mode described above. Others have winding branched roots, which inosculate with each other, and embrace the trunk of a tree in a firm network which effectually prevents its further growth and eventually strangles it. Others again send down to the ground rope-like roots, at first slack and supple, but soon becoming taut and rigid. The way in which a fig supplants a Silk-cotton tree in Jamaica is an admirable illustration of this mode. "A small plant of a fig establishes itself in a rent of the Cotton tree, and throws down a root to the ground, which becomes stretched as taut as a violoncello string, and carries up nutriment to the little plant above, which drops stronger and larger and more numerous roots till it has enveloped the Cotton tree and choked it; and insects do the rest" (Dr. R. C. Alexander in *Hook. J. Bot.* 1850, p. 283). See also a graphic account of the mode of growth of the Banyan fig in Dr. Hooker's *Himalayan Journals*, chap. xxvii.

Forms of Trunks

Were I to unite all my observations on this head, I should be led on to write a complete treatise on the Physiognomy of Plants, which is by no means my intention here; I will therefore sketch briefly such other traits as are most worthy to be noted. The trunks of the trees, except for this occasional dilatation at the base, do not actually depart from the normal tapering cylinder. There was, however, one form at Pará, resembling a clustered Gothic pillar, as if one thick trunk had been formed by the union of several slender ones, whose flowers and fruit I could not obtain; nor did I ever afterwards meet with it, so that I am unable to refer it to its genus, or even family. Another form—a deeply furrowed trunk, here and there positively perforated, so that birds and small monkeys could creep through the holes—I afterwards found to belong to the genus Swartzia, of the leguminous family. Not every species of Swartzia, however, has a perforated trunk, and the most notable example of that peculiarity I have seen was in a beautiful species (which I have called *S. callistemon*) abundant on the Upper Rio Negro.

Everybody knows that the trunks of palms are ringed, each ring being the scar of a fallen leaf; and that the trunks of bamboos are both ringed and jointed, a diaphragm being stretched across the internal cavity at each joint. There are at least two genera of exogenous trees, Cecropia and Pourouma (of the family of Artocarps or Breadfruits), which have the latter peculiarity. On the lower Amazon the cavities are often taken possession of by ants, but in the roots of the Andes by bees, which afford great store of wax to the inhabitants of Maynas.

Varieties of Bark

The bark of the trees is usually smooth, or so nearly so that at a short distance the shallow clefts are undistinguishable; and I have seen no instance in the plains of trunks so picturesquely rugged as those of our old oaks and elms. In many trees the bark peels off in flakes, as, for example, in all arborescent myrtles, in somé Leguminosæ, Rubiaceæ, etc., so that their trunks vary in colour according to the season of the year, being green or olive when they have just shed their old bark, and afterwards turning reddish or brownish.[1]

Some trees have a bark which admits of being split into an almost indefinite number of thin flakes or sheets. Those species of Tecoma (of the order Bignoniaceæ) which have digitate leaves, afford the most perfect examples of this property; and strips of their bark, made up into rolls, are commonly sold in the towns on the Amazon, under the name of Tauarí, as a substitute for paper in the fabrication of cigaritos.

The bark of the Lecythids may be beaten out into a loose mass resembling tow, and is excellent material for caulking seams.

Other barks when beaten out form compact felty sheets, hanging together by the tenacious fibres;

[1] The trunks of large trees, especially near rivers, get sometimes completely encased in the white crust of lichens (chiefly species of Graphideæ); and there is one tree frequent on the banks of the Amazon whose Indian name Miratinga (= White tree) indicates the constant snowy whiteness of its trunk.

[This is probably the origin of the pure white colour of the stems and branches of some of the poplars and plane trees in the United States and Canada, which causes them to look exactly as if whitewashed.—ED.]

they are called Tururí, and are afforded by various figs and Artocarps. I shall have occasion to describe the uses of Tururí when I come to treat of Indian life in the far interior.

Lianas

Of all lianas, rope-plants, or sipós (as they are called in Tupi), the most fantastic are the Yabotím-mitá-mitá or Land-turtle's ladders, which have compressed, ribbon-like stems, wavy as if they had been moulded out of paste, and while still soft indented at every few inches by pressing in the fist. They are usually not more than three or four inches broad, but I have sometimes seen them as much as 12 inches; and they reach two or even three hundred feet in length, climbing to the tree-tops, passing from one tree to another, and often descending again to the ground. They belong to Schnella, a genus of Leguminosæ, and are found all through the Amazon valley. The commonest species near Pará is *Schnella splendens*, Benth.

Lianas of the family of Bignoniaceæ may generally be recognised by their four- (rarely six-) angled stems, the angles being usually obtuse, but sometimes sharp-edged or even winged. At intervals of a few feet the stems have swollen joints, ancient leaf-scars. One of the most gorgeous sights I ever saw was, where a gap having been made in the forest by cutting down some trees, a Bignonia with several parallel stems, which had run lightly over their tops, was left suspended between two lofty trees, 40 yards apart, in a graceful catenary, clad throughout its length by roseate foxglove-like flowers and large twin leaves of a deep green tinged with purple.

Birthworts (Aristolochiæ) are notable for their thick bark, cloven down to the woody axis in six or more furrows. When cut across they give out a strong smell, usually rather fetid, but in some cases pleasantly aromatic. They are scarce in the plains of the Amazon valley, and their singular hooded and often lurid-coloured flowers are difficult to find.

The great majority of lianas, however, have more or less rounded stems; and there is scarcely any family of plants which does not include some members who get up in the world by scrambling upon their more robust and self-standing neighbours. Where two or more of these vagabonds come into collision in mid-air, and find nothing else to twine upon, they twine round each other as closely as the strands of a cable, and the stronger of them generally ends by squeezing the life out of the weaker.

Many lianas are furnished with hooks, which not only aid them in climbing, but are also formidable defensive weapons. The Sarsaparillas (species of Smilax) are the analogues of our brambles, ramping vaguely about, but never up to a great height; and they have either roundish sparsely prickly stems or three-cornered stems whose angles are thickly set with prickles. Sometimes they trail insidiously on the ground, where their presence is only revealed by the wounded foot that treads unwittingly upon them. The Yurupari-piná or Devil's fishing-hooks are leguminous climbers of the genus Drepanocarpus, with broad curved prickles in place of stipules. The Unha de gato or Cat's claw (*Uncaria guianensis*) has long, tough, hooked prickles, capable of sustaining very heavy weights. Both these lianas grow chiefly on the banks of rivers, and are a serious impediment to navigation where the current is strong and canoes must necessarily creep up close inshore. On the rivers entering the Gulf of Guayaquil the Uncaria is still more abundant than at Pará, and there have been instances of a man being caught up by its formidable hooks and suspended in mid-air, whilst the raft on which he was shooting down the stream floated away from under him. But of all river-side lianas, the most to be dreaded are the Yacitára twining palms of the genus Desmoncus, the terminal pinnæ of whose leaves are abbreviated to rigid spines pointing backwards like the barb of an arrow. As the canoe shoots by or under an overhanging mass of Yacitára, woe to the unlucky wight who is caught by its claws, which infallibly tear out the piece they lay hold on, whether it be flesh or garment, or both.

In the virgin forest one not infrequently sees a plant curiously flattened to the trunks of trees, and at a distance looking more as if it were painted than as if it grew thereon. The leaves are from 1 to 3 inches long, closely and symmetrically set on to the stem in two rows, oval, with a rounded apex and a heart-shaped base, of a deep velvety green beautifully netted with the white veins. It is the young state of *Marcgraavia umbellata*, and is so totally unlike the mature state, that it was only by actually tracing the union of the two forms I could satisfy myself of their identity. The stem puts forth here and there clasping roots, which adhere to the tree, or even completely embrace it, if slender; but when it has climbed up into the light it sends out stoutish free branches, clad with long sharp-pointed leaves of a uniform green, and the painted stem-leaves fall away. The common ivy is a familiar example of a somewhat similar mode of growth, coupled with dimorphous leaves.

It was at Pará that I first saw an enormously thick liana, sometimes near a foot in diameter, that wound in a regular spiral up the trees; but often as I tried to trace its upward progress, I always found it intercepted at some height up by an epiphytal Clusia, beyond which I could not distinguish it among lianas of various kinds that had accompanied it from the ground. It was not until I had been accustomed to see it occasionally for some years that I ascertained it to be really the descending axis of the said Clusia, several species of which possess that peculiarity, although the stem or ascending axis never twines, and the whole family of Clusiaceæ or Guttifers has very few true lianas.

Many lianas secrete abundance of fluid sap, usually milky and acrid in Apocynes and Asclepiads (which include a large proportion of all twining plants), turbid and virulently poisonous in some Paulliniæ, but sometimes limpid, sweet, and harmless. The Indians profess to know several lianas whose juice affords a copious and wholesome draught, but I could never trust myself to drink of any but the Dilleniaceæ, chiefly of the genus Doliocarpus. For this purpose it is not sufficient merely to sever the liana, when only a small quantity of fluid would gush out, but it must be cut simultaneously at two points a few feet apart, and the ends of the severed piece held at the same height; then when one end is slightly lowered the liquid runs out in a gentle stream, and may be thus conveniently drunk.

Besides the lianas which twine around and spread from tree to tree, there are others which hang vertically like bell-ropes. These are air-roots of epiphytal Aroids and Cyclanths, and, like the stems of twining lianas, are often armed with prickles or tubercles. As they descend, they send forth rootlets from the point, which finally reach the ground and fasten themselves therein.

Epiphytes and Parasites

This brings us to the consideration of the epiphytes and parasites, which roost in the forks and on the branches of trees, sometimes in such numbers that their foliage and that of the lianas quite hide the leaves of the trees whereon they sit and hang. The common *Arum maculatum* of our hedgerows may give an idea of the aspect of the Aroids, supposing the leaves to be very much magnified, sometimes fantastically jagged or perforated, and in some instances tinged with purple or violet beneath; but some species have long lanceolate or strap-shaped leaves, so as to simulate certain ferns that grow on trees, not unlike our Hart's-tongue fern. The Cyclanths grow, like the Aroids, either in enormous tufts or with succulent creeping stems; but they have broad bifid, or sometimes fan-shaped, leaves. Along with them inhabit multitudes of Bromels, including several species of Tillandsia, with whose aspect our conservatories have made us familiar; and other plants looking exceedingly like the pineapple, but often of gigantic proportions. The viscid sheathing bases of their leaves retain the water of rains, for whose sake these plants are much resorted to by ants; and the stings of these little animals, along with the pungent point and thorny serratures of the leaves, render it by no means agreeable to stumble against a Bromel. On ants' nests, especially those of a sort of termite—large black globose or shapeless masses stuck up in the trees—grow succulent Peppers (Peperomiæ) and a few Gesneriads. Orchids are far scarcer and their flowers usually less showy in the dense forests of the Amazon than in other regions with a similar climate, but where the trees are lower and more scattered. They seem also to avoid trees with pungent or acrid resinous juices, such as the Clusiads, Amyrids, Artocarps, etc., which abound in the Amazon valley. Loranths or Mistletoes, so far as I have seen, absolutely refuse to grow on trees of that kind, which is explicable by their roots not merely clinging to the bark, but actually penetrating the wood and sucking their subsistence thereout. Many of them resemble the common mistletoe in aspect and in their inconspicuous flowers, whose homeliness is, however, often redeemed by their exquisite perfume; but others have showy tubular scarlet or yellow flowers often several inches long. Ferns also of many species—some of them so delicate in texture and so finely divided as to be the most light and graceful of all plants, others rigid and simple in outline—enter into the catalogue of epiphytes; which is closed by mosses, lichens, and a few bark- and leaf-loving fungi.

The Various Forms of Leaves

It will be objected against me that I have described the mere skeleton of the arborescent vegetation, and that the foliage and flowers that adorn those wonderful trees and lianas appear to have been forgotten. In reality, the parasites we have just been considering, and the lianas, do often hide not merely the leaves of the trees, but even their mode of branching, which, when it comes to be examined, is found to be in many cases exceedingly regular and even geometrical, giving rise to a symmetry of outline that can only be appreciated in trees that tower above their neighbours, as in the dome-shaped Silk-cotton trees and some of the taller Nutmeg trees. In looking upward at the fretted leafy arches that span the space between the pillar-like trunks, and are projected on the vault of heaven beyond, the first impression is that the leaves are much smaller than they really are, from under-estimating the height at which they are hung; and, on the contrary, the first sight of the finely divided Mimosa type of foliage at the summit of a lofty tree is apt to exaggerate the apparent height of the latter. But where the foliage can be seen sufficiently near, as on the banks of a river or along a broad forest-path, the general impression to a casual observer would be of massive glossy leaves, intermixed with light feather-like leaves of Mimosas and other similar plants, and with the gigantic plumes of palms; while the botanist would be struck with the

FIG. 1.—INDIA-RUBBER TREE (*Hevea brasiliensis*).
Young trees in a clearing, showing the trifid leaves.

wondrous diversity of forms, two trees of the same species scarcely ever growing side by side. And yet the types of foliage are often very few, rendering the *tout ensemble* exceedingly monotonous; for the leaves of a large proportion of Amazon trees are ovate or lanceolate, leathery, smooth, and entire at the margins. The Laurel type—lanceolate, glossy, entire leaves, with few acute curved veins anastomosing far within the margin, whereof our Sweet Bay may be cited as an example—abounds in the Amazon valley, and includes not only the leaves of all true Laurels, and of scattered species of various other orders, but also the leaflets of many pinnated leaves. It has been sometimes as rare a treat to me to see a deeply-divided leaf as a deeply-furrowed bark; strongly cut, jagged, or sinuated leaves, such as those of our hollies, hawthorns, maples, oaks, etc., being seldom met with on the forest trees. A few trees, however, chiefly of humble or moderate size, such as the Papaws and Cecropias, have enormous leaves, lobed or deeply cloven into finger-like divisions. Others have, like our horse-chestnut, several leaflets (five, seven, or nine) springing from the apex of a common foot-stalk; and these are some of the noblest trees of the forest—Silk-cottons (Bombax, Eriodendron, Ochroma), Bow-wood trees (Tecoma), etc. The leaflets are reduced to three in the India-rubber trees (Siphonia), the Souari-nut trees (Caryocar), and in a multitude of Papilionaceous and Bignoniaceous lianas; although in the latter the middle leaflet is often replaced by a tendril. Lobed or jagged leaves or leaflets are less rare among lianas, especially the herbaceous ones; thus they exist in many Gourd-plants, Passion-flowers, and in some genera of Sapinds (Cardiospermum, Serjania), whose leaflets often recall those of our Lady's-bower. Then there are the pinnated leaves, like those of our ash, walnut, and mountain-ash, possessed by a great proportion of the extensive family of Leguminifers, by Terebinths, Simarubes, etc.; and the bipinnate leaves of several genera of the Mimoseous sub-order, to which the arborescent vegetation of our northern climes affords no parallel. Some genera of Leguminifers have the leaflets reduced to a single pair, so as to look like twin leaves, such as the enormous trees called Yutahí (Hymenæa, Peltogyne), that yield the Brazilian copal; and in some genera (as Bauhinia among trees, Schnella among lianas) the twin leaflets are actually united through part of their length so as to form one cloven leaf, and thus resemble an ox's hoof, whence the Portuguese name Unha de boy.

One of the most marked types of foliage is that of the Melastomes—an order exceedingly abundant in species and individuals, and constituting a large proportion of the undergrowth of all forest, both recent and primitive, but never rising to be lofty trees. These all have opposite leaves—often of considerable size, and sometimes downy or shaggy—traversed by three, five, or seven stout ribs, which are united by transverse, closely-set, parallel veins, giving them a remarkably neat and geometrical appearance. Their near allies, the Myrtles—in some situations almost equally abundant—look very different, from their smaller, glossy, ribless leaves, beset with transparent dots; and, in fact, are always recognisable from their great similarity to the common European myrtle.

There is a type of trees in the Amazon valley—sparingly represented near Pará—with the stems either quite simple or emitting only a few long wand-like branches, which are naked except at the continually lengthening apex, where they bear a few crowded leaves, often of such enormous length as to give them at a distance the aspect of palms. In their season, flowers spring from the naked trunk or branches, generally in clusters, and often noticeable from their size and beauty, as in *Gustavia fastuosa*, which has the large rose-like flowers sometimes 7 inches across. Some of the handsomer Melastomes (Bellucia, Henriettea) are of this type.

Nearly every shade of green is observable in the hues of the foliage, the deepest being usually in the large glossy leaves of Guttifers and the opaque leaflets of Ingas. There are no autumnal tints on the Amazon, for although some leaves turn reddish or brownish with age, the change is very far from being simultaneous in all the leaves of the same tree, and is often entirely hidden from view by the unceasing growth of new leaves. But the absence of that periodical ornament of our northern woods is almost compensated for by the delicate hues of rose and pale yellow-green assumed by the young leaves at the growing point of the branches, contrasting admirably with the deep green of the rest of the tree.

Many leaves are grey or hoary beneath, as in the Cecropias (or Imba-úbas, as they are called by the Indians), and some leaves are clad on the underside with a fine down of a lustrous metallic hue—silvery, coppery, or bronze—especially notable in various Laurels and Chrysobalans. When my boat has been floating lazily on the water, under a burning and dazzling sun, with not a breath of air stirring, I have sometimes—as my eye wandered along the endless forest-margin—given vent to some such exclamation as this: "How tame and monotonous! was there ever seen elsewhere a mass of foliage of such wearisome sameness!" when the coming on of a squall, by simply revealing the glowing tints of the underside of the leaves, has in a moment waked up the scene into life and beauty.

The Flowers of the Tropical Forests

And now a word about the flowers. Were a naturalist to combine into one glowing description all the gay flowers, butterflies, and birds he had observed in any part of the Amazon valley, during a whole year, he might no doubt produce a most fascinating picture; which would, however, utterly mislead his readers, if they were thereby led to suppose that even a tithe of those beautiful objects were ever to be seen all together, or in the space of a single day. Very much depends on seeing any particular site exactly at the time when its most showy plants, insects, or birds are in greatest perfection or profusion; and the effect is always modified by the peculiar tastes of the observer. To the naturalist, the mere fact of an object's being new and strange invests it with a conventional beauty, independent of all æsthetic considerations; and for myself I must confess that, although a passionate admirer of beauty of form and colour, and with a most sensual relish of exquisite odours, I recall with the greatest zest those scenes which yielded me the greatest amount of novelty. But we are again losing sight of the flowers, and, sooth to say, the flowers of Amazonian trees are often so inconspicuous, either from their minuteness or from their green colour assimilating them to the leaves, that none but a botanist ever would see them. There are doubtless many glorious exceptions; but it was not until some years after I had left Pará, and had penetrated to the northern border of the Amazon valley, that I realised my preconceived notion of the loftiest trees of the forest bearing the most gorgeous flowers. At Pará the Leguminifers and Bignoniads, both as trees and as lianas, outshine all other orders in the abundance and beauty of their flowers. Of the former, the taller-growing Cassias and a Sclerolobium are crowned with a profusion of golden flowers; but far more elegant are the large pure white prince's-feather-like flowers of the Bauhinias, coupled with their odd hoof-like leaves. Of the latter order (Bignoniads), the lofty Tecomas are, in their season, one mass of purple or yellow, from the abundance of large foxglove-like flowers, often unmixed with any leaves. The showy white or red gum-secreting flowers of the Clusias and other Guttifers, and their ample glossy, rigid leaves, are sure to attract the botanist's early attention. Some Tiliads (Molliæ sp.) are studded with large star-like white flowers, as striking in their way as the gaudy stars of the passion-flowers that spangle the liana-curtain skirting the rivers. Most novel to the European botanist are the curious leathery, dull-coloured, but often richly-scented flowers of the

Anonads or Sour-sop family; the large white or roseate flowers of the Lecythids or Monkey-cups, notable for the stamens being borne on a large hooded receptacle in the centre of the flower; and the flowers of the Bombaceæ (Eriodendron, Bombax, etc.), with strap-shaped sepals and petals in some instances little short of a foot in length, beyond which hang the still longer bundles of thread-like white or rose-coloured stamens.

Many Myrtles and Melastomes bear a profusion of small white flowers, not equal in beauty to those of our hawthorn, but producing the same general effect. They are remarkable for the suddenness and simultaneousness with which the flowers burst forth and in like manner fade and fall away. On looking over the woods of recent growth in the early morning, they may sometimes be seen studded with patches of white—the flowery crowns of Myrtles and Melastomes—where all was of a uniform green the day before; and a day or two later the patches will have assumed the dinginess of decay.

Of all families of plants—excepting perhaps Leguminifers—Rubiads seem to occupy the principal place in the Amazon valley, from the shores of the Atlantic to the crests of the Andes. They are always easily recognisable by their opposite entire leaves, with interposed stipules; and by their tubular flowers. The latter are often of extraordinary beauty, and, coupled with the great importance to man of the products of many Rubiads—for where else do we get stimulants so precious as coffee and quinine?—render these plants surpassingly interesting to the traveller. . . . We have nothing in our flora comparable to the large tubular flowers of some Posoqueriæ and Randiæ; but the general aspect of the flowers of many species is not ill conveyed by the lilac; others are exceedingly like the honeysuckle; and the small-flowered species of Lonicera have almost exact analogues among Rubiads. Our privet well represents the bushy Psychotriæ, which, along with Melastomes, abound everywhere as undergrowth.

An English botanist misses altogether in Amazon forests most of the familiar forms of his native land; he sees no firs nor yews; no catkin-bearing trees, except a solitary willow; no heaths, roses, berberries, Crucifers, Umbellifers, etc.; but he comes on the relatives of many old acquaintances in an entirely unexpected garb. Violets, for instance, grow to be trees or woody twiners, mostly with small flowers; but their analogues, the Vochysiads, besides being among the noblest of trees, have large richly-coloured and sweetly-scented flowers. Milkworts, whose flowers always bear an unmistakable resemblance to the *Polygala vulgaris* of our moors, are sparingly represented by a few minute herbs; but far more copiously by robust woody twiners (Securidaca, etc.) climbing to the tops of the loftiest trees, and as they descend thence hanging out garlands of purple or white flowers.

Although I have here grouped together some of the more showy types of flowers, I must repeat what I have already said, namely, that the great mass of the trees of the forest, and even many of the lianas, bear inconspicuous flowers; such vast orders, for instance, as Laurels and Terebinths; and others, like Chrysobalans, exceedingly numerous in individuals in the Amazon valley, have scarcely a single example of large or gaily-coloured flowers.

Curious Fruits

Fruits remarkable for their size, beauty, or grotesqueness are perhaps more frequent than handsome flowers. The large pods of the Ingas have been described above, and other Leguminifers bear pods equally large, sometimes containing enormous flattened beans as big as the palm of the hand. The pods of Bignoniads are filled with closely-packed flattened seeds, bordered by a delicate transparent wing often an inch or more in breadth. Globose heavy fruits, like cannon-balls, might seem out of place on the scraggy branches of the humble Cuyeira (*Crescentia Cujete*); but are far less dangerous there than when hung on the lofty Castanheira (Bertholletia), falling from which they often bury themselves in the earth, and would infallibly crack the skull of incautious biped or quadruped that intercepted their descent. The Bertholletia has an exceedingly thick woody shell, without valves or any other natural opening, so that the seeds (Brazil nuts) can only escape from it when it finally rots away; although rodent animals, such as agoutis and pacas, and monkeys often make a forcible entry when it is partially decayed. The fruits of the allied genus Lecythis have, however, a curious convex lid, which comes clean off the cup-shaped capsule when ripe, permitting the ready dispersion of the seeds; hence their Indian name Macacarecuya or Monkey-cups. We have nearly the same thing on a small scale in the pepperboxshaped capsules of our pimpernels. Fruits of many Malpighiads, Polygones, etc., have gaudy red wings to them, so that at a distance they look more like flowers. But the most extraordinary instance of fruits simulating flowers is afforded by the capsules of some Silk-cotton trees, which burst open with valves in a stellate manner, disclosing the beautiful cotton puffed up into a globose mass, so that at a distance they look like large roses or dahlias. Many of the old missionaries, in fact, described this kind of cotton as the produce of a flower; and yet we have seen above that the real flowers of this tribe are often sufficiently large and conspicuous. Most Apocynes and Asclepiads have long spindle-shaped pods, bursting along one side and giving exit to the corrugated seeds, each tipped with a tuft of long silky down. . . . Myrtles and Melastomes are sometimes conspicuous objects from their fruits—yellow, red, or black berries varying in the different species from the size of currants to that of small apples.

. . . . . .

Palms and other Endogens

Nearly all that precedes refers exclusively to exogenous plants, but any description of Amazonian vegetation would be incomplete which should not take into account the palms, whose immense fronds are often as large as some of the smaller trees. The pinnate fronds of the Jupatí (*Raphia tædigera*) and Inajá (*Maximiliana regia*) reach sometimes 40 feet in length; and those of the Mirití (Mauritiæ sp.), etc., are, if shorter, scarcely less bulky, as we have seen at Caripi. But more striking than their size are their graceful forms and wondrous variety—qualities which only a long acquaintance enables us fully to appreciate. The flowers of palms are, it is true, comparatively small; and being usually of a pale yellow colour, are conspicuous only when massed on the large spadices of the taller-growing species; but in their exquisite odour they often yield to no flowers whatever. In many cases the odoür is that of mignonette, but I think a whole acre of that *darling* weed would not exhale as much perfume as a single male spadix of the Caraná palm of the Rio Negro. The flowers of the slender Sangapilla palm of the Peruvian Andes preserve their fine scent for months, even in the dry state; whence the Indian girls wear them in their hair, put them in their beds, and adorn therewith the altars of their household saints.

In some places the lanceolate grassy leaves, suspended from slender wiry branches, of the bamboo mingle with the leaves of exogenous trees and the fronds of palms; although, after having seen the noble bamboo groves of the Andine valleys, the low-growing, intricate, and comparatively inelegant bamboos of Pará pale on the recollection.

Where the ground lies tolerably high and dry, as in the forest at Tauaú, which I have had more particularly in my eye throughout this sketch, the ground-vegetation usually includes but few of the larger herbaceous endogens; but in low moist flats, large-leaved Scitamineæ and Musaceæ give quite a character to the scene by their abundance. There congregate the Heliconiæ, looking like their near allies the Musæ or plantains, but their flower-spikes garnished with showy scarlet bracts; various species of Maranta, Alpinia, Thalia, etc., all having foliage approaching that of the Cannæ now so much cultivated in our gardens; two or three species of Costus, looking like gigantic spiderworts, etc. A sure evidence of a patch of forest being not primitive but of recent growth, and especially of the ground having been at one time devoted to mandiocca, is a carpet of lively green, arising from a compact growth of *Selaginella Parkeri*.

Nor must we omit to mention the roots that creep and cross each other everywhere along the ground, or rise above it in buttresses, arches, or loops, which must be climbed over or under; nor the huge, rotting, reeking trunks—corpses of fallen giants of the forest—partly overrun with mosses, ferns, and lianas. Sometimes a prostrate trunk appears still sound—even the bark is entire—yet it has already been excavated by the voracious termite, so that it yields with a crash when stepped upon, probably prostrating the traveller, and not infrequently disturbing the repose of the snake or toad which has taken up its abode in the cavity.

Fern-Valleys

In traversing the forests of Tauaú, we here and there came unexpectedly on a ravine, stony at the sides, marshy at the bottom, and in some places opening out into a valley, but with no stream of running water. These ravines were perfect fern-gardens. On the stony slopes grew lofty exogenous trees, with a ground-vegetation of several species of Adiantum and Lindsæa. At the bottom was a grove of palms, chiefly of two species, the before-mentioned Assaí and the Paxiuba (*Iriartea exorrhiza*). The latter most singular palm has the trunk supported on, not a tripod, but a polypod, of exserted roots—the spokes of a half-spread umbrella may give a very good idea of them, supposing a few additional spokes to be inserted between the circumference and the axis. Each root or spoke is a rigid cylinder, some two inches in diameter, so beset with hard prickles that it may and often does serve as a grater. The fronds are shorter than in most palms, and have a graceful curl downwards; and the broad leaflets widen gradually to the extremity, where they are obliquely truncate and jagged. The yellow fruits hang in large tempting clusters, like dates, but are too bitter to be eatable—a rare exception among palms. Intermixed with the palms grew noble ferns, species of Lastræa, Litobrochia, Meniscium, Davallia, Gymnopteris, Alsophila, etc. Of the four Alsophilæ seen there, two were decidedly arborescent, having short trunks; thus showing that near the Equator tree-ferns descend almost to the sea-level. On the trunks of the palms themselves grew many species of Asplenium and Acrostichum; as also of Pleopeltis, Campyloneuron, etc., whose scaly rhizomes crept up to a height of 12 or 15 feet, and put forth at intervals lanceolate fronds beset with convex masses of fruit (sori) looking like double rows of buttons; while over both palms and ferns trailed the thread-like stems of various Hymenophylla and Trichomanes, their delicate pellucid fronds varying from a light green colour to a rich rosy brown. Up among the palm-leaves sat perched two or three species of Nephrolepis, with pendulous long riband-like fronds. I have seen in the Andes fern-valleys far more picturesque, with the adjuncts of moss-grown rocks and cataracts; but I do not think I have anywhere found more species growing together, within a small space, than in these palm-swamps at Tauaú.

Other valleys, with a moist (but not swampy) soil, which emitted a dank, disagreeable odour, were occupied chiefly by the Caraná (*Mauritia aculeata*), a fan-leaved palm with prickly stems; and ferns were all but absent.

Although we were in the height of the dry season, a day rarely passed over without rain—usually a smart thunderstorm, beginning a few hours after the sun had passed the meridian. It did not often rain by night, but one night we had a violent storm lasting several hours. The explosions kept up a continuous roll, and one vivid flash was accompanied by a crash so tremendous that I thought it must surely have struck the house, which shook to its very foundations. It was not so, but when I opened the window in the morning, I saw a tall Coco palm only a few yards away, standing without its head, which lay shattered on the ground.

This rain brought out multitudes of toads and frogs; and in walking through the forest the following morning after the sun broke forth we came on a huge toad, nearly as big as a man's head, enjoying a tranquil sitz-bath in a pool of water in the road. I knew not till then of the existence of modern batrachians so enormous. King—a stout fellow over six feet and broad in proportion—picked up a big stone, and with both hands plumped it down on the unsuspecting bather, who seemed at first rather taken aback by the insult, but after a few moments' reflection straightened himself out and marched gravely off, as if nothing in the world had happened.

. . . . . .

Vegetable Products of Pará

I ought not to take leave of Pará without adding a few words on the products of the forest that enter so largely into the consumption and commerce of that port. A complete account of their economic and medicinal uses would, however, require a separate volume; and as many of them, such, for instance, as sarsaparilla, are collected in the far interior, and are only taken down to Pará for sale and for re-embarkation to Europe and North America, I propose to mention only some of the most useful and remarkable as I come across them in the course of this narrative. . . .

One of the objects which most took my attention at Pará was the Maceranduba, Milk-tree or Cow-tree, so called from its bark secreting abundance of drinkable milk. I saw several trees of it at Tauaú, and made trial of the milk, fresh from the tree, both alone and mingled with coffee. The milk flows slowly from the wounded bark; its consistency is that of good cream, and its taste perfectly creamy and agreeable. It retains its fluidity for weeks, but acquires an unpleasant odour. It is extremely viscid, and can with difficulty be removed from the hands or whatever else it touches—a property which renders it an excellent substitute for glue, but a rather unsafe article of diet, and serious cases of constipation have resulted from its being partaken of too freely. When dried it quite resembles gutta-percha, and I have no doubt might be put to the same uses.

Almost every region of tropical South America has its Cow-tree. That of the coast of Venezuela, rendered famous by the researches of Humboldt and Bonpland, and of Boussingault, is an Artocarp, of the genus Brosimum; but this of Pará is a Sapotad, with large leaves, white beneath, close parallel veins, and edible berries, as most others of the tribe have. I afterwards fell in with two species bearing the same native name, and having quite the same habit, on the Casiquiari and Upper Rio Negro. Their milk, however, was scarcely drinkable, although it possessed the other properties of that of Pará, and was in universal use as glue. A hammock which I purchased there for the Museum of the Royal Gardens at Kew has the borders ornamented with beautiful devices in bird's feathers, all stuck on with the milk of the Maceranduba. I gathered flowers and fruit of both species, which proved them to be species of Mimusops, and therefore congeners of the Bully-tree of Tobago, and probably also of the Balata of Demerara.

All the species known to me have a deep, dull red, heavy, close-grained wood, much esteemed for its durability. I have seen a perfectly straight squared log of it, 60 feet long, brought from the

Casiquiari, and fashioned at San Carlos into the keel of a schooner. The Brazilian frigate *Imperatriz*, built at Pará in 1823, chiefly of Maceranduba, was in 1849 still perfectly sound and seaworthy.

At Pará I saw the mode of collecting the native white pitch (Breo branco), which is used there, and throughout the Amazon, for caulking seams. It is yielded by various species of the genus Icica —trees which closely resemble the sumach—and chiefly by one with a tall, clean-growing trunk which was in great request for masts. When the bark of an Icica is wounded, a white milk flows slowly out and coagulates just below the wound, which does not heal up quickly as in most milky trees, but continues to distil for several months or even years. The Indians, therefore, when they come across these trees in the forest, gash them with their terçados, in order that, when they revisit them some time afterwards, they may find a good lump of resin accumulated. Breo branco is brought to market, either in its crude state, packed in baskets lined with leaves, when it is called breo virgem, or in thick cylinders, having been run into moulds of that shape. It is whitish, friable, and exhales a strong agreeable odour. When melted and spread out over a plank or seam, it dries rapidly, and unless a good quantity of grease has been mixed with it in the melting it breaks away; but if that precaution has been taken it adheres very tenaciously, and keeps out the water much better than the black pitch or Oananí, which is obtained from a Clusiaceous tree.

Icíca is the native name for pitch in general; and the white pitch is called by the Indians Icícaritári, to distinguish it from the Yutahí-icíca or copal, which is yielded by the trees called Yutahí (Hymenæa and Peltogyne). Copal distils from cracks or incisions in the bark, and soon congeals into a hard yellowish or vinous mass, not unlike amber. The pods also generally contain pips of it, and large shapeless lumps are sometimes found at the foot of old trees, on or within the earth. It is called Aníme in Venezuela, where it has many uses: it is the best cement for mending broken crockery; an emulsion, with sugar and water, is successfully administered in catarrh and asthma; and it is burnt in churches in lieu of incense, which it much resembles in odour. For this last purpose, the powdered legumes are sometimes used, in the form of pastilles, both in Venezuela and Peru.

CHAPTER II

VOYAGE UP THE AMAZON TO SANTAREM

(*October* 10 *to November* 19, 1849)

About the time of our return to Pará from Tauaú, a vessel arrived from the interior with cargo consigned to the Messrs. Campbell. She was a brig of about 80 tons, called the *Tres de Junho*, and was owned by Captain Hislop, an old settler on the Amazon, resident at Santarem. As the ground from within a hundred miles of the coast upwards was all equally new, and Santarem was 474 miles away, and was, besides, the largest town on the Amazon, it seemed very desirable head-quarters for a campaign. My preparations for the voyage were soon made, the most important items being letters of credit to a merchant of Santarem, and a bag of copper money, weighing nearly a hundredweight, for small change. Of provisions the staple were hard-toasted bread, farinha and pirarucú (large strong-smelling slabs of salted fish from the Amazon, which only necessity and much practice can bring any one to relish). Besides these, I took a small stock of taínha, a smaller and well-tasted fish caught in the Pará river; eggs, coffee, sugar, and other lesser matters. I provided myself also with a sort of canteen, called a patuabalaio, at that period an indispensable article for a traveller. It had compartments for stowing plates, knives and forks, etc., and especially for frascos—large square bottles of the capacity of about two quarts—to hold molasses, spirits, vinegar, etc.

We embarked in the *Tres de Junho* on the 10th of October, at 9 P.M. Our course lay at first westerly, trending a little south, across the bays of Marajó and Limoeiro—the latter at the mouth of the Tocantins; then nearly west for about sixty miles, along a channel narrower than those bays, but still of considerable width, and with numerous islands on its southern side at the mouths of several tributary rivers. Still keeping the isle of Marajó on our right, we entered a narrow channel called the Furo dos Breves, on which stands the small village of Breves. Our course began now to trend a little northerly, and after crossing a deep lake called the Poço (well), we entered another channel (Canal de Tagipurú) which, after a long winding course, brought us finally into the Amazon.

The Poço was a great rendezvous of floating aquatics, detachments of which made excursions a little way up the Tagipurú with the flood-tide, then back again and a little way down the Furo dos Breves with the ebb-tide. The Tapuyas called them all Mururé, but they were made up of plants of widely distinct families, the most abundant being the common *Pistia Stratiotes*, whose foliage is not unlike that of our broad-leaved Ribgrass, although the plant is really cryptogamic and closely related to the ferns. Another Mururé was the singular *Pontederia crassipes*, which bore short spikes of pale blue flowers springing from among the roundish leaves, whose stalks became inflated and filled with air, so as to serve as floats. Another and a handsomer plant of the same family —a species of Eichhornia—with large spikes of violet flowers, has the same property; but both plants, when thrown on the muddy shore, take root there, and the swollen petioles disappear, being no longer needed.

In the wide bay of Marajó the wind blows and the waves roll almost as much as in the open sea; but in the narrow channels of Breves and Tagipurú, and amongst the islands that precede them, there is either unbroken calm or brief and uncertain winds. Then the mariner has no aid but from the tide, and, if his vessel be too bulky to be propelled by oars, must either lie by between tides or creep along by espia (the Indian word for cable) in this way. A boat, having in it a large roll of cable, one end of which is tied to the prow, or to the foremast, of the vessel, rows ahead until the cable is nearly all paid out, when its other end is fastened to a stout overhanging branch by the river-side; and the sailors, reunited on board the vessel, draw in the rope until they reach the point where it is tied. The process is then repeated, and thus a slow progress is kept up during the hours of ebb. . . . The passage of the Tagipurú occupied us five days. I went on shore thrice during that time, but found little in flower that I had not already seen at Caripi and Tauaú. The Bussú palm (*Manicaria saccifera*) abounded on both banks, and the

FIG. 2.—BUSSÚ PALM (*Manicaria saccifera*).

houses at Breves and in the scattered sitios were mostly thatched with its fronds, which are almost unique among palms in consisting of a single piece, like those of the plantain (Musa), and not of distinct leaflets; so that each frond forms a long tile reaching from ridge to eaves.

. . . . . .

We were at length fairly in the Amazon, whose muddy waters, varying from a dull yellow colour to that of weak chocolate, according to the light in which they were viewed, ran too deep and strong to be turned back by any tide, so that we must thenceforth depend entirely on our sails. We were, however, in only one of the channels, or paraná-miris, of the King of Rivers, not exceeding two miles in breadth. The land on our right was a long island, beyond which lay another channel, and another (or perhaps other two) beyond this before reaching the true northern shore of the Amazon. All through the 20th we were sailing with a fair wind up the paraná-mirí, which kept about the same average width; and having passed the first island we came to a second, from which it was separated by a narrow furo. The wind failed us before sunset, but after dark got up again in strength, and about midnight brought us round the point of the second island into a wider channel. Throughout the dry season the easterly wind—a continuation of the trade-wind of the ocean—blows up the Amazon, at least for several hours every day, and sometimes day and night without remission, especially in the months of September and October. Early in the morning we passed Gurupá, a village on the right bank where there is a strong fort, generally attributed to the Dutch during the brief period they had possession of the Amazon; although Baena says it was built by Bento Maciel Parente, Capitaõ-Mor of Pará, after he had expelled the Dutch in 1623. Our course lay then through narrow channels, among islands at the mouth of the Xingú, where the current was not so strong as in the main Amazon, although they were beset with shifting sandbanks, which made the passage not devoid of danger. The islands were mostly densely wooded; but one of them (probably of recent formation) presented the appearance of a beautiful meadow, being clad with long grass, sprinkled with low trees, with here and there a clump of arborescent Aroids; and begirt by a natural fence of *Salix Humboldtiana*, a graceful willow, notable for its long, narrow, yellow-green leaves, and for its being distributed, in varying forms, along the banks of rivers of *white* water (but not of *black*) throughout equatorial America.

. . . . . .

All through the following night it blew a perfect gale of wind, but fortunately in the right direction. . . . At daybreak we came out into the main channel, and for the first time got a sight, across an intervening wooded island, of the real north shore of the Amazon, which rose abruptly into a ridge of hills, called the Serras d' Almeirim, apparently nearly a thousand feet high. . . .

Some way higher up we came in front of the more extensive and picturesque Serras de Parú. . . . Our course lay still along the southern shore of the river, which continued as flat as ever; but the ground stood mostly high out of the water, and was clad with lofty forest, including very few palms, so that it was probably not inundated even when the river was full. . . . To westward of this appeared the higher hills of Monte Alegre (*i.e.* the Delectable Mountains), and at their foot the town of that name, formerly called Curupatuba, from the river which enters two leagues higher up. We rarely saw the whole breadth of the Amazon without any intervening island; and the broadest of these intervals of clear water, a little higher up than the Velha Pobre, was barely six nautical miles across.

. . . . . .

In the rebellion of 1835, to be unable to speak Lingoa Geral and to have any beard were crimes punished with death by the Cabanos, who carefully extirpated any vestiges of hair from their own faces; but in 1849 the fashion had entirely changed. Such of our Tapuyas as rejoiced in a few straggling hairs on the chin and upper lip, and especially two or three of them who might have a drop of white blood in their veins, were never weary of admiring themselves in the glass, and of making believe to comb out their beards. Many of them had guitars (called violas) of Lisbon manufacture, costing in Pará six or eight milreis each, and would spend hours together in strumming them to the same melancholy tune, consisting of some eight or ten notes, and nearly always in a minor key. In the evening they would sometimes dance, the performers being one, two, or three—the step a sort of quiet heavy shuffle, varied by occasionally lifting a leg, and by sundry snaps of the fingers and slaps on the thighs. I learnt afterwards that both music and dance were modified from the Landúm, one of the national dances of Portugal.

On the 27th, about midday, we reached Santarem, at the junction of the river Tapajoz with the Amazon, and came to anchor in front of Captain Hislop's house, which stood at the eastern extremity of the town, on a grassy terrace sloping down to a broad sandy beach. At the back rose a morro or low rounded hill, which hid the rest of the town from view, and was crowned by a fort, for merely looking at which, in 1837, Lieutenant Mawe had been made prisoner, and sent down to Pará under guard. We were cordially received by Mr. Hislop, who invited us to dine with him, and sent out to seek a house for us. We found him a sturdy, rosy Scotsman, who had in his younger days followed the sea, but had been settled on the Amazon no fewer than forty-five years. He had at one time traded extensively to Cuyabá, the capital of the mountainous province of Matto Grosso, which is reached by ascending the Tapajoz nearly to its source, and passing thence by a short portage to one of the head-streams of the Paraguay, whereon Cuyabá is situated. The staple produce of Cuyabá was diamonds and gold-dust, and Santarem could offer in exchange guaraná, the produce of plantations in the immediate neighbourhood, and salt, brought from Portugal: two articles of the first necessity to the miners of Cuyabá, and at that time scarcely to be had except from Santarem. Mr. Hislop had for some years back nearly relinquished the Cuyabá trade, having suffered serious losses in it from the roguery of his agent and the failure of some of his creditors there, and now limited himself to the trade with Pará. He was a devoted reader of newspapers, of which he kept large files, to be perused and reperused; indeed, he assured me that he read a newspaper that had been laid by six months with greater zest than when recent. . . . Of books he read but two—Volney's *Ruins of Empires* and the Bible; and by combining their contents had framed for himself a creed of very motley complexion. Whenever he indulged in a few extra glasses of port after dinner, he was certain to favour his guests with a dissertation on the character of Moses, whom he affirmed to have been "a great general, and a great lawgiver, but a great impostor"! Combine with these oddities the frank and hearty bearing of a sailor, and it will be understood how I found in the old captain an amusing companion and a valuable friend during my sojourn at Santarem.

A house having been obtained, we removed to it the same night, and Mr. Hislop lent us one of his slaves to cook for us until we could get a cook of our own. The house, which was a fair sample of the average at Santarem, was of only a single story, but the rooms were airy, the roof tiled, and the floor of bricks, instead of mother earth as in houses of inferior class; and there was a small yard at the back, with kitchen and other necessary offices. True, it did not contain a single article of furniture, but there were rings in the walls wherefrom to suspend our hammocks; Mr. Hislop lent us a few chairs, and some cedar-planks, out of which with the aid of piles of bricks we extemporised shelves for our parcels of plants and other effects; and we obtained the loan of a large table from Mr. Jeffries, an Englishman settled and married at Santarem, to whom, and to his relative Mr. Golding, we were indebted for many little services. To the utensils brought up from Pará I had only to add by the purchase of a large waterpot and a lamp, and our simple *ménage* was complete.

. . . . . .

The local name of the Tapajoz at Santarem is

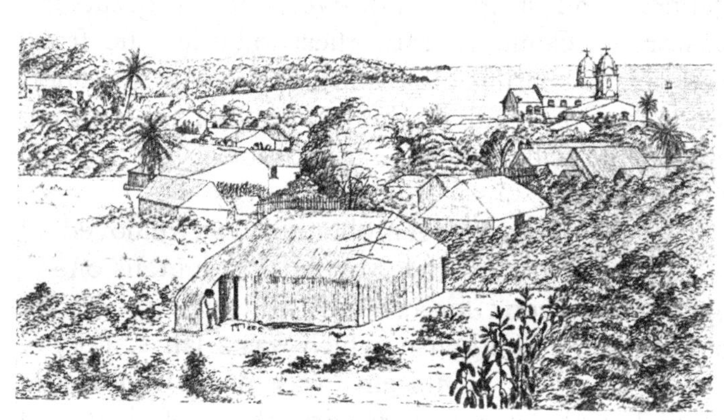

FIG. 3.—SANTAREM.
From a hill near the Fort. (R. Spruce.)

"Rio Preto" or Black River, but the real colour of its waters is a deep blue. When I first saw it, in the dry season, the blue water extended down the southern side of the river for several miles below Santarem, before being absorbed in the muddy expanse of the Amazon; and there was a broad firm beach of white sand stretching about a league in that direction; while up the river it was continued for full five miles, following the sinuosities of the coast; but when the Amazon became swollen by the rains to its winter level, its waters dammed back those of the Tapajoz, and not a bit of blue water or of sandy beach was to be seen in the mouth of the latter. The town of Santarem extends up the Tapajoz about a mile, and fronts to north by west. The eastern half of this frontage, with two parallel streets at the back, constituted the town, properly so-called, and was occupied by the more aristocratic portion of the population: it contained a neat and spacious church, ornamented by two towers. The western half, called the aldea or village, was the residence of Indians and other free people of colour, who inhabited huts with mud walls—or with no walls at all, but bare posts in their stead—and roofs of palm-leaves. The population of both villa and aldea would at that time scarcely exceed 2000.

. . . . . .

Instead of the forest-clad plains and artificial pastures of Pará, I found at Santarem natural campos or savannahs sloping gently upwards from the banks of the Tapajoz, and at the back rising into picturesque but not lofty hills—apparently of 500 or 600 feet, but I had at that time no barometer to measure them. The soil is mostly a loose white sand, but the hills are strewed with volcanic scoriæ, and towards their summits appear volcanic blocks of considerable size. A brook of remarkably clear water, the Igarapé d' Irurá, takes its rise at the foot of the most distant hills, among lofty forest, and runs along the eastern base of the nearer hills, where it is from 3 to 5 feet deep in the dry season; then across the western side of the campo to enter the first bay on the right bank of the Tapajoz, precisely where the sandy beach of Santarem terminates. A narrow belt of lowish forest marks the course of this igarapé, and it is in many places almost impassable from the dense growth of a stemless palm, called Pindóba (*Attalea compta*, Mart.), and of a tall Heliconia.

A similar, but rather larger stream, the Igarapé de Mahicá, has its source near that of the Irurá, but runs in a contrary direction to join the Amazon a league below Santarem. The lower part of its course is across an extensive flat of grassy marshy land, flooded so deeply in winter that canoes traverse it in every direction, and doubtless at no very ancient period a permanent lake.

The vegetation of the upland campos reminded me of an English pleasure-ground. It consisted of scattered low trees, rarely exceeding 30 feet in height, and here and there beds of gaily-flowering shrubs, with intervening grassy patches and lawns. The grass in the dry season looked rather dreary, for it consisted of but one species of Paspalum, growing (like many tropical grasses) in scattered tufts, whose culms and bristle-like leaves were hoary with white hairs; so that it differed widely from the dense green turf of an English meadow. Among the trees then in flower, the Cajú or Cashew-nut (*Anacardium occidentale*, L.) was exceedingly abundant; and an old Cajú, with its rough bark, its branches touching the ground on every side, its young leaves of a delicate red-brown, and its numerous pear-like yellow or red fruits (more properly enlarged fruit-stalks), each tipped with a kidney-shaped knob (the real fruit), is a picturesque object, notwithstanding its humble size. With the Cajú grew the Caimbé (*Curatella americana*, L.), a small tree not unlike a stunted oak in habit and in the sinuated leaves, which are, however, so rough that they are used in lieu of sandpaper by the carpenters of Santarem. It is one of the very few trees of the hot plains that have deeply-furrowed bark, which accounts for the name Alcornoque (Cork tree) I afterwards heard given to it on the llanos of the Orinoco, where also it is a common tree. But the finest of these trees was the Suca-úba (*Plumiera phagedenica*, Mart.), an Apocyneous tree which grows to about the size of the common holly, and has long coriaceous leaves of the richest green, with terminal clusters of white flowers the size of primroses, but very fugacious, followed by curious spindle-shaped twin pods full of winged seeds. The milky juice of the Suca-úba has great repute as an anthelmintic. Another tree of similar growth, but of the family of Rubiads (*Tocoyena puberula*), had wider rugose leaves, and ochre-yellow flowers with a tube 4 inches long. A Murixí (*Byrsonima Pöppigiana*, A. Juss.), with numerous racemes of pretty yellow flowers, and another (*Byrsonima coccolobæfolia*, H. B. K.) with similar racemes of pink flowers, and leaves like those of the Cajú, were both very ornamental. With these grew here and there *Hylopia grandiflora*, St. Hil., an Anonaceous tree, notable (like many others of the order) for its pyramidal mode of growth, for its two-ranked, rigid, lance-shaped leaves, and especially for the thick leathery petals (six, in two rows), being of a fine rose colour within, instead of the yellow or green prevailing in the flowers of most of the order; although they want the fine fruit-like odour of the flowers of many of the Anonas—of the Chirimoya, for example.

Among the branches of the trees sat or hung many sorts of mistletoe, some of them with bunches of long yellow or scarlet, and often sweet-scented, flowers. But more admirable than these were the lichens, encrusting the old trunks with patches of vivid yellow and red, and bearing fruits of the most curious and novel character. On some trees, lichens of the family of Graphideæ prevailed, their fruits resembling cabalistic or oriental characters strongly written in black or scarlet on a white or greyish ground, and of types that seemed wonderfully various to one who was familiar only with the Opegraphæ of Europe.

Of the shrubs the most striking was a Rubiad (*Chomelia ribesioides*), which in habit and in its numerous pendulous racemes strongly resembled a currant-bush, only the yellow downy salver-shaped flowers were much prettier. There were also several Myrtles and Melastomes; the former bore abundance of black berries, the size of sloes, considered eatable by people who did not object to a strong flavour of turpentine.

Over the bushes twined and climbed various lianas. Bignoniads and Apocynes with bell-shaped white, yellow, or purple flowers; Diocleæ, looking like and really closely allied to our scarlet-runners, but with long compound spikes of purple or violet flowers; Serjaniæ, with compound leaves of three or nine deeply-toothed leaflets, spikes of whitish flowers and capsules of three membranaceous white or roseate wing-like pieces (follicles), each with a small black globose seed at the apex; and *Davila Radula*, Mart., allied to the Curatella, and like it with rough leaves, but bearing very large panicles of flowers, whose yellow, two-valved, indurated, and persistent calyces looked not unlike half-split peas. Over bushes and lianas trailed the thread-like entangled stems of *Cassytha brasiliensis*, a leafless herb like our Dodders, but in the structure of the minute white flowers so closely allied to the Laurels that it must perforce be classed along with them, notwithstanding the contrast in external appearance between one of the humblest herbs and some of the noblest trees in the world.

Some parts of the campo had been fired early in the dry season, and the burnt ground had become partially clad with beds of two curious hoary plants, from 1 to 3 feet high, the one a Leguminifer (*Collæa Jussiæana*, Bth.), with leaves of three leaflets clad with white down, and small purple flowers; the other a Menisperm (*Cissampelos assimilis*, Miers) with roundish excentrically peltate woolly leaves, and blackish corrugated capsules resembling a dead caterpillar curled head to tail.

The annual burnings are in some places followed by a dense growth of a fern, *Pteris caudata*, which no subsequent burnings, or anything but fairly stubbing it up, can thenceforth eradicate. This fern is scarcely to be distinguished as a species from the common bracken (*Pteris aquilina*) of our heaths, although the longer, slenderer, and drooping points of the divisions of its front give it a rather different aspect.

Near the aldea, where the soil was rather less

sandy, small plots of ground were under cultivation, but produced little more than water-melons, pumpkins, and plantains of poor quality. Many such plantations, long since run to waste, had got grown up with caapoera, comprising a dense growth of small trees, shrubs, and twiners, mostly different from those of the campo. The commonest plants in flower at that time were species of Casearia and Lacistema, two genera often found growing together, and with much external resemblance, but very widely separated in character. The former are not unlike our hazel bushes, and have large, two-ranked, toothed leaves, axillary clusters of greenish or whitish flowers, and three-valved capsules very like those of violets. The latter, with a similar but rather more rigid habit, have small axillary catkins. There were also species of Erythroxylon—a genus rarely absent from caapoeras in the Amazon valley. They grow to small trees, not unlike our plum trees or sloes, though usually more rigid in their ramification and the texture of their leaves, with almost the solitary exception of *E. Coca*, whose thin submembranaceous leaves are as indispensable a stimulant to the inhabitants of the Peruvian Andes as those of tea to the Chinese. In the environs of the town, especially along the beach, two oriental trees, the Tamarind and the Azedarach, had become naturalised; as had also the handsome *Cæsalpinia pulcherrima*, brought perhaps originally from the Antilles. I have since seen these three plants, growing in the same way, and still more abundantly, in the plains of Guayaquil.

The lowland campos on the Mahicá had again a different vegetation. There were many sorts of grasses which kept fresh and green all the year round, some of them bearing spikes of feathery flowers; others (species of Panicum) had long distantly-jointed stems, simulating slender bamboos, and supporting themselves on the branches of the trees that margined the campo, climbed to a height of 15 or more feet. Wherever the soil was turfy, there were cushion-like patches of the Mahicá, a small monocotyledonous herb, whose densely-set, deep green, bristle-like leaves give it quite the aspect of *Polytrichum juniperinum*, one of the common mosses of our moors, from which it is widely separated by the pretty flowers of three petals, red in one species (*Mayaca Sellowinna*, Kunth) and white in another (*M. Michauxii*, Endl.). It gives its name to the campo and igarapé, and is a singular instance of even an insignificant herb having the same name (with the difference of a letter) in the Amazon valley and in French Guayana, where it was first found and made known to science by Aublet.

Along with the Mayacæ, and elsewhere on the campo where there was little grass, trailed a delicate Rubiad (*Sipanea ocymoides*), so like, in its opposite lanceolate leaves and pink flowers, to the European *Saponaria ocymoides*, which I had seen a few years before ornamenting crumbling schists in the Pyrenees, that at first sight I could hardly believe it was not the same. Besides these, the few plants in flower on the campo were two or three Jussiææ, *Coutoubea spicata*, Aubl. (of the order Gentianeæ); *Peschiera latiflora*, Benth., an Apocyneous shrub only a foot high, but with large jessamine-like flowers; and a few annual Melastomes.

On that side of the campo nearest Santarem, it was fringed by a dense growth of tall prickly palms, interspersed with a few trees of *Simaruba versicolor*, St. Hil., conspicuous for their glossy pinnate leaves and bunches of green flowers; and with a Rubiaceous shrub (*Palicourea riparia*, Benth.) which I have met with again in similar situations all through the plains: it has slender forked stems, green bark, opposite lanceolate leaves, lax panicles of which all the branches are red, and waxy yellow flowers the size and shape of those of the lilac.

The above sketch of some of the plants in flower at the time of my first visit to Santarem may serve to give an idea of the aspect and character of the vegetation in the month of November. I should add that along the shore of both the Amazon and Tapajoz there was a rather dense growth of trees of the gapó, mostly of humble stature, but becoming loftier as one descended the Amazon. Very few of them were in flower in November, and I obtained them all afterwards in perfect state, so that I need not now further particularise them. There were also many small lakes near the rivers, with very little water in them at that season, and containing only a scanty, sickly vegetation.

At Santarem I had the pleasure of meeting Mr. A. R. Wallace, of becoming acquainted with the paths across the campo under his guidance, and of his animated and thoughtful conversation in the evenings; although, after a hard day's work, we both of us found it difficult to keep our eyes open after 8 o'clock, for it was not until I had been some time longer in the country that I got into the

Fig. 4.—The Victoria Regia
As it grows in the Amazonian lakes and side channels.

way of taking a short siesta in the heat of the day, which enabled me to enjoy the evenings more. He had lately returned from an interesting trip to Monte Alegre, and was preparing a boat to ascend to the Rio Negro. At Monte Alegre he had fallen in with the famed aquatic *Victoria amazonica*, and had brought away a fragment of a leaf quite sufficient to show that there was no mistake about the plant. During my voyage from Pará I had learnt from the Tapuyas that in lakes around Santarem there was a water-plant called in Portuguese the Forno or Oven, in Lingoa Geral Auapé-yapóna (the Jacaná's oven), from the resemblance of its enormous leaves to the circular oven used for baking farinha, and from the little river-side birds called Jacaná or Auapé being frequently seen upon them. Captain Hislop and other residents at Santarem confirmed this report, which pointed plainly to the Victoria. Having obtained precise directions to one of its localities, Mr. Jeffries was so kind as to lend me a boat and men, and to accompany myself and Mr. Wallace to see the Forno. We crossed the main channel of the Amazon to what appears from Santarem to be its northern shore, but is really the north side of a very long island, called Ananarí; and then went a little way up a creek to a sitio called Tapiíra-uarí. A walk thence of about two miles across the island brought us to a paraná-mirí, in which we had the satisfaction of finding a patch of the Victoria about 10 yards in diameter. There was barely 2 feet of water where it grew, rooted into nearly an equal depth of mud. The leaves were packed as close as they could lie, and none of them exceeded $4\frac{1}{2}$ feet in diameter. I wished to obtain proof as to whether its duration was annual or perennial, but was unable to decide, although the evidence seemed in favour of the latter. I found no prostrate submerged trunk, but a thick central root penetrating so deep that we could not dig to the bottom of it with our terçados. This root, notwithstanding its size, might be annual; but then every one who knew the plant assured me that the Forno was never wanting all the year round in that and other localities; in which I afterwards found them to be correct; not so, however, in their statement that, when the lakes and creeks rose to their winter level, not only did the petioles lengthen out to keep pace with the rising waters, but the floating leaves went on increasing proportionately in diameter, until they sometimes attained a breadth of 12 feet. I found, in this and other instances, that the measuring-tape was needed to correct the illusions caused by the exaggerated statements of others, or even by the apparent evidence of my own senses.

The Water-lilies I have since seen in South America are certainly all of them annual; and one which springs up on the savannahs of Guayaquil, when the winter rains transform them into lakes, takes only from two to three months to attain its full dimensions and ripen its edible seeds.

CHAPTER III

AN EXCURSION TO OBYDOS AND THE RIVER TROMBETAS

(*November* 19, 1849, *to January* 6, 1850)

[THE Journal of this excursion, as written out by Spruce for publication, has here been considerably reduced by omitting most of the ordinary details of a traveller's daily life, while retaining all those descriptions of the vegetation and the general aspects of nature which are of permanent interest, as well as some of the more eventful incidents of the journey. I have also omitted some long geographical discussions, as well as a detailed account of Cacao culture in various parts of South America, thus reducing the narrative portion of this chapter by about one-half. It was natural that on his first exploring journey into a new district, Spruce should have kept a very full Journal; but I think that had he lived to publish his whole Travels, he would himself have found it necessary to excise and condense his MSS. as vigorously as I have been obliged to do, in order to reduce the whole work into a moderate compass.]

On our voyage up the Amazon from Pará, we had had at first no rain beyond an occasional short thunderstorm; but for the last two days on board, and for three or four days after landing, there was almost continual drizzle—a break in the dry season such as is always expected at Santarem towards the end of October. I considered myself indebted to it for the burst of flowers on the bushes of the campo, of which I had not failed to profit. It was followed by dry sunny weather, and as I was told I might still expect near two months of summer, I resolved to put in execution a project I had formed at Pará of visiting Obidos and the river Trombetas. Having obtained berths on board a batelaõ or cattle-boat, bound for Obidos and Faro, we embarked on the 19th of November, and after a tedious voyage of nine days—the distance from Santarem being only 70 miles—arrived at Obidos towards night of the 28th. . . . Had the vegetation of the south bank, along which our course lay, been more interesting, I would not have demurred at the delay, for I was able to get on shore every day when the vessel was anchored or lying to for a wind; but nearly the whole coast, to a considerable breadth, was clad with plantations of Cacao (called cacoals in Portuguese, cacoales in Spanish); for it is in this part of the Amazon that Cacao cultivation is most extensively carried on. The cacoals either reach to the very margin of the river or have an intervening narrow fringe of such weeds, shrubby and herbaceous, as grow commonly on inundated river-banks. A few of these were new, but they were nearly all of insignificant aspect.

After doubling Ponta Paricatuba—the north-western extremity of the island or peninsula of that name—we entered on a wide bend of the river to southward, whose coast was mostly a steep cliff, at that time, when the water was at its lowest level, rising to a height of near 200 feet, and with a good deal of stratified rock exposed at its base. Here we found a few interesting plants, especially a handsome Rubiad (*Calycophyllum coccineum*, D. C.), with long rampant stems, rufous bark peeling off in thin flakes, large opposite leaves, and flowering peduncles one to two feet long, thickly beset with cymules of small yellow flowers, the outermost flower of each cymule subtended by a large leafy bract, near 3 inches long, scarlet above, red beneath, and with its stalk so united to the calyx as to seem a continuation of one of the teeth of the latter. Some parts of the cliff appeared from below in a perfect flame from the abundance of these gorgeous bracts, to which the plant owes its Indian name Corusé-caá or Sun-leaf. With it grew a fine Bignonia, with downy flowers of the deepest purple; and on the top of the cliff, under the shade of trees, there was good store of a fern (*Gymnogramme rufa*, Desv.), whose pinnate fronds were marked on the underside with numerous close reddish streaks (rows of capsules).

. . . . . .

The cultivation of the Cacao was far more interesting to me than the indigenous vegetation. The Cacao tree (*Theobroma*—Food of the Gods) has been so often described, as have also these very plantations of Santarem and Obidos, that it is useless to describe them further; but as I have since seen the Cacao plantations of Guayaquil—perhaps the most important in the world—a comparison of the latter with those of the Amazon will not be uninteresting. And first I would remark on what seems to me a defect in their management, in both localities, namely, the overcrowding of the plants, which (if I may trust to my memory) is more excessive at Guayaquil than on the Amazon. It is notable to see solitary trees near houses, and rows of trees adjacent to streams and roads, heavily laden with corpulent fruits; whereas in the centre of the plantations, where the trees stand so close together that their branches interlace, and the broad leaves completely shut out the sun's rays—where there is no circulation in the dank, mouldy-smelling air that hangs over the ground—a large proportion of the flowers drop off without being fertilised, and the few fruits that do reach maturity are more slender, and the seeds smaller and thinner, than on those trees to which light and air have had free access. A well-grown Cacao tree, in fact, affords of itself sufficient shade to its trunk and principal branches, whereon (as is well known) the flowers and fruits chiefly grow; and there is no need to hem it in so with other trees as to cut off the small portion of air and light that would otherwise penetrate under its drooping branches. The planters themselves have not failed to note the greater yield of the trees along the skirts of the plantations, but without attributing it to the true cause; and they think it necessary to go on planting the trees at just the same distance apart as their forefathers did.

. . . . . .

As we approached Obidos we saw before us a steep cliff, rising to perhaps 150 feet above the river, along whose northern bank it stretched away for a couple of miles. It was composed of variously-coloured earths and clays, and in some places a coarsely-grained sandstone, like that of Paricatuba, peeped out at its foot. On a plateau towards the eastern end of this cliff stood the town of Obidos, of which we could see nothing from the river except the church tower and the roofs of two or three houses; but on climbing up to it we found it a considerable place, nearly equal to Santarem, although by no means so regularly and neatly built. I had letters of recommendation to the Commandante Militar, Major Joaõ da Gama Lobo Bentes, who installed us in a room which we shared with his son, a dissipated young man, who divided his time between his hammock, his viola, and his cachimbo or pipe. We had therefore scant space for indoor work, and the range of our outdoor operations was also rather limited, in consequence of there being no broad paths leading far into the interior, as there were at Santarem. We found, however, a moister climate and a more vigorous vegetation, although flowers and fruits were, from that very circumstance, less accessible. Virgin forest came up to the very skirts of the town, and the Guaribas (Howling monkeys) used to serenade us from thence about daybreak.

. . . . . .

The small collection of plants made during my brief stay at Obidos, and at an unpropitious season of the year, represents its vegetation very inadequately. The beach alone was gay with flowers, chiefly of annual Phaseoli, Euphorbiads, and Composites. On the cliff grew a few humble Melastomes, mixed with a very fine Gentianad (*Lysianthus uliginosus* var.) with the habit of a Campanula, and with similar bright bluebell-shaped flowers. Here and there hung large bunches of *Lycopodium cernuum*, with gracefully curling branches, and fruit-bearing spikes like those of the common club-moss. With it grew another fern, *Gleichenia glaucescens*, with long rampant stems (like all its congeners) repeatedly pinnate, the pinnæ often reduced to a single pair, so that the type of division might seem to be forked. *Gymnogramme calomelanos*—its much-divided fronds of the deepest green above but beneath covered with a white pruina, as if strewed with flour—appeared wherever wet trickled from the cliff. It is perhaps the commonest of all the ferns of tropical America, and struggles even up to the cold paramos of the Andes, where, although dwindled from 3 feet (its size in the plains) to as many inches, it preserves all its characteristic features.

On the eastern side of the town a small rivulet ran down a valley from the northward, and before entering the Amazon expanded into a lake, known as the Lago de Obidos. The gentle descent to this rivulet was sandy, and the forest of rather humble growth. Of the few trees then in flower the most striking were a wild Cacao (*Theobroma Spruceana*, Bern.), 40 feet high, with a crown of leafy branches at the summit, and bunches of flowers all the way up the straight slender trunk; a fine Chrysobalan (*Licania latifolia*, Bth.); a Guarea with pinnate leaves and long racemes of small white flowers; various species of Inga, Cupania, etc. Under the trees grew a nightshade

(Cyphomandra) bearing abundance of pure waxy white flowers; and a Melastome (*Tococa scabriuscula*) notable for a bladdery dilatation of the leaf-stalk, in which active stinging ants take up their abode. The slender bramble-like stems of *Acacia paniculata* climbed to the tops of the trees, and all the way up put forth their panicles of minute cream-coloured flowers gathered into globose heads.

Near the lake the ground was marshy—evidently in the rainy season laid under water—and the vegetation was peculiar. Very abundant was a low bushy Euphorbiaceous tree (Peridium), which diffused a strong scent of honey from its numerous red flowers, that consisted each of a pair of hemispherical cups (like a bullet-mould) enclosing the minute florets. Other trees of humble growth were species of Mayna, Burdachia, Cybianthus, etc. The lake itself was fringed with sedges, chiefly species of Hypolytrum, quite like our Carices in habit, and in the spikes of beaked fruits; and with a pretty fern (*Nephrodium Serra*) very like the *Lastræa Oreopteris* of our moors. On its waters floated *Salvinia hispida*, which is also a fern (in the widest acceptation of that term), but from its ovato-reniform olive-coloured leaves looks quite alien to that family; besides a Water-lily (*Nymphæa Salzmanni*) rather like our English species, but not near so pretty.

From the opposite shore of the lake rose the Serra d' Escamas, clad with lofty trees, and with a dense undergrowth among which I found some fine flowering shrubs.

The weather continued to be much broken, so

that we sustained frequent drenchings in the woods, and had great difficulty in preserving our collections from rot and mould. It seemed as if I should have to renounce my project of exploring the Trombetas, but Major da Gama assured me that when the rains set in thus early, the weather generally took up again after Christmas Day, and the whole of January was tolerably dry, forming what in Spanish America they call a "Verano del niño," or we in England might call a Christmas summer. He offered, too, to lend me his own igaraté[1] or galiota for the trip, and to send for Indians to man it from the Trombetas itself. I gladly accepted his offer, and the Indians were sent for. They should have been five, but only three responded to the call. With these we had to content ourselves, as none were to be had at Obidos; but the Major gave us an order to embark the two recreant Indians on the way, if we could only catch them. Even the other three had not come with a good will; they would rather, poor fellows, have been in their forest-homes, hunting, working, or playing as they listed, than plying their paddles all day in the hot sun or the pelting rain. Two of them were stalwart fellows, apparently over thirty; the

[1] *Igára*, a canoe; *igára-té*, a great canoe. An igára, which is merely a trunk hollowed out and fashioned like a boat, is made into an igára-té by adding ribs to it and nailing thereto one or more planks on each side, so as to enlarge its capacity. A flooring of boards or palm-stems is laid in the stern and dignified with the name of tolda (quarter-deck), and it is sheltered by a toldo or awning, much like the cover of a gipsy's cart, only that on the Amazon it is made of palm-leaves and not of canvas; but about Guayaquil the latter material is often employed, and the cabin is called a ramada. The toldo is usually closed behind, but sometimes it is open at both ends, which are protected when needful by yapás or mats. As Major da Gama's boat had a cabin made of boards, instead of palm-leaves, it was dignified by the name of galiota. A small light canoe, fashioned in the shape of a skiff, is called a montaria.

third, who was to be pilot, would be nearer sixty; he had ascended high up the river, and was familiar with its navigation. My desire was to go, if possible, as far up as to where the river began to have rocks in its bed and hills on its banks; and from the pilot I learnt that at a few days' navigation up the Trombetas, a large tributary, the Aripecurú, entered it on the left, by ascending which I should find what I sought much sooner than by keeping up the main stream. So the caxoeiras or rapids of the Aripecurú were fixed on as our goal, and I laid in a stock of the indispensable pirarucú and farinha for food on the way.

We got off on the 17th of December, at about 10 A.M., and it was $3\frac{1}{2}$ P.M. when we reached the mouth of the Trombetas, although only six miles away in a direct line from Obidos. The Trombetas is there about a mile across, including a small island.

At $8\frac{1}{2}$ P.M. we reached the mouth of an igarapé and lake called Quiriquirý, where our pilot's brother had a sitio, in which we were glad to take refuge from the rain and the carapanás. It was decided to remain here the whole of the following day, in order to make yapás or mats wherewith to shelter the fore-part of the galiota, where our provisions were stowed, for the rain had wetted them considerably.

Dec. 18.—This day (I quote now from my Journal) on the igarapé and Lake Quiriquirý. Our host, Elisardo, is a carpenter and a very ingenious fellow. He is also something of a farmer, and a luxuriant meadow of Canna-rana, bordering the lake, enables him to fatten a few young cattle. He keeps three or four apprentices and assistants, and seems comfortably off. In the morning he took us across the lake, and we remained until near night in a valley, traversed by a feeder of the lake, where the Tapuyas cut palm-leaves and wove their yapás, and I searched about for plants. On entering the forest next the lake, I was startled at seeing what seemed to be two snakes lying across the path—they were leaf-stalks of an Aroideous plant (Dracontium), and were mottled with white, green, and black (or brown) exactly like the venomous Jararáca, whence their name, Jararáca-tayá. I found a few growing plants which had a bulbous root, like that of *Ranunculus bulbosus*, but flatter beneath. It is edible, like the roots of many other Aroids, but the acridity which pervades it has to be got rid of by maceration, or by throwing away the first water in which it is boiled.

During the dry season the waters of the lake had receded so as to leave a broad beach, with a bushy border next the forest. On the beach grew several annual grasses; an undescribed Sensitive-plant (*Mimosa orthocarpa*); and a pretty shrubby Leguminifer (*Tephrosia nitida*), clad with silky down, like our Alpine Lady's-mantle, and bearing numerous purple vetch-like flowers; it is called Ajarí, and the leaves are used for stupefying fish, the same as those of *Tephrosia toxicaria*, Pers., a much less handsome species, which I afterwards saw cultivated for that purpose at Santarem and in Peru. The bushes consisted of various species of Croton, Büttneria, etc., but especially of *Gustavia brasiliensis*, Mart., which is known as Arvore de

Chapelete or Little-hat tree, from the fruits being likened to a miniature hat, of which the five large persistent calyx-segments, spreading horizontally from the margin of the disk, represent the brim or flaps. Gayer than any of these were the climbing plants—Malpighiads, Asclepiads, and, above all, *Stenolobium cœruleum*, Benth., which has leaves of three rhomboidal leaflets, and bright blue flowers in panicled spikes: I afterwards found it to grow as a weed all through the plains. In the forest, but within reach of inundations, grew a good deal of *Licania Turniva*, a tree of 50 feet or more, with minute green flowers in pinnate panicles: it is called Caraipé das agoas, and its calcined bark is used in the fabrication of pottery, in default of better, for it contains a very small proportion of silex. It is commonly remarked among the Indians that the products of trees of the gapó—whether bark, timber, fruits, or resins—are inferior to those of other trees, their congeners, growing on terra firme or beyond the reach of floods.

I wandered a long way up the gently sloping valley, but the forest became very dense, and neither trees nor shrubs were to be seen in flower. When I returned, the Indians were just finishing their yapás. The broadish leaflets of the fronds of the Pindóba (whereof the yapás were made) are nearly contiguous, and are set on to the rhachis or midrib with great regularity at an angle of about 45°, so that when two fronds are laid side by side, one half-covering the other, the leaflets cross at right angles, and are readily woven together. Six fronds being thus laid in two layers, or nine fronds in three layers, the whole were interwoven into a compact mat or yapá, impervious to any rain. When required for thatch, the midrib of the frond is split nearly through along its whole length, and the two halves turned over to one side, when the leaflets of the one half fall over the interstices of the other half. The fronds are then spread out to bleach and dry, when they become perfectly white or light straw-colour; nor do the leaflets curl up in the least; so that a house newly thatched with Pindóba has a very neat and pretty appearance.

. . . . . .

We had now to leave the Trombetas and turn into the Aripecurú, which we found to have two islands in its mouth. We took the channel between the islands; towards its farther extremity it was nearly choked up with the Luziola (an aquatic grass), through which we had some difficulty in pushing with poles. Other islands succeeded, and we went on threading narrow channels, walled in by lofty trees which were festooned with climbers from base to summit, until after midday, when we emerged into water clear of islands, where the river was more than 500 yards wide. At this point there was a little sandstone rock exposed on the banks, resembling that of Obidos. Sandbanks began to peep out, and some way farther on the river was so obstructed by them that we had difficulty in finding a passage. Towards evening we came on a very long beach, about 200 yards wide, standing high and dry out of the water, of which there was only a narrow strip along its western side. This is known as the Playa grande de tartaruga or Great Turtle-bank; and we drew up alongside it as night closed in. Our men lighted a fire, and spread out the sail to sleep on; and by searching about found a few turtle's eggs; although, from the multitude of shells lying about, it was plain that most of the young turtle had already taken to the water.

. . . . . .

[In three days more the first cataract on the river was reached. The course of the stream was generally to the north, though occasionally winding considerably; the banks became steep, and after the first day's journey hills of considerable height began to appear, some estimated at 1000 to 1500 feet. On the night of the 23rd the Indians and Mr. King slept on a sandbank by a large fire, and in the morning the tracks of a jacaré (alligator) showed that one of these dangerous beasts had come out of the water and passed close by them, unheard by any one. On the morning of Christmas Day the river became narrower, the stream swifter, stratified rocks appeared on the banks, which soon became low, vertical, dripping cliffs, above which the steeply sloping banks were clothed in the richest foliage. Here and there slender rivulèts poured in cascades over the cliffs with the musical sound now heard by the travellers for the first time since leaving England. (Condensed by Editor.) The Journal then continues:—]

At length the current becomes too furious to be stemmed by either poles or paddles. The Indians leap on shore and cut strong sipós, stems of a Bignonia, fasten them to the prow, and two of them yoke themselves thereto to haul alongshore. The pilot takes the helm, which requires all his force to manœuvre; and the fourth man stands in the prow with a long pole, to guard the galiota from being dashed against the rocks at the side or under the water, but is not dexterous enough to prevent it from receiving some pretty hard thumps.

It was midday when we moored the galiota in front of the first caxoeira, having reached the limit of the navigation of the Aripecurú. We landed on the left bank, on a small beach skirted by numerous myrtle bushes, which, being covered with snowy blossoms, resembled so many hawthorns, and emitted as delicious a perfume. Here we cooked our breakfast, or dinner, and mingled our cachaça with the water of the caxoeira to drink a "Merry Christmas" to our friends in England, who, whilst enjoying their roast turkey and plum pudding over a blazing fire, were perhaps pledging the travellers in choicer beverages.

Thus far the weather had favoured us, for we had experienced no heavy rains, and I was in hopes that it would keep dry long enough to enable me to make a large collection of plants. I wished to erect a rancho on the beach, but the Indians declared themselves fatigued, and put off the task until the morrow, contenting themselves with making a fall-to roof with the yapás. The two following days and nights were rainy, with violent thunderstorms at brief intervals, making the want of a hut severely felt, and yet serving as an excuse to the Indians, who could not (they said) cut palm leaves in the midst of rain and drag them through the wet forest. On the 28th the sky was perfectly clear at daybreak, and seemed to promise a fine day; so that I was tempted to try to reach the Serra de Carnaú, and even to ascend it if there were time. We could not see it from our station, but the last view I had had of it on our way up had satisfied me that it rose directly from the eastern bank of the river. Leaving one man to guard our encampment, we took the other three along with us to open a track through the forest. The sun had barely risen when we started, and my advice was to follow the river-bank; but with the view of getting round the heads of some igarapés, whose mouths we could see at some distance up the river, the Indians struck into the forest to eastward, ascending hills and descending into valleys choked with bamboos and Murumurú palms, the latter bristling with prickles of several inches in length. We had gone along thus for some hours, when they appeared doubtful which way to steer. Three several times they climbed lofty trees to look out for Carnaú, but could see neither mountain nor river. At noon, having been on foot six hours, we stopped to deliberate on the probable direction of our goal, when two of the men, without saying a word of their intention, set off to retrace their track to our camp. My experience of forest travelling was as yet very slight, and I knew not how essential it was to never lose sight of my Indian guides. I supposed (erroneously as it proved) that we were at no great distance from the river, and that we might easily reach it by tracking the course of one of the numerous igarapés. . . . So, with the Cafúz Manoel, who was the one left with us, as pioneer, we sought about for an igarapé, and having found one, began to descend along it—no easy task, for its course, where not densely beset with bushes and lianas, ran through flats of entangled bamboos and cutgrass, which were passable only on our hands and knees. The day was excessively sultry—not a breath of air stirring—when suddenly the sky became overcast, and the solemn stillness was broken by a soughing in the forest, soon deepening into a roar, and a terrible thunderstorm burst upon us. In the midst of this, King stopped to break open the shell of a castanha, and got left behind. The torrents of rain so obscured the air, and the incessant roll of thunder and the pattering of the raindrops on the leaves so deadened every other sound, that for some time we did not miss him, nor hear him calling out to us, as he afterwards told us he had done. We thought he would surely soon rejoin us by following the downward course of the igarapé; but by halting for him I lost sight also of Manoel, and half an hour elapsed before we found each other again. I then made him climb a lofty tree, and I from its foot, he from the top, called on our companion until we were hoarse. I bade him look out also for the river, but he declared he could see naught but tree-tops. It had got to 3 o'clock, when, to our very great joy, we heard King's voice, and he shortly afterwards came up with us. After picking his chestnuts out of their shell, he had by mistake gone *up* a tributary of the igarapé, and the rise was so slight that he did not suspect his error until he had gone about a mile, when by floating two leaves he ascertained which way the water ran, and immediately retraced his steps.

The igarapé seemed endless, and we were beginning to fear it would end in some palm-swamp, when, at about 4 P.M., and just as the rain was passing off, we were gladdened by the sight of the river—whose aspect, however, was quite strange to us, still and tranquil as a lake—and the very mountain we had been in quest of close at hand to northward. At some distance to westward, another stream came rushing down over rocks to join the one by which we were standing; and there was a peninsula of rude granite blocks piled up to a great height at their junction. We were plainly a long way from our camp, and our only thought was to reach it as speedily as possible. We started, therefore, down the river, but it was impossible to follow the very margin, for there was no beach, and the forest was denser and more entangled there than at a little way inland. I found that Manoel could get along much more rapidly than we could, and when the sun was getting low I sent him ahead, with instructions to cook something when he reached the canoe, and await our arrival—another error on my part, for Manoel's terçado had greatly facilitated our progress through the forest.

We continued to struggle on until a little past sunset, when it became too dark to allow us to proceed; for although the moon was only just past the full, it was some time ere she rose above the tree-tops. We sat down at the foot of a large tree, in the angle between two sapopemas; but both tree and ground were very wet, and we ourselves were thoroughly soaked, for, even after the rain ceased, every bush we pushed through, every liana we cut, brought down on us a shower of drops. Our situation was no enviable one, for we had no arms save King's terçado and my lichenological hammer, and no materials for lighting a fire. We had in a bag a little roast pirarucú and farinha, and although the latter had been transformed into a glutinous paste by the rain, we made a scanty meal on them. After a while we began to feel chilly and drowsy; but had we given way to sleep under such circumstances, we might have awaked too stiff to move; to say nothing of the risk of being assaulted by jaguars, which we had been told abounded in the forests of the caxoeiras. We resumed our march, but the night was cloudy, and scarcely any of the moon's light penetrated the dense forest. However, we scrambled on—now plunging into prickly palms, then getting entangled in sipós, some of which also were prickly. Even by day the sipós are a great obstruction to travelling in the untracked forest; what must they be, then, by night! One's foot trips in a trailing sipó —attempting to withdraw it, one gives the sipó an additional turn, and is perhaps thrown down; or, in stooping to disentangle it, one's chin is caught as in a halter by a stout twisted sipó hanging between two trees. At one time we got on the track of large ants, which crowded on our legs and feet and stung us terribly, and we were many minutes before we could get clear of them. . . .

Bewildered and exhausted, we sought the riverside, and scrambled to some granite blocks standing high out of shallow water. There we lay down and waited until the moon approached the zenith, when we again plunged into the forest, with just light enough to enable us to select the thinnest parts, but not to show what stones, stumps, and sipós lay in our way. With cautious steps and slow we persevered, keeping the river always within hearing, and now and then crossing an igarapé, either by wading through the water or by passing along some fallen slippery trunk which bridged it over; and at 1 o'clock in the morning reached our camp—sadly maltreated and wayworn.

The effects of this disastrous journey hung on us for a full week. Besides the rheumatic pains and stiffness brought on by the wetting, our hands, feet, and legs were torn and thickly stuck with prickles, some of which produced ulcers. In comparison with these, the annoyance caused by the bites of ticks large and small and the stings of wasps and ants was trifling and transitory.

I have been thus minute in my account of this adventure, in order to give some idea of what it is to be lost or benighted in an Amazonian forest. . . . Let the reader try to picture to himself the vast extent of the forest-clad Amazon valley; how few and far between are the habitations of man therein; and how the vegetation is so dense that, especially where the ground is level, it is rarely possible to see more than a few paces ahead; so that the lost traveller may be very near to help, or to some known track or landmark, without knowing it. I have heard an Indian, recently established in a new clearing, relate that, having gone out one morning to cut firewood, he had wandered about the whole day before he could find his hut again, although, as he ascertained afterwards, he had never been more than a mile away from it. . . .

In making one's way through the forest, it is advisable not to cut entirely away the intercepting branches, but to cut or break them half through and bend them forward in the direction of one's route; and this is especially necessary when there are several persons in company, and the turning of a large tree may completely hide the leader from view, although only a few paces ahead. In the excitement of gathering new plants, or of the chase of wild animals, one often forgets to mark the way properly; and it has several times happened to myself, when deep in the forest and quite alone, to be unable to find my track when I wished to return along it. It is a rather painful moment when one becomes convinced that the way is irrecoverably lost, and stouter nerves than mine would probably not be entirely unmoved by it. There are no trees all leaning over in the direction of prevailing winds, no mossy side to the trunks, as in the forests of the temperate zones. My plan has been to sit down and patiently watch the sun through the tree-tops until I ascertained his course; then to calculate carefully my own course therefrom, and to follow it unswervingly; by which means I have always come out safely. A pocket-compass is no doubt a very good companion in such emergencies, but it requires to be carried in a waterproof case or pouch, for the bush is almost constantly wet, however clear the sky may be overhead.

To return to my narrative. As my main object had been to reach the mountain, I did not delay our progress by herborising much on the way, and I gathered only two plants worth noting; the one an anomalous plant, allied to Ebenads, which Mr. Bentham has proposed as a new genus, under the name of *Brachynema ramiflorum*. It is a small tree, not unlike the Cacao, and with similar long veiny leaves, tapering at both ends, the lower ones borne on very long stalks. The flowers, which grow in clusters on the naked stem and branches, have the tubular corolla mottled with brown and yellow, and its segments rolled back like ram's horns. The enlarged calyx forms a cup to the fruit, like that of an acorn. The other plant is a Calathea, with large Maranta-like leaves, and yellow Crocus-like flowers from the root: it covered the top of a sandy hill, under the trees, where the cutías or agoutis had burrowed extensively.

Everywhere grew noble trees—Bertholletiæ, Lecythides, Icicæ, Licaniæ, etc., and above all various Lauraceæ, including the Itaúba (*i.e.* Stone tree), which yields the hardest and most durable wood for shipbuilding. Scarcely any of them, however, were in flower; but near our encampment I got a few of the humbler trees in good state. Very frequent was *Nonatelia guianensis*, Aubl., a handsome Rubiaceous tree, with ample opposite leaves; tubular flowers, red at the base, yellow upwards, and rather shorter than those of the honeysuckle; and small many-ribbed fruits, looking externally not unlike those of the hemlock. It is found scattered all through French and Brazilian Guayana. *Swartzia grandifolia*, Boug., the Mïrá-pishuna or Black tree of the Brazilians, we had seen all along the banks of the Aripecurú. It grows to a large tree, and its dark-coloured and durable wood is much esteemed for cabinet-work. The leaves are pinnate, with the midrib winged between the leaflets, like that of an Inga. The flowers, growing on the naked trunk, like those of so many Amazon trees, consist of a single large yellow petal, and of numerous declinate stamens, yellow above, violet below; and they are followed by legumes like those of the horse-bean.

Among the climbing plants, *Norontea guianensis*, an odd Guttifer, shot forth from the mass of its dark green foliage as it were jets of flame—spikes of 2 feet long, bearing each some two hundred curious pouch-like bracts of the finest rose-colour, accompanied by minute purple flowers. A Combretum was very showy from its cylindrical spikes of flowers, each consisting of a tubular calyx, with minute yellow petals stuck just within it, and long thread-like stamens, of a deep red, hanging out of it. *Drepanocarpus ferox*, Mart., bore panicles of pretty purple vetch-like flowers, not, however, to be plücked without risk, on account of the strong hooked prickles of the stem.

But the most curious plants grew on the rocks of the caxoeira, where they were kept constantly moist by the foaming waters. They were Podostemeæ—a family in which are strangely blended polypetalous flowers with foliage resembling that of seaweeds or lichens, or sometimes of Jungermannias. They were in great abundance, and had eaten the hard rock into holes, reminding me of the way in which our chalk cliffs in England are eroded by a minute moss (*Weissia calcarea*) and by certain lichens (Verrucariæ). I gathered three kinds, the handsomest being a new species (*Mourera alcicornis*), with pale violet flowers, and fronds recalling those of Iceland moss (*Cetraria islandica*).

In sandy places among the rocks flourished a

small herb of the Violet tribe, *Ionidium oppositi-folium*. Various species of the same genus grow in other parts of Brazil, where their roots afford a sort of Ipecacuanha, quite equal as an emetic to those of the true Ipecacuanha (Cephælis), but not so mild in their operation. I did not, however, again fall in with any Ionidia until many years afterwards, when I came upon them in the Andes of Quito at 9000 feet elevation.

Shady rills, that came down the declivity on the right bank of the river, nourished a good many ferns on their banks, but no very noticeable species.

Of palms rising to the height of trees there were seven or eight kinds, all of which I had seen also on the Amazon; but there were several palms of humbler growth, species of Bactris and Geonòma, which I had not noticed before.

Damp shady hollows, where the vegetable mould lay deep, were often overspread with *Helosis brasiliensis*, Mart. (of the natural order Balanophoraceæ), one of the lowest forms of flowering plants, looking quite like the young state of some fungus (Agaricus or Polyporus), until what seems to be an unexpanded cap is found to be a solid oval head, of a reddish-brown colour, studded with minute flowers of the most rudimentary structure. I have seen it at several points in the Amazon valley, and it reappears near the coast of the Pacific, at the western foot of the Andes.

The following additional observations on the caxoeiras or cataracts of the Aripecurú are all I could make during the four rainy days of my stay there.

The first caxoeira is a distinct fall of a few feet

when the river is low, but in the time of flood it is probably a mere rapid. The rock seemed to me to be clay-slate, of a purplish-grey colour, rarely reddish. The strata dip to S.S.E. at about 10°, and the sections of the principal planes of cleavage run E.S.E. and N.E. The uppermost strata, as seen in adjacent declivities, are thin, shaly, and arenaceous; and they are overlaid by a soft sandstone, in thick strata, whether conformally or not I could not ascertain. On the top of a sandstone hill to west of the fall are strewn a few dioritic blocks, quite like those seen elsewhere in the Amazon valley.

A little above the first fall, granite rocks begin to appear on the left bank, and from thence upwards there is no other rock, the second and all the upper falls being over granite. The rocks, whether of slate or granite, over which the water falls are coated with a black varnish, in some places with a lurid yellow tinge. I have since seen apparently the same kind of deposit at the cataracts of the Orinoco, where it had previously been seen and described by Humboldt. He supposed it to be peculiar to rivers of white or muddy water, founding his opinion on the absence of any such deposit on the granite rocks in the black waters of the Rio Negro. But the Aripecurú has as clear water as the Rio Negro; and at the cataracts of the Huallaga, whose waters are still whiter than those of the Orinoco, there are no varnished rocks. I suppose, therefore, that the deposit is owing to some mineral held in solution (not merely in suspension) in the white waters of the Orinoco and the black waters of the Aripecurú.

From the Serra de Carnaú downwards I counted six caxoeiras. In the intervals the river spreads out wide, and is sprinkled with small islands, some of them wooded, others mere heaps of naked granite blocks. In the same space, seven igarapés enter the river on the left bank—how many on the right I could not tell—and several others come down the steep banks of the narrows below the first caxoeira.

We saw and heard a good many monkeys and curassows (mïtúns) in the woods. My thoughts ran so entirely on plants, that I had neglected to take my gun with me from Santarem; and a pair of pistols which I had taken were useless for shooting birds and monkeys. The Indians carried two guns, and I gave them of my fine powder; but they were bad marksmen, and did not shoot a single head of game throughout the voyage. They found once a jabotím or tortoise in the woods; and this was the only variation from our fare of pirarucú and farinha we enjoyed at the caxoeiras.

There was a little bird which interested me exceedingly by its song, although I did not get a sight of it. It is called Uirá-purú (which means merely Spotted bird), and is saïd to be about the size of a sparrow. As Senhor Bentes had told me I should certainly hear it at the caxoeiras, adding that "it played tunes for all the world like a musical snuff-box," I was constantly listening for it; and at length one day, just after noon—the hour when birds and beasts are mostly silent—I had the pleasure of hearing it strike up close at hand. There was no mistaking its clear bell-like tones, as accurately modulated as those of a musical instrument. Its "phrases" were short, but each included all the notes of the diapason; and after repeating one phrase perhaps twenty times, it would suddenly pass to another—sometimes with a change of key to the major fifth—and continue it for an equal space. Usually, however, there was a brief pause before a change of theme. I had listened for some time before I bethought me of writing down its song. The following phrase is the one that oftenest recurred:—

Simple as this music was, its coming from an unseen musician in the depths of that wild wood gave it a weird-like character, and it held me spellbound for near an hour, when it suddenly broke off, to be taken up again at so great a distance that it reached my ear as no more than a faint tinkling.

The only other animal that took my attention was a beautiful frog, frequenting moist shady rocks and the roots of trees. The belly and legs were of the deepest indigo blue; the back blackish, with a green band on each side, beginning at the nose and running the whole length of the body; and the toes were papillate.

Except on the day of our excursion towards Carnaú, we scarcely ever saw the sun. Thermometrical observations made at midnight and daybreak gave every day the same results, viz.—

Temperature of air at	0 A.M.,	75°.
,, ,,	5 A.M.,	75°.
,, ,,	6 A.M.,	73°.
,, of water at	6 A.M.,	83½°.

I rose several times in the nights for star-observations, but so cloudy was the sky that I got only a single meridian altitude of α Eridani, which gave for latitude 0° 47′ S.

The 29th of December was cloudy and showery, and it seemed probable that the Christmas summer had been put off until another year. I found that the reason why our men had erected no rancho was that they hoped thus to prevent my making a prolonged stay. They began now openly to express their discontent—the sound of the waterfall, they said, was "muito triste," and, with the excessive cold, prevented their sleeping—and I saw plainly that if I did not move at once they would take French leave of me. On the 30th, therefore, at 7 A.M., we started on our return voyage. The river had risen much, and we shot rapidly down. Towards night we drew up at the second turtle-bank, of which we found only a very small space left uncovered by the water.

. . . . . .

[The return journey from this unfortunate and unproductive expedition occupied eight days, the Journal of which is chiefly occupied with hearsay geographical details as to the river Trombetas, now superseded by later information. A day was passed at the farm of Senhor Bentes, near the mouth of the Aripecurú, to dry their soaked clothes, mats, and sails before continuing the journey, allowing Spruce to observe and collect a few more plants; and the notes on these, as well as an interesting account of the different trees called "cedars" in various parts of the

Amazon valley, are of sufficient interest to be given entire.]

The rain did not clear away sufficiently to permit me to enter the woods until 10 o'clock, and very few trees were to be seen in flower. On ground inundated by the Caipurú grew *Parkia discolor*, a handsome Leguminous tree, with leaves of the Mimosa type, *i.e.* twice pinnate, with very numerous close-set, scimitar-shaped leaflets; and with purple flowers, gathered into large pendulous heads exactly like tassels, having a knob of male flowers at the base, and an apical fringe of long thread-like styles. *Cynometra Spruceana*, growing along with the Parkia, and belonging to the same family, is notable for the fruit being not a legume, but a drupe, resembling a wheat-plum.

On ground beyond the reach of floods I saw a few of the trees called Cedros or Cedars, and had one of them cut down. The timber of the Cedro is to the inhabitants of the Amazon what deal is to us at home, being more abundant and more easily worked than any other. It is also more accessible (and this is a great consideration), for, of the large trunks seen floating in the Amazon, by far the greater part are Cedros; so that all that is necessary is to catch them as they float down in the time of flood, and tow them to wherever they may be needed. The trees grow chiefly by rivers, on alluvial barrancos, which, although too high to be inundated, are being continually undermined, and portions of them precipitated into the water. The northern tributaries of the Amazon do not produce much Cedro; but the great rivers which flow from the southward through alluvial valleys, viz. the

Madeira, the Ucayáli, and the Huallaga, bring down vast abundance of it.

On the voyage from Santarem to Obidos, I measured a cedar-trunk left by floods on the beach, and found it 110 feet long, although its top had been broken off a little above the first branches, and where its diameter was still above 3 feet. It had four sapopemas at the base, which measured each 9 feet across. . . .

The Cedros of the Amazon valley belong to the genus Icica (Amyrideæ), some species of which yield the white pitch of Pará, as we have already seen; but whether any of them be identical with the Cedar of Demerara (*Icica altissima*) I am unable to say. They are widely removed from the Conifers, to which the Cedars of the Old World belong; yet the colour of the wood, its grain, and particularly its scent, are so like those of true Cedars, that it is no wonder the Spanish and Portuguese settlers called them Cedros. Colonists are very apt to bestow names of the old country on the trees and herbs of the new, wherever they find any resemblance, either in the aspect or products, to the familiar plants at home. The Cedro of the hill-forests of the Andes consists in part of a species of Cedrela, perhaps *C. odorata*; but what is called Cedro in the central valley of the Quiterian Andes is a Euphorbiacea (*Phyllanthus salviæfolius*, H. B. K.), whose branchlets are crowded at the extremity of the branches, and are so closely beset with two-ranked leaves that they look quite like the long pinnate leaves of an Icica; so that even a botanist might have some difficulty in deciding that they were really branches, and not leaves, unless they bore flowers, which spring in clusters from each leaf-axil.

To return to my diary. Donna Cesaria treated us well—gave us arrowroot for breakfast, wild pig for dinner. She and all her people were very curious about the object of my collections. I explained it to them as well as I could, but the Senhora was not satisfied, and seemed to have convinced herself that they were intended as patterns for fabrics of cotton and silk, England being associated with woven goods in the minds of most South Americans. I showed her through my lens some beautiful lichens covering the surface of a leaf. "O Deos!" exclaimed she to her women who were standing around, "em Inglaterra todo esto vai ser pintado em chita!" ("in England all this will be painted on calico!"). Her parting command to me was to send her one of the handsomest prints our manufacturers should devise from the materials I had collected on her farm.

A meridian altitude of the sun gave for the latitude of Caipurú 1° 37′ S.

. . . . . .

After crossing the Amazon on January 6, I landed on the island opposite the cliffs of Paricatuba at daybreak to gather specimens of the Arrow-reed, *Gynerium saccharoides*, a magnificent grass, which grows in broad masses on the inundated shores and low islands of the Amazon, often accompanied by *Salix Humboldtiana* and two species of Cecropia. It is called in Portuguese Arvore de frêcha; in Tupi, Uíwa; both names signifying "Arrow tree." It grows here to 15 or 20 feet high, and the stout, solid jointed stems, as thick as the wrist, are leafless almost up to the point, where they bear a fan-shaped crown of large sword-shaped leaves, closely set in two ranks. The terminal smooth, shining, taper peduncle, 3 to 5 feet long, is the material of which the Indian forms the shaft of his arrows; and it is surmounted by an ample panicle clad with myriads of minute purple-and-silver flowers, turned to one side, and waving gracefully with every breath of wind.

Having gathered my specimens, we started, and the swift current bore us rapidly onward. In an hour we turned into the Igarapé Açú, and it was barely 6½ A.M. when we reached Santarem. In crossing the Tapajos I obtained a better view of the town than I had previously done, and I was struck with the beauty of its site. The newly-risen sun illumined the lines of white houses, stretching parallel to the river, where numerous vessels of all sorts and sizes were anchored or moving about; and at the back the shrubby campos swelled into bare hills, backed by distant blue wooded ridges.

CHAPTER IV

RESIDENCE AT SANTAREM: OBSERVATIONS ON THE VEGETATION AND ON THE INHABITANTS

(*January* 6 *to October* 8, 1850)

WE now settled down at Santarem for the winter or rainy season, which, having set in there about Christmas Day (as we have seen that it did also on the Trombetas), continued with unrelaxing severity throughout the first four months of the year, without any of those fits of sunny weather in January and February such as the residents affirmed to be the rule. Violent thunderstorms were frequent, and the heaviest rains were generally by night, while from 10 to 3 of the day there was often bright sun, and invariably intense and oppressive heat; for the trade-winds, that blew daily for many hours together during the dry season, were now partially dormant—sometimes for several successive days—and when they did get up in strength, rarely lasted for more than an hour or two. The rivers and the small inland streams rose rapidly, gradually narrowing the range of our excursions. Ilhas de Caapím, *i.e.* Islands of Grass, floated down the Amazon in vast numbers, and sometimes an Ilha would make its way through the Igarapé Açú into the Tapajoz, and encumber the port of Santarem. These floating Grass-islands are a sure indication of the river beginning to rise, and they merit a particular description here, from being a remarkable and indeed unique characteristic of the Amazon and of its tributaries with white or turbid water, but not of those with blue or black water, nor indeed of any other rivers in the world that I have seen or read about. The rafts of driftwood on the Orinoco, described by Humboldt, and seen there more lately by myself, have their counterparts on the Amazon, the Mississippi, etc.; but the Grass-islands of the Amazon are totally different things: they are compact masses of grass, in a growing state, varying from 50 yards in diameter to an extent of several acres. What kind of grass they consist of, and how they came there, I will now try to show.

Along low shores of the Amazon, especially in deep sheltered bays, there is often a broad belt of Caapím (the Tupi name for grass, in general); and the same feature, more strongly marked, is seen in some of the still paraná-mirís, and in lakes that communicate with the river by a short channel. This Caapím consists chiefly of two species, the Canna-rana or Bastard-cane (Echinochloæ sp.) and the Piri-membéca or Brittle-grass (*Paspalum pyramidale*)—amphibious grasses, for whose production *white* water is essential, as is proved by their absence from the Tapajoz and Rio Negro throughout their entire course, and from the Trombetas above the Furo de Sapuquá. Lakes, it is true, have mostly clear water in the dry season, but the lakes into which white or turbid water enters during the rainy season are the only ones which produce those two grasses, and sometimes in such abundance that they become periodically choked up. The same thing happens also in some of the paraná-mirís. Whilst the waters are falling, the belt of Caapím extends inwards, wherever it finds that shallow water in which it most luxuriates; and thus increases vastly in breadth. But when the next flood comes, the earth is gradually washed away from the roots of the Caapím, until, having no longer anything to retain it in its place, the loosened mass is detached from the shore and floats down the stream. In some cases the lower part of the stem is actually decayed, and thus has so slight a hold on the ground as to be readily dislodged by the swelling stream; and as the stems are much entangled, it is only in masses they can be liberated. The circular Grass-islands are mostly the product of lakes, whose outlet has become silted up during the ebb of the river, and is not reopened until the waters, having already risen considerably, burst the barrier and rush like a cataract into the lake, liberating the Caapím, whirling it round and round, and finally carrying it off to the Amazon. I have been in no small peril from the irruption of the Amazon into one of these closed channels, as I shall have occasion to relate shortly.

Grass-islands are often of immense thickness. One which I examined on the upper Amazȯn consisted entirely of *Paspalum pyramidale*. After many futile attempts, I succeeded in drawing up an entire stem of the grass, which measured 45 feet in length and possessed 78 nodes; so that, making all allowance for the tortuosity of the stems, the island could scarcely be less than from 20 to 30 feet thick. All the nodes, save three or four of the uppermost ones that were above water, sent out rootlets, doubtless to extract subsistence from the water; and several of the lowest internodes were dead and half-decayed; yet nearly all the stems bore vigorous panicles of flowers, so that at a short distance the island resembled a luxuriant meadow. Floating on the water, and kept in by the grass-stems, were several minute plants: an Azolla, two Salviniæ, a small Pistia, and an undescribed Frogbit (*Hydrocharella chætospora*, gen. nov.); besides several small molluscs.

Sometimes the voyager finds refuge from a squall by forcing his canoe into the yielding mass of a Grass-island, which breaks the shock of the waves; but when the river is rising rapidly, floating islands oblige the pilot to keep a sharp look-out, especially by night, and in the wet season no vessel anchors in the Amazon: the least evil that could result from such imprudence would be the dragging of her anchor by the onslaught of a Grass-island. From what has been said above of their bulk, and also taking into account that the winter current of the Amazon is at the rate of four or five miles per hour, some idea may be formed of the effect of their meeting a vessel stemming the stream, or even anchored in it, and there have been instances of vessels getting half-buried, and sometimes swamped, in the floating mass. In 1836, the year following the rebellion of the Cabanos, five sloops of war were sent from Pará to receive the submission of the various towns on the river, and whilst lying at anchor in the port of Santarem, a Grass-island of some acres in extent found its way into the Tapajoz, and coming full upon those vessels, tore them all from their anchorage and carried them bodily down the river. A strong body of soldiers, blacks and Indians, amounting to some hundreds, were dispatched to liberate them, and it cost many hours' labour with axes and terçados to effect it, for the island was several yards in thickness. Numbers of snakes (Anacondas), and even some cow-fishes (called Peixe-boys), were found in it and killed.

When I ascended the Amazon to the roots of the Andes, and saw floating islands of grass quite as abundant there, in proportion to the breadth of the river, as I had seen them 1500 miles lower down, I could not help asking myself what became of that immense quantity of grass which was every year carried out to sea. I cannot learn that much of it is cast ashore on the islands in the mouth of the river; but when the floating islands meet the tide they must get broken up, and the grass is probably soon decomposed by the salt water. The fate of the floating trunks and branches of trees, met with in great numbers throughout the Amazon, must often be far more protracted.[1] Many a log, grown on the eastern slope of the Andes, is conveyed by the waters of the Amazon to the ocean, then, by the continuation of the current of the same river, into the Gulf Stream, by which it may finally be deposited on the coast of Ireland or Norway, or even of Spitzbergen!

[1] A little below the mouth of the Huallaga I came on a palisada (as Spaniards call an accumulation of driftwood) stretching across nearly the whole breadth of the Amazon, and had some difficulty in passing it in my canoe.

INUNDATED LAND AND ITS EFFECTS

Nobody at Santarem could recollect the Amazon and Tapajoz rising so rapidly as they did in 1850. They attained their maximum the preceding year on the 12th of June; but this year they had risen above the flood-mark of 1849 by several inches as early as the 15th of April, after which date they maintained the same average height—now rising, now falling a few inches—until early in June, when they began to subside. Many of the cacoals between Santarem and Obidos were inundated, and the people who resided on them were driven into the towns, in the outskirts of which they erected temporary habitations of palm-leaves. Our countryman, Mr. Jeffries, had a plot of mandiocca on a small river (the Aripixuna) which enters the wide bay of the Tapajos, and being alarmed by the sudden rise of the waters, had set all his hands to work to get up the roots, dress them, and bake the farinha. This took them several days, and on their last day it was near midnight when they withdrew from the oven the last batch of farinha. The next morning the oven and the whole of the field were laid completely under water! We ourselves suffered in the matter of provisions; for the milch-cows were flooded out of their pastures, and strayed away into the forest, so that often no milk was forthcoming at our breakfast—a great privation. The rich low meadows opposite Santarem, on the spit of land called the Ponta Negra, between the two rivers, were transformed into a lake; so that of the cattle kept thereon to fatten for the Santarem market some were starved, others drowned, and not a few of the younger ones fell victims to alligators, thus rendering our supply of beef as precarious as that of milk. I was told, but cannot vouch for the fact, that those rapacious monsters (the alligators) thread their way in the water, concealed by the gigantic marsh-grasses, and thus approach unperceived their unconscious victims, whom they first stun with a blow of their tail and then speedily crush in their enormous jaws.

About the same time there was a great mortality—a sort of murrain—among the alligators in lakes lying to north of the Amazon, a day's journey from Santarem; but it fell short of what Captain Hislop recounted to me as having occurred many years before, when it was computed that no fewer than a thousand alligators died in the Tapajoz, and floated down to Santarem, where so great was the stench of their decomposing carcasses that the principal merchants had all their boats and men employed for some weeks in towing them down the river to a safe distance below the town.

When the waters were at their highest, I visited the meadows of the Ponta Negra, principally with the object of procuring seeds of the Victoria. It grew there in two small lakes, to attain which we had to push our canoe through a thick grove of grasses, which stood out of the water to a height of from 2 to 5 feet, besides having at least an equal length of stem buried in water and mud. These grasses formed an elegant fringe, with their nodding plumes of purple-and-green flowers, to the little round lakes, in each of which grew a single plant of the Victoria, each plant with a single flower rising from among its gigantic leaves. In the lakes, and among the tall grasses, were several small floating plants, chiefly cryptogamic, such as a Riccia, an Azolla, and a Salvinia; but there was also a curious and beautiful Euphorbiad (*Phyllanthus fluitans*, sp. n.), with two-ranked, roundish, heart-shaped leaves, of a pale green colour tinged with rose; a fascicle of white radicles from the base of each leaf, and two to four small white flowers in each axil. Though wide as the poles asunder from the Salvinia, it was so like it in external aspect that I could hardly believe my eyes when I found it to be a flowering plant. This is one out of many instances that have fallen under my notice of plants, widely different in the structure of their flowers and fruits, becoming assimilated in habit and in the form of their merely vegetative organs from being subjected to the same conditions of existence. That this is *one* cause, I cannot doubt; but there are probably others, lying deeper than we have hitherto been able to penetrate, which, as in the analogous case of what has been styled "mimicry" in insects, have aided in originating these startling and unexpected simulations. It was strange, also, to see great quantities of a floating Sensitive-plant, *Neptunia oleracea*, whose slender tubular stems were coated with cottony felt of an inch in thickness, as buoyant as cork, serving to sustain completely out of water the heads of pale yellow flowers, and the delicate bipinnate leaves, which shrank up at our approach. The same plant occurs here and there, in shallow waters, throughout the Amazon valley, and also on the western side of the Andes, on the borders of the Pacific; and it reappears in China, where, indeed, it was first found and described by Loureiro.

About the middle of April we were horrified by the news that yellow fever had broken out at Pará with extraordinary virulence. Above half the population, it was said, were ill at one time, and many people of distinction fell victims to that dreadful malady, including Her Britannic Majesty's Consul, Richard Ryan, Esq. Yellow fever had never before invaded the shores of the Amazon, and great was the alarm it created, even at Santarem. The good people of Santarem are not ordinarily remarkable for attention to religious observances, except at Christmas and other festivals, when there is a pious display of rockets, crackers, and balloons, and of processions of a very dramatic character; but when we were in daily fear of the dreaded fever reaching us, we had vespers every night in the church, and those families who were happy enough to possess a rude daub of some saint assembled round it on their knees at stated times, and recited a number of prayers taken *ad libitum* from the breviary. A more amusing process was the dragging a couple of field-pieces through the streets, and discharging them at short intervals, with the object of clearing the atmosphere, and so preventing the entrance of the threatened pesta. With the same intention lumps of the odoriferous white pitch were fastened on poles, stuck up at the crossings of the streets, and set fire to after sundown, thus illuminating the whole town, and emitting a perfume by no means disagreeable. But the most efficacious precaution of all was considered to be the kissing a small wooden figure of St. Sebastian, which was nightly exposed at the foot of the altar, during the novenas of Whitsuntide, to receive the homage of such as feared the pest and trusted to secure the saint's intercession against it, including every man, woman, and child in the church, with the exception of the estrangeiro, whose omission did not fail to be remarked on; but, as he contributed his mite towards the expenses of the feast, his crime was considered venial.

Although Santarem happily escaped the plague, for that time, it was for several months unusually unhealthy. Almost everybody had attacks of constipaçaõ and slow fever—I myself did not escape—and a good many cases resulted fatally; while in the villages up the Tapajoz ague of the worst kind was rife, and above four hundred people fell victims to it. I attributed these maladies, in part, to the unprecedentedly rapid rise of the rivers, and the consequent premature inundation of the lowlands. Nearly all the tributaries of the Amazon, but especially those of clear water, are either aguish throughout their course or have known aguish sites or districts. In the case of the Tapajoz, the inhabitants ascribe it to the insalubrity of the water at certain seasons, and this is doubtless one, although not the chief cause. As the annual rise of the Amazon is somewhat higher than that of the Tapajoz, and as the latter begins to ebb a little earlier, the waters of the Tapajoz are dammed back by those of the Amazon, and are thus rendered nearly stagnant for several weeks, about the time of highest flood and beginning of ebb. During that period they become unfit for culinary purposes, in consequence of the admixture of a quantity of yellowish-green slime, called limo. I examined the latter with the microscope and found it to consist chiefly of decomposed Confervæ, with a very few Diatoms intermixed. It originates in small lakes and sluggish igarapés, whose mouths, connecting them with the Tapajoz, become dried up in summer; and, when they are reopened by the swelling rains, the limo which had accumulated in them whilst stagnant is discharged into the river. No doubt this slimy water is very unwholesome, and those who are obliged to make use of it filter or strain it as well as they can; but at Santarem those who have boats and men send out for Amazon water, which is always wholesome, and apparently grows sweeter the longer it is kept; whereas that of the Tapajoz, when at its best, is apt to acquire a sickly smell if kept a few days. I have seen a similar effect from a similar cause in the river Atabapo, a tributary of the Orinoco.

Allowing its due weight to the cause thus briefly sketched, there is another and more important one, first pointed out by Humboldt, to account for the healthiness of the rivers of equatorial America which run east and west, and the unhealthiness of those whose course lies north and south, namely, that the former alone are accessible to the full force of the easterly or general trade wind. On the main Amazon, especially in the lower part, ague does not occur as an epidemic once in thirty years, thanks to the prevalent easterly wind; yet even there we had sometimes, about new or full moon, a day or two of what is called vento da cima or "wind from up river" (*i.e.* westerly), and it is justly esteemed a vento roim or "noxious wind," for it brings with it neuralgic pains, colds, and fevers. So that we may apply to the equatorial regions, in the western hemisphere, the English adage reversed, and say—

When the wind is in the East,
'Tis healthier for man and beast!

I did not keep any meteorological register at Santarem, but the heat in the wet season seemed to me to surpass what I have felt anywhere else on the Amazon. Neither sweltering heat nor soaking rains ever caused me to intermit my labours, and I went on collecting all through the wet season. But I found it very difficult to preserve my specimens, which I prepared in large quantities, and therefore needed every day to dry great piles of damp paper. To this end I made an agreement with a French baker who lived near to have the use of his oven every morning after the daily bread had been withdrawn from it; but the paper never got half so well dried in this way as it did when spread out on the sand under a broiling sun, where there was free evaporation.

When the flooding of the lowlands reduced my excursions by land to very narrow limits, I used to explore the coasts of the rivers and igarapés by water whenever I could get a boat and men. Boats I could have at any time, but the gente to man them were difficult to catch. Should I need men but for a day, I must ask the Capitaõ dos Trabalhadores (Captain of the Workmen) for them, and then wait perhaps a fortnight before I could get them; for in all probability a detachment of soldiers would have to be sent into the interior, to beat them up at their sitios. These delays were so annoying, that I preferred the chance of hiring three or four men from the crew of any vessel that happened to be laid up in the port, which was but seldom.

We had for some months been unable to get into the hills, on account of the intervening Igarapé d' Irurá having widely overflowed its banks; but when, in the month of June, the rivers began plainly to ebb, I was desirous to see how the igarapé was affected. We visited it one day with this intent, and were well satisfied to find it fordable by wading up to the middle. The ground on the opposite side, though still plashy, was not impassable, and we saw that the foot of the hills could be reached without difficulty. On a slightly rising ground a little beyond the igarapé were the ruins of a cottage, half of the walls and roof of which had fallen, and had got so overgrown with rank grasses as to quite hide the beams and rafters from the eye. In passing over these ruins, Mr. King had the misfortune to tread on a large nail, which was sticking in a rafter, point upwards, and having (like myself) only india-rubber shoes on, which are a protection against naught but wet, he was severely wounded in the broad part of the foot. As the wound was very painful, I thought it better that he should return to the igarapé and wash it, and there await my return, as I wished to penetrate a little farther. Having gone far enough to satisfy myself that there was no obstruction from water, I was retracing my steps, and expected I had already passed the dangerous ground, when I felt myself pierced in the left foot, and was immediately thrown forward with violence. On withdrawing my shoe, my foot was bathed in blood; a nail had entered the narrow part of the sole and pierced through a little below the ankle. How we reached Santarem I hardly know. We cut sticks wherewith to aid our faltering steps, but the excruciating pain obliged us every now and then to throw ourselves on the ground; and it took us three hours to drag ourselves over the three miles. On reaching home I had poultices applied to our swollen feet, and as I knew rest to be the best of all remedies in such a case, we did not attempt to leave our hammocks for three days. In a week's time we were able to get about again; but a year afterwards my wound broke out afresh and caused me much suffering.

It was a singular coincidence that the builder of the cottage at Irurá had come to his death *by a nail*. This man, a Portuguese, was pursuing a runaway slave along a narrow track in the forest; the slave, who was armed with a musket, ascended a tree, and as his master passed underneath it, shot him in the forehead with a nail.

We had another adventure in the same valley two months later on. Nearly south-west from Santarem there is a small lake called Maracanámirí, communicating with the Tapajoz by a short channel. In November 1849, when the rivers were at their lowest, it was a walk of an hour and a half to reach this lake, by the broad beach of the Tapajoz; but they ebbed so slowly in 1850, that in the middle of August the mouth of the Irurá was unfordable, being still near half a mile wide; so that to reach the lake we had to cross the igarapé about two miles up, and then penetrate the forest extending along its banks, to reach an open campo which stretches away to the shores of the lake. We crossed the igarapé, and then attempted to pierce the forest; but the track by which we entered it ceased after we hád followed it a while, and we had then to cut our way through entangled lianas and Pindóba palms, steering by compass in the direction of the campo. While thus progressing slowly and with difficulty, I heard a distant roar, very much like that of a jaguar; but as I had seen several cattle on the Santarem side of the igarapé, I was willing to suppose the sound might have come from one of them. Shortly afterwards it was repeated, and a little nearer; and in a few minutes more it was repeated, so loud and near, that it brought us both to a standstill. King had heard the two former growls, but, like myself, he had not spoken. We were armed only with terçados, and had barely arranged our plan of defence when we heard a tremendous crash among the underwood. After this, however, we heard no more. When we afterwards recounted the adventure to some Indians, they told us that the crash we had heard was undoubtedly the tiger, either springing on some deer, of which he had been in chase, or, arriving in sight of us and doubting his capacity to overcome us, betaking himself to flight.

Rarely are jaguars met with so near Santarem; yet a few years before an engagement took place between three men and a jaguar, in the very same valley. One of these men was armed with a musket, another with a terçado, and the third—a tall powerful man—was quite unarmed. It was upon the last that the jaguar made his first attack, springing upon him out of a bush; and he had fortunately sufficient activity and presence of mind to seize the jaguar by the fore-paws, one of which he secured by the wrist, and the other lower down, and consequently less firmly. They struggled until the jaguar released this paw, and made a claw with it at the man's head, tearing his scalp completely over his eyes. At the moment of the attack the man who had the musket was some distance in the rear, but the one with the terçado flew to his companion's assistance, and the jaguar, leaving the latter, turned on his new assailant, whom also he succeeded in wounding severely. He then sat down midway between them, eyeing first one and then the other, and looking, I daresay, as amiable as a cat might between two disabled mice, uncertain which to devour first. At this critical conjuncture the third man came up, and the contest was renewed, resulting in the death of the tiger, but not until he had wounded all his assailants. The man who had been scalped was living at Santarem in 1850, and constantly wore a black skull-cap, his head being still very tender.

It is fortunate for me that Mr. Bates's much longer residence in Brazil, and consequent more intimate acquaintance with the people, have enabled him to give a far more complete account of their manners, morals, and customs than I could pretend to do. He lived long enough among the Brazilians to learn to like them, which I confess I hardly got to do. My impressions, however, derived from my own personal acquaintance with so remote and fragmentary a portion of Brazil as the Amazon valley, should by no means be taken as applying to the whole of that vast empire. The Portuguese race have hung together wonderfully in South America, and if they continue to do so who can doubt that they have a great destiny before them?

If I cannot say much in favour of the Amazon folk, as a whole, I retain a pleasant and affectionate remembrance of many individuals among them, both natives and foreigners. I do not know that I have anywhere in the world met with a more gentlemanly, well-educated, and honourable man than Dr. Campos, the Juiz de Direito at Santarem. Thoroughly urbane, both in his public and private capacity, he was yet well known to be inaccessible to a bribe; whereas his predecessors in office had been notorious for the opposite quality. Community of taste brought us together as much as my very limited leisure would allow. He was an ardent student of mathematics, and my familiarity with some branches of that science enabled me (he said) to render him valuable aid. In the course of our conversations on general subjects I found him well acquainted with English and French literature, and from the original sources.

I had no other friend among the Brazilians so intimate as Dr. Campos. Among the foreign residents there was no pleasanter fellow or better friend than Abraham Bendelak, a Jew of Tangier, who sought us out soon after our arrival at Santarem, was ever ready to render us a service, and often accompanied us on our excursions. At the time of my visit to the Amazon there were a good many Moorish Jews settled in the principal towns; only temporarily, however, for many of them, like Bendelak, had left wives and families in Morocco, and intended to return thither as soon as they had scraped together a few thousand dollars.

Even a small place like Santarem had its dangerous classes, and they were chiefly free people of mixed race. The slaves, especially the pure blacks who had been brought when young from the African coast, were mostly civil and humble, but merry withal, and pleasant to deal with; and the mulattoes, although apt to be proud and restive, were tractable enough when held properly in hand. The free people of colour, however—except the cross between pure white and Indian, whose worst property is usually laziness and "shiftlessness"—were too often bad citizens and dangerous neighbours; and there, as elsewhere in South America, the Sambo or Cafúz—the mongrel bred between the Negro and the Indian—was accounted the most vicious of all the cross-breeds. In Venezuela I have heard it asserted that nine-tenths of the really atrocious crimes were committed by Zambos. I know not if the proportion were as great in Brazil, where a good many Sambos called themselves "Mulattoes," and it was rare that a man would own to the title of "Cafúz." The towns on the Amazon were affirmed to be much freer from crime than many others in Brazil—such, for instance, as Pernambuco—but it was difficult to get at correct statistics on this head; for, on account of the defective organisation of the police, and the repugnance of judges and juries to admit circumstantial evidence, a criminal not taken in the act was almost secure of escaping conviction.

During my sojourn at Santarem an incident happened of which I copy here the full relation from my notebook, to serve as an illustration of what I have just stated.

On the 2nd of August, we had made a long excursion by land, skirting the base of the hills, and then striking the Igarapé d' Irurá near to its source. We did not reach home again until long after nightfall, and I was so much fatigued that when I lay down in my hammock I found it impossible to sleep—a thing that always happens to me after over-exertion. About midnight I was startled by a loud rattling at the door, and by the Porteiro or town-crier calling out my name. I inquired what he wanted. "The Delegado has sent to call you," said he. I repeated my question, and got only the same answer. "What conspiracy am I about to be involved in now," thought I; "do they want to make a second Lieutenant Mawe of me?" My disturbed imagination and aching head suggested I know not what medley of plots and false accusations, and I was about to tell the Porteiro that if the Delegado wanted me he might come and fetch me himself, when he dispelled my apprehensions, but at the same time gave me a greater shock by calling out that some one had stabbed the "Capitaõ Inglez," that is, my merry old friend Captain Hislop! At that I sprang from my hammock, and as King was by that time awakened by the tumult, we huddled on our clothes and sallied forth into the dark streets. Arrived at the house, we found the Delegado and a number of other people collected before it, on the terrace, whose base is laved by the Amazon. I could get no certain information of what had happened from them, and at the Delegado's request I entered the house, not without considerable misgivings of finding my poor friend in a deplorable state. He was, however, sitting on a sofa, moaning very much, but still quite able to hold himself up; and they had already stanched the bleeding and bandaged the wound, which was in the lower part of the chest. I remained until I saw a bed made up for him on the sofa, and everything as comfortable as circumstances would admit. I would have stayed with him all night, but there seemed no imminent danger, and he himself did not think it necessary. The weapon had been aimed at the heart, but the point had glanced upwards, penetrating the base of the breast-bone, and making a wound three inches deep.

The occurrence had quite stunned the old captain, although the wound itself gave him little pain; and it was not until some days afterwards that he could so far recollect what had happened to him as to give a tolerably connected account of it. He had sat up late, reading, and after locking the outer door and another door leading into the kitchen—as was his wont—he went into a little back room to undress. In this room there was a large native leaf-mat set up leaning against the wall; he had no occasion to disturb it, and there is no doubt that the assassin was concealed behind it—how he could get there unobserved we shall see presently—and that he had passed from thence into the captain's sleeping-room by the door, which was always left open. The captain was in the habit of promenading on the terrace in front of his house, from the time of leaving the dinner-table—that is, from 5 or 6 o'clock until about 8—for the sake of the fresh breeze up the Amazon; and this whether the moon were shining or not. On that night there was no moon, and as it took him some minutes to walk from one end of the terrace to the other, at his easy pace, and he used never to look behind him, a person might enter the house, and even take anything out of it, without his perceiving it. To continue. He put out the light, lay down in his hammock, and went to sleep. About half an hour afterwards, as he supposed, he was awakened by a noise near his hammock, and fancied that cats must have got into the room. He had been stabbed whilst asleep, but on first waking up he did not feel the wound. He rose up in the hammock and felt himself in contact with a man's arms; they grappled him and he shook them off. Then he felt himself attacked on the opposite side; he tried to cry out, but his assailant held his head down and covered his mouth. After a struggle of a few moments, he disengaged himself sufficiently to be able to call out, whereupon the other released him, caught up a trunk which was standing in the corner next the hammock, and was making off with it when, having nearly reached the door, he fell over a chair—trunk and all; but instantly regaining his feet, he escaped through the street door, which it seems he had taken the precaution of setting wide open before commencing his deadly operations.

Meantime a woman who slept in an upper room had heard the captain's cries—three times he called "Ah Jesus!"—and she shouted to the mulatto cook (who happened to be the only other person in the house, the captain having sent off all his people the day before up the Tapajoz in quest of produce), "Joaquim! Joaquim! o patraõ está gritando! Levanta-te depressa!" ("Joaquim! the patron is crying out—get up quickly!"). At this moment she heard the noise made by the trunk and chair falling, and called out louder, at the same time running to the window, from which she saw by the dim starlight a man running at full speed down the sandy shore of the Amazon. The cook struck a light, and the two went together round by the street door to Mr. Hislop's room. They found him still in a sort of waking dream, and, astonished to see them enter by the door which he had left locked inside, he called out to ask them how they had got in. He did not even know that he was wounded—he had felt his hands wet, but thought it was with the sweat of the man he had grappled with; but they told him that the blood was running from him profusely, and that his hammock and the mat underneath it were covered with blood. The cook immediately ran to call the surgeon and an apothecary, and on his way roused the neighbours and the Delegado de Policia.

The assassin would seem to have thrown away the knife as soon as he had inflicted the wound, thinking probably that the latter (like Mercutio's) would serve; for he did not repeat the blow, as he might easily have done, whilst Mr. Hislop was struggling with him. The knife was afterwards found on the floor, and it had been fashioned for the purpose out of a piece of old hoop, beaten into the shape of a dagger, with a sharp point three inches long. Some blacksmith had evidently been at work, and the handle was an old file-handle, such as the smiths of Santarem were wont to use. The magistrates examined all the smiths in the town, but to none of them could the file-handle or the making of the knife be traced. Suspicion attached to a young blacksmith—a mulatto or Zambo—who was undergoing a term of imprisonment in the fort for an assault, and was in the habit of bribing the sentry to allow him to go out of a night and visit his wife. He was out on the night of the attempt on Mr. Hislop's life, and was met in the streets early in the morning by a policeman, who took him into custody. There was other circumstantial evidence of his having been both the fabricator of the knife and the assassin; but it was considered insufficient to authorise even his being apprehended on suspicion, and so the affair was allowed to drop, although all Santarem was convinced of his guilt.

The circumstances that in all probability led to this murderous assault are briefly these. Not many days previously Captain Hislop was paying a mulatto girl for sweets she had sold him the sum of one milreis (2s. 4d. sterling). He wished to give her a bank-note of that value, and he took out of a trunk a little tin box containing his paper-money, wrapped up in small parcels—the one-milreis notes separate from those of higher value. He set the box on the table, and taking out of it a parcel of notes, went up to the window to see if they were of the value he wanted, for it was in the dusk of evening. While his back was turned, the wench put her hand into the box and gripped at a venture a quantity of notes, which proved to be of 20 and 50 milreis, to the value of 470 milreis (nearly £55 sterling). He did not discover his loss until some days afterwards, and even then would have taken no steps to redeem it, fearing the well-known vindictiveness of the Brazilians, and especially of the mistress of this girl, who had been drummed out of Obidos not long before on account of her iniquitous conduct; but one of his friends mentioned the matter to the Delegado, who immediately had the girl brought up and flogged until she gave up what still remained in her possession of the money, viz. 270 milreis. Twenty milreis more were subsequently recovered from another mulatto girl, who had received them from the actual thief. This affair was the talk of the town, and the story of the captain's square trunk, containing (it was supposed) untold sums of money, excited the admiration of every one, and the cupidity of probably not a few. It is plain, however, that the thief had not known which was the trunk that contained the money-box, for the one he actually attempted to carry off contained only old clothes. It was every one's belief that he had been instigated to the crime by the mistress of the sweetmeat girl; but if it were so the accomplice escaped conviction as the principal had done.

I was told of a similar case of stabbing which had happened about three years previously. Our friend Luiz, the French baker, had a quintál—half yard, half orchard—at the back of his house. There his oven stood, under a tiled roof, supported by posts.

but without any walls; and his two sons—youths, the one of thirteen, the other of seventeen years—used to sling their hammocks between the posts in the dry season, when there was neither rain nor mosquitoes, for the sake of sleeping *al fresco.* They were reposing thus, on a dullish moonlight night, when the younger of the two, happening to wake up about midnight, saw a man stealing gently about the quintál and approaching his brother's hammock. The robber, for such he was, noticed that the lad was awake, and to frighten him into silence drew his knife across the elder brother's throat. The younger, at sight of this, gave vent to a shriek of terror, whereupon the villain sprang upon him, buried the knife in his body, and fled out of the quintál. The wound was in his left side, and though deep, had fortunately not pierced any vital organ; yet it was sufficiently severe to confine him to his bed for two months. As regards the assassin, there is the usual tale to be told—the police failed entirely to make him out. The lad had not seen his face distinctly—he had only noticed from his hair and the colour of his skin that he was a mulatto. A free mulatto, who had been twice imprisoned for theft, was suspected. This man embarked shortly afterwards for Pará, and he had not been there long when he was detected in some crime, for which he was put in prison, and there died.

I do not tell of these crimes because I consider them unusually atrocious, or because when I think on them I thank God that we Englishmen are not as other men are, especially as those Brazilians. On the contrary, I acknowledge that the records of our police-courts show how habits of drunkenness, or a generally reckless course of life, may render an Englishman's heart as black as any Zambo's, and lead him to the commission of equally atrocious crimes.

CHAPTER V

A GEOLOGICAL SKETCH OF THE LOWER AMAZON, AND AN ACCOUNT OF SOME CURIOUS PHASES OF VEGETABLE LIFE AT SANTAREM

I SHALL preface what I have further to say of the vegetation of Santarem by a brief sketch of the geology, and an attempt to connect the latter with my observations elsewhere in the Amazon valley. Geology, however, had no place in my programme, for my previous studies had not prepared me for working it thoroughly, and the apparently entire absence of fossils from the rocks of the Amazonian plain (for Messrs. Wallace and Bates had no more than myself been able to find any) took away from the pursuit any interest which it might otherwise have had for me.

As I have already stated, the unmistakably volcanic character of much of the rock at Santarem was the most remarkable geological feature. Along the shore of the Tapajoz, but especially in the hills lying south and east from it, misshapen blocks, glazed and honeycombed, quite resembling the slag from a smelting furnace, and often of enormous size, were strewed about, or confusedly heaped up. Were these deposited on the spot, or, if not, how did they get there? If I recapitulate all the data

I accumulated, bearing on this point, it may assist abler physicists than myself to decide the question. I have spoken before of a conical peak, called the Serra d' Irurá, looking from afar like enough to a small volcano. It lies S. 37° W. and four miles away from Santarem, and I suppose it may rise to 300 feet above the Tapajoz. This peak is overstrewn with scoriæ of the kind described, but the top is rounded and there is no semblance of a crater. Beyond it are ridges, strewn with similar blocks, and with intervening hollows; but the latter are not crater-like, and there is no trachyte or basalt or volcanic rock of any kind beside those boulder-like blocks; nor, although there has been much denudation, as we shall presently see, is there any remarkable tilting up of such stratified rocks as are still *in situ*.

I may here enumerate all the sites in the Amazon valley where I have seen these volcanic boulders, beginning at the coast and proceeding westward. The first is Caripi, near Pará, in lat. about 1½° S., long. 48½° W.; (2) Santarem, long. 54° 40′ W.; (3) the cataracts of the Aripecurú, lat. 0° 47′ S., long. 56° W.; (4) Villa Nova, long. 57° W.; (5) on the Paraná-mirí dos Ramos, long. 57° W.; (6) Serpa, long. 58° W.; (7) on the Rio Negro, at various points for a short distance up, long. 60° to 60½° W.; (8) Manaquirý, on the Upper Amazon, lat. about 4° S., long. 60½° W. This is the most westerly point at which I have observed them myself, although I have been told of their existence at several intermediate points, and also at Coari, still farther west. It may well be supposed that when I reached the foot of the Andes I looked to find them much more abundantly;

but although I have traversed the eastern base of the Andes, from about 7° S. lat. nearly to the Equator, along the rivers Huallaga, Pastasa, and Bombonasa, nowhere did I see the volcanic boulders of the Amazon reproduced. Even among the volcanoes themselves of the Andes, I have never seen any scoriæ so perfectly vitrified on the surface as those of Santarem, nor any lavas so completely fused as those of Etna and Vesuvius. The tufas of Cotopaxi have, in fact, been boiled rather than fused. But if we go still farther to the west, beyond the Andes and the American coast, we come on a group of volcanic islands (the Galapagos) lying upon the Equator, where glazed scoriæ like those of the Amazon abound.

Before pursuing the considerations to which these facts lead, let us return to Santarem and see how far my scanty observations will aid us in ascertaining what has happened to the stratified rocks, and how the bed of the Amazon may have been excavated therein. Due south from the Serra d' Irurá, and with three lowish intervening ridges, there is a curious isolated table-topped hill, the abrupt and naked southern side of which looks at a distance like the ruins of a Gothic castle, from its being cleft into masses resembling towers. On examination it is found to consist of white sandstone, in horizontal layers; and the thinnish top layer being much harder than the next subjacent layers, it has resisted atmospheric or other decomposing agencies to a greater degree, and projects all round like the coping of a wall. The edges of similar thin compact layers form projecting rims here and there on the vertical surface. There was little vegetation beyond a few scattered grasses on the summit, and a few minute ferns and Selaginellas under the shade of the projecting cope. A similar hill, with a much broader flat top, apparently of the same elevation, rises on the other side of the valley of the Irurá, to eastward, or N.E. ½ E. of the first. From the summits of these hills there was a good though distant view of the Serras of Monte Alegre, among which might be distinguished many table-topped summits, some of them apparently much higher than those of Santarem. On referring to Mr. Wallace's account of his visit to Monte Alegre, I find the following description of one of these hills: "We now saw the whole side of the mountain, along its summit, split vertically into numerous rude columns, in all of which the action of the atmosphere was more or less discernible. They diminished and increased in thickness as the soft and hard beds alternated, and in some places appeared like globes standing on pedestals, or the heads and bodies of giants." And of a cave which he went to explore: "The entrance is a rude archway, 15 or 20 feet high; but what is most curious is a thin piece of rock which runs completely across the opening, about 5 feet from the ground, like an irregular flat board. This stone has not fallen into its present position, but is a portion of the solid rock harder than the rest, so that it has resisted the force which cleared away the material above and below it."

There is precisely the same kind of white sandstone, with alternating hard and soft strata, some hundreds of miles higher up the Amazon, in a low table-land extending parallel to the left bank of the Rio Negro, from its mouth to I know not how far up, and upon which several small tributary streams of the Rio Negro take their rise. I have ascended nearly to the head of two of these, the Igarapé da Cachoeira and the river Tarumá, both of which run in the upper part of their course over a white sandstone, usually so soft that it crumbles under the foot, but with interposed layers of marble-like rock, which cause those streams to descend in a succession of falls, a few miles apart, each fall being over a slab or sheet of this harder rock. I have visited three falls of the Igarapé da Cachoeira having this character, the lowest of them being no more than 12 feet high, while the first fall of the Tarumá (the finest cataract in the Amazonian plain) is over 30 feet, and the slab of white stone from which it leaps projects so far that one may walk on a ledge a few yards lower down without getting a drop of water from the fall. This structure quite corresponds to what I observed at Santarem, and Wallace at Monte Alegre. Moreover, on the Rio Negro, as on the Amazon and Tapajoz, the white sandstone reposes on the gritty sandstone of Pará.

Several of the hills below Monte Alegre, namely, those of Parauaquara and Paru (see Plate in Bates's *Naturalist on the River Amazons*, chap. vi.), are table-topped, but some of those covered with forest appeared to be round-backed. I hardly doubt, however, that all are of the same formation, and that they vary only in the material being of a more or less yielding nature; for there are a few bare flat summits which I suppose to have a similar hard coping to that of the table-mountains of Santarem.

The bed of the Amazon has plainly been excavated, by whatever means, in this white sandstone and in the subjacent Pará grit.[1] At abrupt points or capes, on the Amazon, the grit crops out to view, as at Paricatuba, Serpa, Poraquecoara, etc.; and even where the shore is low and alluvial, and there are subsidiary channels at the back of the main stream, on penetrating far enough inland one is sure to come on the ancient rocky margin of the Amazon. Thus at Manaquirý, lying S. of the Amazon about fifty miles above the Rio Negro, where there is a perfect labyrinth of lakes and channels, we find at the back of all a wall of nearly horizontally stratified Pará grit rising 30 feet above the present high-water mark, and often so hidden by vegetation as to seem from the water a steeply-sloping bank; but I traced it for many miles continuously, and found it to be here and there indented with deep bays or cirques; and I cannot doubt that anciently the main Amazon laved this rocky wall, and in time of flood rose to a level with its top. There is said to be a similar wall a little farther westward at Menacapurú on the northern side of the Amazon. A few volcanic blocks, mostly isolated, repose on the wall at Manaquirý.

. . . .

At Caripi I thought I saw indications of the sandstone rock on which the scoriæ reposed having been affected by heat at the point of contact, but I sought in vain for such indications elsewhere.

. . . .

[1] [The American geologists consider the Pará sandstone to be a more recent deposit.—ED.]

RECENT GEOLOGICAL WORK ON LOWER AMAZON

[The preceding remarks were intended to have been rewritten by the author before publication in order to incorporate the results of later geological research. This is shown by a pencil note written soon after Spruce's return to England. I have thought it better, however, to give the account as I find it, because it is very clear and precise, and embodies facts which I cannot find in any of the descriptions by the American geologists who have investigated the geological history of the Lower Amazon. My friend Professor Branner of the Stanford University, who has travelled all over Brazil as the successor of Prof. Hartt, the former Government geologist, referred me to papers by Mr. Derby and Prof. Hartt in the *Proceedings of the American Philosophical Society* and in the *Boletim do Museu Paraense* for the best account of the geology of the Lower Amazon. These papers show that the valley lies in an elongate basin of Palæozoic rocks in narrow belts of Silurian, Devonian, and Carboniferous age, the outcrops of these rocks forming the cataracts of the various tributary rivers. On the north side of the valley these rocks are met with at less than 50 miles from the river, while on the south side they are from 100 to 150 miles distant from it; beyond them extends the great granitic region of Guiana and Brazil. In the Silurian strata at several points a rich molluscan fauna has been found closely agreeing, often in the very species, with those of corresponding age in North America, so that their position in the geological series is perfectly well established.

We now come to the series of flat-topped hills extending for about 150 miles on the north bank of the Lower Amazon to beyond Monte Alegre, and a few of precisely similar form and structure on the south side, as described above by Spruce. These consist of horizontal beds of sandstones and clays, and are often about 1000 feet high. They are isolated and have suffered a large amount of circumdenudation; but as no fossils have been found in them, their exact age is unknown, though they undoubtedly belong to the Tertiary formation.

In the lower land behind, and sometimes between, these there is exposed a large extent of coarse massive sandstone, with intercalated beds of shale. These rise into rounded hills farther inland, and also near the river in the Serras of Ereré; and they are all more or less inclined and disturbed, besides being traversed in various directions by trap dykes, which in some parts are very numerous, while in others the volcanic rock seems to occur in intrusive layers. In some places these dykes stand above the surface like ruined walls; in others they have been denuded more than the adjacent rock so as to form sunken channels. These sandstones contain fossilised wood and abundance of dicotyledonous leaves fairly well preserved. Hence it is concluded that they cannot be older than the Cretaceous age; while their being always more or less disturbed and penetrated by trap dykes shows that they are much older than the overlying softer sandstones, which are *always horizontal* and *never penetrated by the dykes.* It is therefore considered that they were either of Cretaceous or Eocene age, probably the former.

The preceding summary of the geological structure of the Lower Amazon valley, as described by the American geologists, enables us to understand the probable origin of the country. The highlands of Guiana and Brazil were evidently in existence in Archæan times, and from their denudation the Silurian, Devonian, and Carboniferous rocks were successively formed in the seas around them. The upheaval of these deposits must have extended across the intervening valley, unless the sea there was very deep, otherwise we should find some indications of secondary rocks formed during the enormous lapse of time between the Palæozoic and the Upper Cretaceous formations, though it is possible these may exist below the extensive Cretaceous and Tertiary deposits. In either case it seems certain that the central and chief portions of Guiana and Brazil have been continuously dry land since the close of the Palæozoic period, while considerable portions must always have been above water to furnish the source of the early Silurian and other sedimentary rocks. During this whole period denudation must have been continuously at work, the results of which are to be seen in the numerous isolated ranges and mountains of the vast Amazon-Orinooko plateau—the huge domes of granite or gneiss, and the great blocks or ridges of palæozoic or metamorphic rocks, the plains around which must have been all once buried under vast masses of superincumbent strata many thousand feet thick. Denudation has reduced this to what the American geologists term a *pene-plain*, from which now rise the denuded and weathered domes of granite, cubes or ridges of sedimentary rocks, and those strange rock-pillars which here and there rise above the forest towards the sources of the Rio Negro.

The very interesting work of Professor Hartt and his colleagues appears to have been concentrated on the north side of the great river, while the less extensive hills of Santarem receive the most meagre notice. Mr. Derby, towards the end of his careful paper, says: "The Tertiary beds of the southern side of the valley are, in the Santarem region, considerably lower than those of the north. The highlands behind Santarem are 400 feet high. . . . In a bed of blue clay exposed on the slope of these highlands, I found worm-tubes, the only fossils that the Tertiary beds of this region have yet afforded." He then goes on to say that the coarse sandstone beds of the plains about Pará and in Marajo "are certainly more modern and belong to the later Tertiary or the Quaternary." It is quite certain, therefore, that he could never have visited the remarkable rounded hills, almost buried in scoria-like masses, described by Spruce and cursorily examined by myself, these being not "behind" but 3 or 4 miles to the south-east of Santarem; while those farther inland, as described by Spruce, are exactly like the table-topped Tertiary hills of Monte Alegre. If Spruce's observation at Caripi (near Pará) of "trap rock penetrating into clefts of the sandstone which it had actually fused" be correct, it would seem to indicate that the Pará grit is, as Spruce supposed, older than the horizontal sandstone of Ereré and Santarem.

The only other fact bearing on this point is in a recent letter from Prof. Branner. He says: "I was at Santarem and saw dark scoria-like rocks, probably the same as those mentioned by Spruce. They closely resemble volcanic rocks, and were so compact that they broke with a glassy fracture; but those I saw were sandstones cemented with iron and silica. Similar rocks occur about Pará, and also over plains north and east of Macapa, where they cover large areas." These, however, I cannot think refer to the same rocks as those of the Serra of Irurá and adjacent low hills, which correspond much better with those of the early Tertiary or Cretaceous formation described by Mr. Derby and Professor Hartt. The former writer states that throughout these beds "diorite is very common, forming immense dykes, and sometimes apparently forming sheets between the strata of sedimentary rocks." And again he says: "The surface of these dykes is always decomposed, presenting a scoriaceous appearance, and enclosing crystals of quartz and fragments of the adjacent sedimentary rocks." Professor Hartt says that these dykes are often "so decomposed and eaten away that it is difficult to say what they originally were."

These descriptive phrases will apply well to the scoriaceous rocks observed by Spruce and traced by him over so large an area, and I think they prove that on the north as well as on the south of the river both the newer Tertiary and the older Cretaceous rocks occur adjacent to each other, the former characterised, in both areas, by loftier hills of horizontal strata and table-topped outline, while the latter are lower and often rounded, disturbed and penetrated by abundant dykes of trap or diorite often of very large size or even in intrusive layers. These latter hills have been more denuded, and those near Santarem are often thickly covered with the scoriaceous remains of the volcanic dykes which probably occur in or upon them. These volcanic blocks almost covering the slopes and summits of these hills are so overgrown with shrubs, grasses, and other herbaceous vegetation that the subjacent rock on which they rest was apparently not visible. But no doubt a little systematic search would discover exposures of it. There is evidently here an interesting problem for the next geologist who may visit Santarem. It may possibly be that this "conical" hill, so strangely covered with volcanic débris, may really be the fragmentary remains of the plug or core of one of the old Cretaceous or Tertiary volcanoes, the more massive and harder blocks of its débris having protected its more friable portion from complete degradation.—A. R. W.]

Aspects of Vegetation at Santarem

I proceed now to complete my account of the vegetation of the mouth of the Tapajoz, and to show how it was affected by the change of seasons, from wet to dry, in the year 1850.

The first effect of the rains was to bring out a luxuriant crop of grasses—tall, rank, and succulent on the banks of the rivers and in swampy ground, slender and wiry in the groves and thickets on the campo. Some idea of their variety may be obtained from the fact of my having gathered ninety species at Santarem. Sedges were less numerous, both in species and individuals, but included some pretty things, especially the species of Dichromena, which have the heads of flowers subtended by parti-coloured tracts, green below, white above. For the first three months in the year little was to be seen in flower but these grasses and sedges, and a few weedy plants in the neighbourhood of habitations. The trees—instead of being revivified by the rains, as in some other parts of tropical Brazil, where they lose their leaves during the dry season, or, in other words, *æstivate*—looked every day more and more dingy; and it was not until well on in the wet season, or even at the beginning of the dry, that most of them pushed forth new leaves and threw off the old ones. A very few shrubs, however, on the arid campo, that had seemed withered up at the end of the dry season, were clad with new verdure under the influence of the rains. One of these, *Connarus crassifolius*, sp. n., with leaves of three leaflets, like the Laburnum, but much thicker and stouter, and bearing a profusion of snowy flowers, was very handsome; it belongs to a small order (Connaraceæ) which trenches closely on some of the outlying members of Rosaceæ and Leguminosæ. As the ground became saturated with moisture, bare sandy and gravelly places on the campo got spotted over with patches of vegetation, some white, others green—the former composed of *Polycarpæa brasiliensis*, a pretty weed, rather like the Spurrey of our cornfields; the latter of a grass

(*Chloris foliosa*) with rigid tufted leaves barely half an inch long, which, after I had watched it nearly four months, at length put forth culms bearing at the apex six or seven silky, feathery spikes. These and a few other herbaceous plants made the upland campos look fresher than at any other time of year; but I never again saw them so gay and flowery as when I first visited them in the month of November.[1]

The great burst of foliage and flowers was, however, at the time when the rains and floods began to abate, and it was most obvious on the river margins. It was marvellous to see the myriads of minute annuals—or I might almost call them *ephemerals*—which sprang up on the shores of still bays and at the mouths of creeks of the Tapajoz. Following in the wake of the receding waters, they sprang out of the sand, flowered, and ripened their seeds; and by the time the sand had got quite dry—*i.e.* in a few days at most—they had quite withered away. And notwithstanding their humble size and transitory existence they were all pretty things, many of them with showy white, yellow, or pink flowers; and nearly all proved to be quite undescribed. They comprised two fairy water-plantains, resembling the *Alisma ranunculoides* of our English brooks in miniature; several Eriocaulons, Utricularias, a Hyris, a Herpestes, some slender annual sedges (Eleocharis and Isolepis), and a few other plants.

[1] I did not again see a Chloris until I reached the coast of the Pacific, south of the Equator, where in ground somewhat similar to that at Santarem, but usually for many months or even years together unmoistened by any shower, the rains of 1862 clothed the desert with a verdant carpet, wherein several species of Chloris were conspicuous.

One of the Utricularias (*U. Spruceana*, Benth.) was surely the simplest in structure of all its tribe, and may serve to give an idea of the general aspect of these ephemerals. Stems of the size of an ordinary sewing-needle, fixed into the sand by a little cone of rootlets, no leaves, but a minute tubular two-lipped bract a little below the flower, which is white and comparatively large, complete the description of its outward aspect; but then it grew in such abundance that patches of sand of many yards in diameter were white with it. The plant, however, that most interested me was an Isoetes (*I. amazonica*, Mgg.), exceedingly like the *I. lacustris* which inhabits our northern lakes. It was the first of its tribe that had been found near the Equator. A second species I found afterwards in nearly the same latitude on the cold paramos of the Andes at an elevation of 12,000 feet.

These ephemeral plants on the beaches of the Tapajoz are a most remarkable feature of its vegetation, and I have seen nothing like it elsewhere, except on inundated islands in the cataracts of the Uaupés. Certainly vegetation is on a most gigantic scale in the Amazon valley, not only as regards the vast size attained by some of the species, but also in the range of magnitude from the enormously large to the extremely minute. Compare, for instance, the lofty Eriodendrons and Caryocars with these lowly Utricularias and Alismas.

Whilst the beach was thus being bedecked with pretty but transitory flowers, the more permanent vegetation of its sandy or stony outer margin was also putting on a flowery garb. Low bushy trees, averaging 20 to 30 feet high — with here and there a taller, but never a lofty, tree intermixed—and most of them bearing showy blossoms, fringed the beach of the Tapajoz. Where the shore rose abruptly inland, the fringe of gapó was narrow; but where it was nearly flat, as at the mouths of rivulets, there was a great breadth of that peculiar arborescent vegetation which flourishes only where the plants are wholly or in part submerged during some months in the year, being to them a sort of hibernation. The gapó vegetation of the Tapajoz has quite the same character as that of the Rio Negro, where I afterwards found several of the identical species of the Tapajoz, especially certain Leguminifers, such as *Campsiandra laurifolia*, Benth.; *Outea acaciæfolia*, Benth.; *Leptolobium nitens*, Vog.; and a Chrysobalan, *Couepia rivalis*, sp. n. The first of these is a low spreading tree or shrub, bearing a profusion of flowers, white within, rosy without, not unlike those of the peach or almond, but grouped in large corymbs. On the extreme edge of the gapó it sometimes forms a continuous fringe of miles in length, especially by the Rio Negro. The flowers are followed by pods containing large flat beans, which little Indian boys find suitable for making ducks and drakes with; and their mothers grate down and (having got rid of the bitter narcotic principle by straining and baking) make passable farinha thereof; but this is only when mandiocca runs very scarce.

[The accompanying photographic print of the gapó vegetation of one of the tributaries of the Lower Amazon near Pará shows conspicuously two species found here by Spruce. The small tree in the foreground with conspicuously mottled trunk is that just described, and the colour of the bark is due to variously-coloured lichens, as already referred to (p. 27). Almost behind the Campsiandra is the Jauarí palm, the lower portion of the stem being thickly clad with long spines directed downwards. As the tree gets older these fall off, as shown in the two specimens to the right, which exhibit the scars of the fallen leaf-stalks in beautifully regular pale rings. Both these trees are common in the clear-water tributaries of the Amazon and Rio Negro.—Ed.]

Other trees of the Tapajoz were species of Terminalia, Genipa, Tecoma, etc. But the most ornamental tree there was *Pithecolobium cauliflorum*, Mart., a Mimoseous tree of moderate size, with a gnarled tortuous trunk on which and on the main branches grow the flowers, consisting almost wholly of long thread-like stamens, lake-red above, white below; and they are so densely packed as to give the trunk the appearance of being enveloped in toucan's feathers, thus producing, along with the green, leafy, but flowerless crown of the tree, a striking and novel effect.

Among the very few palms at Santarem, one, the Jará (*Leopoldinia pulchra*, Mart.), grows gregariously by the Tapajoz; and it reappears on the Rio Negro in such abundance as to be one of the characteristic plants of that river. It is of humble growth, rarely exceeding 12 to 15 feet, and its most marked feature is the rigid leaf-sheaths, split into finger-like divisions, which remain clasping the stem like so many gauntlets after the leaves themselves have fallen away.

[Another of the interesting palms found by Spruce in the caapoeras of Santarem is the small

FIG. 5.—IN THE GAPÓ.
Campsiandra laurifolia. Astrocaryum Jauary.

Mumbáca palm, which grows from 8 to 12 feet high, and has a slender prickly stem, beautifully regular pinnate leaves, and small red or orange-coloured fruits. It is rare in the virgin forests, but more abundant in second-growth thickets near Pará and Santarem. The photographic print (on page 155) shows a group of these palms in the undergrowth of the forest. It was taken near Pará, but is equally characteristic of the places where Spruce met with it, as described in his paper on "Equatorial American Palms," in the *Journal of the Linnean Society* (vol. xi. 1869).—ED.]

The vegetation of the shores and islands inundated by the turbid waters of the Amazon was almost entirely diverse from that of the blue Tapajoz. Enormous figs, often with tortuous deformed trunks, and sometimes sending down props like the Banyan; Silk-cotton trees; India-rubber trees (*Siphonia Spruceana*, Benth.); and the Itaúba-rana (*Ormosia excelsa*, sp. n.), a fine tall timber tree, with hard discoloured wood, and panicles of lilac flowers, were conspicuous among the trees of the gapó. But more abundant than any of these, and (as I afterwards found) extending along the banks of the Amazon to the very roots of the Andes, was the Pao Mulatto or Mulatto tree, so called from the colour of its bark, which is continually peeling off and being renewed. It grows 60 to 100 feet high, and branches in such narrow forks that its top is usually in the form of a reversed cone; which peculiarity, along with its shining reddish-brown skin, and (in the season) its corymbs of flowers resembling those of the hawthorn in colour and scent, render it everywhere a striking object. It is closely allied to the Chinchonas or Peruvian Barks, and Mr. Bentham has made it the type of a new genus under the name of Enkylista.

Several small permanent lakes communicating by short channels with the Tapajoz—as well as flats and hollows which had become lakes during the rainy season—brought forth many curious plants in their waters and along their borders in the months of July and August. It was notable there, as in Europe under similar circumstances, how those aquatics which rear themselves erect, and thus bear the flowering part of their stem well out of water, have the submersed leaves set round the stem in whorls, quite alien to the habit of their congeners growing on terra firme. Thus a Jussieua (*J. amazonica*) had the narrow submersed leaves so closely whorled as to quite resemble the Mare's-tail of our pools; while the emersed ones were solitary, as are all the leaves in the other species of the genus. *Sipanea limnophila*, sp. n., had many-leaved whorls under water, while the leaves just out of water stood four together, and the uppermost were merely opposite; whereas in the other species of the genus (which in habit, and in their pink or white flowers, resemble our Soapworts and Campions, although their affinity is really with Madders and other Rubiads) all the leaves are either opposite or rarely three in a whorl. These plants, and others that grew along with them and showed the same peculiarity, were all, strictly speaking, *amphibians*, the water wherein they had first vegetated being completely dried up ere they had ripened their seeds. The true aquatics—such as passed

FIG. 6.—MUMBÁCA PALM (*Astrocaryum Mumbaca*).

the whole term of their existence in the water—had all of them some contrivance for sustaining their flowers high and dry until fertilisation had been effected, or indeed until the fruit was fully ripe. A Utricularia (*U. quinqueradiata*, sp. n.) deserves especial mention; it is a small species, with submersed finely-divided leaves bearing numerous bladders; but the flower-stalk, which is about two inches long, has midway a large involucre of five horizontal rays resembling the spokes of a wheel; this floats on the surface and keeps the stalk always erect, and the solitary flower well out of water; the whole recalling a floating night-lamp, especially as the large yellow flower may be considered to represent the flame.

The aquatics bred in the turbid waters of the Amazon have already been described in my account of the Ponta Negra (p. 113).

On the margins of lakes, and elsewhere in moist sandy grounds, grew several small plants, distinct from those already mentioned of the shores of the Tapajoz; such as several Milkworts (Polygalæ) and Xyrides, the latter looking like miniature Daffodils. *Polygala subtilis*, H. B. K., and *Burmannia capitata*, Mart., two fairy little plants, both having nearly leafless stems and heads of cream-coloured flowers, but otherwise extremely unlike in their structure, reappeared in a similar site on the Rio Negro, and again on the savannahs of the Orinoco. *Pectis elongata*, H. B. K., a Composite herb with a strong Tansy-like smell, abounds in the same places, and still more on the savannahs of Guayaquil.

The vegetation of the upland campos has already been sketched as it appeared in November, after the autumnal rains. It was then in its prime, as that of the river-margins was in July, August, and September. A few of the small trees, such as the Anacardium and the Plumiera, flowered more or less all the year round; but the taller trees that grew scattered about the campo flowered chiefly from July to August. The finest of these were two Leguminifers; the one, *Bowdichia pubescens*, Benth., had bright blue or violet flowers; the other, *Lonchocarpus Spruceanus*, Benth., had long compound spikes of red-purple flowers; both were very ornamental, and yet, from growing dispersedly, they nowhere produced the effect that a great mass of gay colour does, as seen in our fields of flax, clover, etc. *Vochysia ferruginea*, Mart., a very handsome tree, bearing spikes of yellow flowers which exhaled a most delicious odour, was common in the low grounds; and I saw it again in similar sites on the Casiquiari and Orinoco, and in the roots of the Peruvian Andes.

The volcanic hills proved to have a very meagre vegetation, although I explored them most sedulously, and devoted several fatiguing excursions to them. Some of the slopes were clad with a dense growth of stout reedy grasses, which, together with the rough stony ground, made the ascent sufficiently painful. Two of these grasses, however, were very handsome; *Paspalum pellitum* from its sharply-folded Iris-like leaves, and *P. pulchrum*, Mart., from its spikelets—closely set on six digitate spikes—being each surrounded by a row of golden-yellow bristles. Of trees there were very few, and those mostly solitary—rarely gathered into groves. One of them was a Euphorbiad, *Mabea fistulifera*, Mart., notable, like many of its congeners, for its long twiggy fistulose branches, which are in common use on the Amazon as tubes for tobacco pipes, under the name of Tacuari. The same species had previously been gathered by Pohl and Martius in the provinces of Minas, Goyaz, etc. There was one very fine tree in the serras, and also in stony places on the campo, towards Santarem—*Salvertia convallarioides*, St. Hil., a Vochysiad. It grew 30 to 40 feet high, and the leaves and branches being arranged in whorls, six or seven together, gave the tree a symmetrical, candelabrum-like aspect, which was rendered more striking by the branches being upturned at their extremity and bearing each a panicle of large white hexapetalous flowers. These had the delightful scent of the Lily of the Valley, so that in walking through a grove of Salvertias in flower, I was constantly reminded of that charming though lowly plant. In drying, they assumed the still richer odour of the Violet.

In stony valleys grew *Lafoensia densiflora*, St. Hil., a small tree with large curious flowers not unlike those of the Pomegranate, but white instead of red. The Mabea, the Salvertia, and the Lafoensia grow all through the hilly campos of tropical Brazil. I did not see them elsewhere in the Amazon valley, any more than *Strychnos brasiliensis* and other South Brazilian plants gathered by myself only at Santarem.

Although I did not visit the hills of Monte Alegre, yet, as they are evidently a continuation of those of Santarem and have quite the same character, and as I was told by Mr. Wallace that he found their ascent obstructed by a dense growth of coarse grasses quite similar to those of Santarem, I should expect equally to find there *Curatella americana*, *Mabea fistulifera*, *Salvertia vochysioides*, and other plants of Central and Southern Brazil. And if there are, as the natives assert, other bare hills to northward of those of Monte Alegre, then there is probably a break right across the Amazonian forest, in longitude 54°-55° W. from the granitic region of Central Brazil to that of the frontiers of Dutch Guayana, of open hilly ground, wooded in some intervening valleys and hollows, but of a quite different character from the remainder of the densely forest-clad Amazon valley.

Lofty primeval forest was rare near Santarem. To reach any such I must penetrate by land to the sources of the Irurá and Mahicá; or go by water a few miles down the Amazon and then up some igarapé. I shall mention only a few of the forest trees which are notable for their products, and conclude this chapter with a notice of some of the edible wild fruits of Santarem.

Itaúba, *i.e.* Stone tree, so called from the hardness of its wood, which is more esteemed for shipbuilding than any other on the Amazon, is a noble tree of the family of Laurels, which was undescribed until my specimens of its flowers and fruits afforded materials for its determination. There are two varieties of it, the *preta* or black (*Acrodiclidium Itauba*, Meissn.) and the *amarella* or yellow. The former attains a larger size, and the wood is a deep dull purple—the heart-wood nearly black; while that of the latter is paler and yellowish. Itaúba wood is somewhat heavier than water, so that a canoe made of it infallibly sinks when full of water, as I have found to my cost; but for the construction of a large boat there is no timber on the Amazon equal to it. The Laurél amarilla (*Ocotea cymbarum*, H. et B.) of the Casiquiari and Alto Orinoco is the only American tree superior to it, for the wood is equally hard and imperishable, and it is lighter than water. The Greenheart of Demerara belongs to the same family. The Itaúba bears an oblong black berry, of which the pellicle is studded with glandular dots and has on it a bloom like that of a plum; it contains a single large almond-like seed invested with pulp, an eighth of an inch thick, which is good eating, spite of its strong resinous flavour, and is sometimes made into wine like the pulp of the fruit of the Assaí and other palms. The Brazilians compare them, and justly, to a small variety of olive of which large quantities are imported from Portugal.

Cumarú-rana or Bastard Tonga-bean (*Andira oblonga*, sp. n.), a Leguminous tree growing in woods beyond a site called Urumanduba, is notable for its flowers and fruits, having a fine odour of orange-peel and balm (Melissa), approaching to that of the seeds of the true Cumarú (Dipteryx); whether they possess the same properties I know not.

Cupa-úba or Balsam Capivi tree (*Copaifera Martii*, Hayne). Of this there were a good many specimens on the wooded slope intervening between the upland campos and those of Mahicá; but they were said to yield so very little oil (or balsam) as not to be worth tapping. The habit of the tree was, however, quite the same as that of other species of Copaifera from which capivi is obtained in great quantities along various tributary streams of the Amazon. All the species have the small flowers closely set on the branches of a rigid pinnate panicle, the flattened pink ovary standing out beyond the four or five white petals and the free stamens (eight to eleven); and the leaves consist of two or more pairs of deep green leaflets beset with pellucid dots. In old trees the trunk becomes hollow at the core, and there the oil accumulates and is extracted by boring with an auger. But on the Casiquiari I saw the trunks tapped by cutting out a wedge near their base, deep enough to reach the deposit of oil.

Pitómba (*Sapindus cerasinus*, sp. n.), a shrub 6 to 10 feet high, with pinnate leaves and white flowers, grows on stony slopes, at Cape Mapirí and elsewhere, on the Tapajoz. It bears a yellow fruit the size of a cherry, and has something of the same taste. The thin pulp envelops a single seed, which, on tasting, I found to have a pleasant flavour of black currants, and therefore ate several of them; nor did any ill consequences result, but when I told my Santarem friends of it, they said they had never known of the seeds being eaten, and that I had acted imprudently, for the plant belonged to a poisonous family. I knew, however, that the seeds of the nearly-allied Guaraná were wholesome, and I afterwards found the seeds of most of the Sapindaceæ are at least harmless, notwithstanding the deadly properties of the stems and roots of such plants as *Paullinia pinnata*.

Tapiribá or the Tapir's fruit (*Mauria juglandifolia*, Bth.). A tree belonging to the Anacardiaceæ, like our ash in its leaves, but not attaining so great a size, and it bears an oblong yellow subacid drupe about the size of a wheat-plum.[1] The tree is frequent throughout tropical South America, and is known in Venezuela by the name of Jovo or Hobo, and in Peru by that of *Ciruelo amarillo* or Yellow Plum; but I do not know that I have anywhere seen it truly wild. It is extremely tenacious of life, so that a stake cut from it nearly always takes root in the ground, and (if allowed) grows to be a tree; on which account it is much used for fencings of corrals on the Orinoco, and of cane-fields, etc., about Guayaquil. At Santarem, rows of stakes of Tapiribá, which had been stuck by the roadside leading out of the town to the cemetery on the campo, about a year before my arrival, were already acquiring leafy heads, and promised to form soon a shady avenue.

Apiránga (*Mouriria Apiranga*, sp. n.). A small tree about the size of the plum, belonging to a curious genus intermediate between Myrtles and Melastomes. It bears a pleasant-tasted red berry with three stony seeds.

Araçá (*Psidium ovatifolium*, Berg.). This is a Myrtle—a sort of small Guayaba, rather more acid than the common kind. The acid fruits of several Eugenias are also called Araçá.

Tapiíra-guayaba (Belluciæ sp.). A Melastome, with a slender—often unbranched—trunk, reaching 50 feet high, and bearing a few large leaves at the top and a profusion of white rose-like flowers on

[1] [This is an old Yorkshire plum, so named from being ripe at harvest-time, when pies were made of it. It was not of very good quality and is now superseded by the Victoria plum.—ED.]

the naked trunk. The fruits are like small apples to look at, and partly to the taste, but they are mawkish, like all their tribe; and in reality they are berries, divided into twelve cells, and enclosing numerous minute seeds.

Yenipápa (*Genipa macrophylla*, sp. n.—Cinchonaceæ—and other two new species)—*Genipa americana*, L., the most widely distributed species of the genus, I have seen wild in many places across the whole breadth of South America. In Peru it is called Huítu; in Ecuador, Jagua. Its fruit (a large olive-green berry) affords a permanent black dye, and is in universal use by the Indians for staining their skins; it is also pleasant eating, when allowed to become over-ripe, having then the consistence and much of the flavour of the medlar. Three kinds of Yenipápa grew along the shores of the Tapajoz, and proved to be all undescribed. One of them has leaves full 18 inches long, and globose fruits as large as a swan's egg. All have the same properties as *G. americana*.

Uirarí-rana (*Strychnos brasiliensis*, Mart.). A small bushy tree, with twiggy, decussate branches, growing in the outskirts of Santarem, and bearing red three-seeded fruits, whereof the pulp is edible though insipid. I met with a second example of the occasional harmlessness of the fruits in this deadly genus on the river Uaupés, where the wild turkeys eat the berries of *Strychnos rondeletioides*, sp. n.

S. brasiliensis does not climb—at least I saw no example of it at Santarem—although the twiggy branches seem apt for it, should need occur. I did not meet with the plant elsewhere, but it is frequent farther south.

A long list might be made of plants having an edible pulp to the fruit, although the seeds and every other part of the plant may be virulently poisonous. The example of the fruit of the Yew is familiar to every one.

CHAPTER VI

VOYAGE FROM SANTAREM TO THE RIO NEGRO BY WAY OF THE FLOODED AMAZONIAN FOREST

(*October* 8 *to December* 10, 1850)

I HAD every reason to be satisfied with my collections at Santarem; but when I had nearly exhausted the Flora accessible within a day's journey, I began to long for new fields, and I fixed on the mouth of the Rio Negro for my next centre of operations. Untoward circumstances had prevented my making any long excursion from Santarem, besides that to Obidos and the Trombetas. I had planned an expedition of a month up the Tapajoz, in the rainy season, in a small vessel of Mr. Hislop's, and had made every necessary provision for it, when I was struck down by fever on the very eve of starting. I had not at that time any boat of my own, nor in all probability could I have got sailors to man it at Santarem, where every free man of colour was in debt to the resident merchants, who would have exacted payment of the debts before allowing the men to embark on a voyage. I had hoped to get a passage up the Amazon for myself and my companion on board a schooner from Pará, belonging to an Englishman named Bradley, and which actually passed Santarem on its way up in July, but so heavily laden, and already so crowded with passengers, that there was no room left in it for us. We finally left Santarem for the Barra do Rio Negro—now called the City of Manáos—on Tuesday the 8th of October, in an igaraté belonging to Monsieur Gouzennes, a French gentleman who had been many years settled at Santarem, and was accustomed to send vessels up the Amazon every year to procure salt fish, turtle oil, Brazil nuts, and other produce, in payment for goods advanced the previous year. Our vessel was a very small one, of little more than 3000 arrobas (= 9600 lbs.) burden, and my baggage half-filled it. For want of room we were put to much inconvenience in preserving such plants as we could collect on the way; and, what was still worse, the palm-leaf toldo or cabin was so ill-constructed that every heavy rain penetrated it, and gave us afterwards much trouble in drying our soaked clothes, papers, and eatables. However, there was no alternative, and for this conveyance, wretched as it was, I had waited nearly three months.

Our crew consisted of but three men: the Cabo or captain—a fine young fellow named Gustavo, eldest son of the French baker—and two mariners, the one a Mamaluco or half-breed, the other a pure Indian of the Yuma tribe, which inhabits the lower part of the Madeira. As it was calculated that we might still have before us two months of dry weather and brisk easterly breezes, this scanty crew was considered sufficient; but, as it turned out, the weather was broken, rainy, and either calm or squally, from the very beginning of our voyage, and winter had fairly set in before its close. M. Gouzennes himself, with his family, accompanied us in a much larger vessel, called a cuberta, as far as Villa Nova.

. . . . . .

Rarely is there perfect silence on the banks of the Amazon. Even in the heat of the day, from 12 to 3 o'clock, when birds and beasts hide themselves in the recesses of the forest, there is still the hum of busy bees and gaily-coloured flies, culling sweets from flowering trees that line the shore, especially from certain Ingas and allied trees; and with fading twilight (6½ P.M.) innumerable frogs in the shallows and among the tall grasses chaunt forth their Ave Marias, sometimes simulating the chirping of birds, at others the hallooing of crowds of people in a distant wood. About the same hour the carapaná (mosquito) begins its night-enduring song, more annoying to the wearied voyager than even the wound it inflicts. There are, besides, various birds which sing, at intervals, the night through, and whose names are uniformly framed in imitation of their note; such are the acuráu, the murucututú—a sort of owl—and the jacurutú, whose song is peculiarly lugubrious. A sort of pigeon, which is heard at 5 o'clock in the morning, is called, and is supposed to say, "*Maria, já he dia!*" ("Mary, it is already day!")—a name which reminded one of "Milk the cow clean, Katey!" a Yorkshire appellation of the stockdove. Among the birds which most amused me with their note by day were the "*Bem te vi!*" ("I saw thee well!") and the "*Joaõ corta páo!*" ("John, cut the stick!").

I one night much amused the sailors by inquiring what *bird* it was that was making a croaking noise in an opposite cacoal. It was not a bird, they said, but a small quadruped, the size of a rat, which had its abode in the cacoals and lived on the fruit. It is one of the animals specially resorted to by the Indian payés or wizards, and great importance is attached to its replies, which are merely a repetition of its note—written *Torô* by the Brazilians, but sounding almost like the French *trou*—for an affirmative, and perfect silence for a negative. I was in my turn diverted by one of the crew holding a conversation with the Torô, whereof what follows is a nearly literal translation.

"Your worship sings very sweetly all alone by night in the cacao tree!"—"*Torô! Torô!*"

"Your worship seems to be enjoying your supper on the delicious cacao!"—"*Torô! Torô!*"

"Will your worship tell me if we are to have a favourable wind in the morning?"—Torô respondeth not.

"Your worship, do me the favour to say if we shall arrive at Obidos to-morrow?"—Again no reply.

"Your worship may go to the devil!"—An insult of which Torô taketh not the least notice; and so ends the dialogue, the Indian being too angry to interrogate further.

When lying-to for a wind I obtained a few plants unobserved, or left ungathered, the preceding year; and when slowly beating against the strong current in the Strait of Obidos, I twice swam on shore to gather a stout Mimoseous twiner that adorned the banks for miles with its thick spikes, a foot long, of minute pale yellow flowers.

Obidos seems unlucky for travellers. Here Spix and Martius had been delayed to repair their helm, thirty years before; and we ourselves had scarcely embarked, early in the morning of the 15th, when our boat took the ground in a stony place, and the ironwork of the helm was broken by the shock. It took a smith the whole day to repair it, and it was not until 10 of the following morning that we got it fastened on and again set forth on our voyage.

Above Obidos, we began to meet with vast numbers of alligators. When anchored on the night of the 16th in the still bay at the mouth of the Trombetas, we were surrounded by them, mostly floating nearly motionless on the water and only distinguishable from logs by the undulations of the back. Their grunt is something like what a pig might make with his mouth shut; our people imitated it, and thus drew several of them quite near us, but I did not care to waste powder and shot on them. The following morning, coasting slowly along a low muddy shore, we saw a multitude of them—large and small—put off from land into deep water at our approach.

The female alligator of the Amazon piles up her tough-coated eggs, the size of swan's eggs, to the number of from forty to sixty, and covers them with dead leaves and other rubbish, so that the pile or nest looks like a small haycock. One morning the people of M. Gouzennes's cuberta went on shore to collect firewood, and as they ran along in Indian file they passed close to an alligator sitting on her nest. She took no notice of them, but the hindmost man called out to M. Gouzennes on board the cuberta, which was lying close inshore, and pointed her out to him. M. Gouzennes then fired twice on her with ball; but although hit each time the only effect was to make her turn round on her nest and look very angry. This is the only instance that has fallen under my own observation of the alligator incubating, but I have often heard it spoken of as a fact.[1]

It took us ten days to reach Villa Nova from Obidos, although the distance is only 95 miles; for there was seldom any wind beyond the squalls preceding thunderstorms, and those rarely blew in the right direction; and it was very slow and very hard work for a couple of paddles to make head in our heavy craft against the current of the Amazon. The river was at its lowest, and had receded from the forest-margin, leaving in some places a bare sandy or muddy beach, so broad that I could barely walk to the farther side of it and back whilst our breakfast or dinner was being cooked; and then I rarely found anything in flower besides *Mimosa asperata*, and two or three common river-side Ingas and Myrtles. All we could do, therefore, was to lie under the toldo and doze or read our books and old newspapers.

. . . . . .

The whole coast from Obidos to Villa Nova is flat and uninteresting, until a little below the latter town its tameness is somewhat relieved by a lowish wooded ridge, called Os Parentins, running close

[1] In ascending one of the rivers of Guayaquil in a boat with two men, we came to where a bit of earth-cliff had fallen in and had left exposed at the top a deposit of alligator's eggs. One of the men picked out a couple of eggs and dashed them on the water. They broke with the shock, and out of each darted a fully-hatched alligator and dived out of sight. I never saw a finer example of instinct, or inherited reason, being called into play at the very moment of a creature's coming into the world.

by the right bank. This ridge is the eastern limit of the new Province of Amazonas, which extends westward to the Peruvian frontier. According to Baena (*l.c.* p. 230) the Jesuit missionaries founded a village of Indians of the tribe Parentins on the flat top of the hill; but it did not last long, for the neophytes revolted against their teachers, burnt all the houses, razed the church, and buried the bells. Local tradition asserts that those subterraneous bells may still be heard to ring every Christmas Eve. It was late in the evening of the 24th when we reached Villa Nova. We found it a miserable-looking town, the houses going sadly to ruin; and there was but a single small vessel in the port. It stands on a small bay, skirted by a lowish cliff, upon which are piled blocks of diorite, like those of Santarem. We went on shore and visited the Vicar—"Padre Torquato," the celebrated story-teller of Prince Adalbert's *Voyage up the Xingú*. We found him a young man—certainly under forty—good-looking and rosy—exceedingly courteous in his manners, but delighting wonderfully to hear himself talk, and therefore not unlikely to be led into the relation of marvellous tales, *as true*, although himself sceptical respecting them. He seemed highly flattered to hear that the Prince had made mention of him in his travels.

[See also my *Travels on the Amazon*, pp. 109-10 and 266, and Bates's *Naturalist on the Amazon*, pp. 147-48, for further examples of Padre Torquato's character and of his universal kindness to European travellers. I may mention here the unfortunate habit of altering names of places so prevalent in Brazil. At the time here referred to "Villa Nova" was the universally known name of the little town at the entrance of the great Paraná-mirí dos Ramos, which extends for more than 200 miles to a little above the mouth of the Madeira, its official name being "Villa Nova da Rainha," as given on the excellent maps of the Society for the Diffusion of Useful Knowledge in 1852. But in all modern maps I have seen, including the large map of Brazil issued by the "International Bureau of the American Republics," no such name appears, but instead we find the old Indian name PARINTINS in capitals, with (Villa Bella da Imperatrice) in brackets as the former name—so that the town has had four distinct names in about half a century. In the memorial edition of Bates's book, published in 1892, it is termed "Villa Bella" only, so that the place where he resided twice and made some of his most interesting collections may be looked for in vain! Similar cases occur everywhere in Brazil, so that it becomes almost impossible to follow the route of any of the older travellers on a modern map.]

Where Villa Nova now stands was formerly the "Mission of Tupinambarána," established by a certain José Pedro Cordovil in 1803, who gathered together several Mauhé and Mundrucú Indians, and induced them to settle there. The name he gave the mission was to indicate that the people were not *true*, but *spurious* Tupinambás or Tupís. It was not raised to the rank of "Villa" until 1818. This explains why on many maps a broad strip of country along the right bank of the Amazon appears as "Ilha de Tupinambaránas"—not that there was originally any nation bearing that name, nor is it known at all to the actual inhabitants, except as the name of a river forming the eastern boundary of the so-called "island." The southern boundary of the island—or, more properly, series of islands—is a long winding channel called the Furo de Urariá, or sometimes Paraná-mirí dos Ramos, which, leaving the Madeira in about 4° S. lat., runs parallel to the Amazon through about three degrees of longitude and joins that river near Villa Nova, in $2\frac{1}{2}$° S. lat. . . . The region between the Urariá and the Amazon is literally sown with lakes, which communicate by short channels, some with the Amazon, others with the Urariá. About midway of the Urariá a channel branches off to the Amazon, and this is considered the upper mouth of the Ramos; while the Urariá thence to the Madeira is often called the Furo de Canomá, from the principal river that enters it. . . .

The Urariá bears some resemblance to the Casiquiari—the celebrated channel uniting the Orinoco to the Rio Negro—not only in its length and other dimensions, but in some of its other features; and as I shall have to describe them both, my readers will be enabled to make the comparison for themselves.

Our little vessel was destined to traverse so much of the Urariá as bears the name of "Ramos," to look up some of M. Gouzennes's creditors there; while that gentleman himself proposed proceeding up the main river as far as to the upper angle of the Madeira and Amazon, where there is another great region of lakes and channels called the Uautás.

We had passed below Villa Nova at least three outlets of the Ramos; but we entered it, a few miles above the town, by a channel called the Limoẽs, along which there was a scarcely perceptible current towards the Amazon. We expected to be detained only a few days in the Ramos, whereas, as it happened, we spent an entire month there. In that time I might have made many interesting observations respecting the great country of Guaraná and Pirarucú, had I not unfortunately been taken ill soon after entering it. We passed the night of October 29 at a great bend of the river, and being wishful to determine its position by an astronomical observation, I lay all night outside the cabin for that purpose—a thing I had done many times on the Amazon without taking any harm from it; but the night was very cloudy, and so much dew was deposited that in the morning my blanket was soaked, which brought on an attack of fever; and although it abated in three or four days, I did not fairly recover from it until I got out again into the broad Amazon. During our stay in the Ramos, we had constantly heavy night-dews, whereas on the Amazon the dews were none or scarcely perceptible. This is doubtless owing to the breezes which sweep *up* the Amazon, but in narrow channels like the Ramos are reduced to light puffs blowing in no settled direction. I say "narrow," compared to the breadth of the main river, although I estimated the breadth of the Ramos at from 400 to 600 yards. . . .

As we slowly ascended the Ramos, we came at every few miles on a sitio or clearing, consisting of one, two, or three houses, tenanted by people of mixed race. The adult males were nearly all absent, either fishing in the lakes or collecting turtle oil on the Amazon. Many of these people were debtors of M. Gouzennes, but his agent, Gustavo, could get nothing from them but promises to have the payment ready by the time of his return voyage. On the 3rd of November we reached a sitio called "As Barreiras" (The Cliffs), consisting of two houses perched on the top of a cliff of sandy clay on the right bank. Here the men were actually occupied in catching and preserving fish for M. Gouzennes in a lake called Lago das Garças, lying on the opposite side of the river; and we had to wait until it should be dry enough to be embarked, which detained us for twelve days.

The lakes which lie thickly on both sides of the Ramos are all richly stored with fish. In the height of the dry season, when the water of the lakes is low, numbers of fishermen resort to them for the purpose of taking pirarucú; including not only all the available population of the Ramos, but also fishing-parties from places as far distant as Pará and Macapá. When I had somewhat recovered from my sickness, I managed to reach the Lago das Garças; to do which I had to thread a narrow track above three miles long through thick forest, consisting chiefly of wild Cacao trees, Bertholletias, and Urucurí palms (*Attalea speciosa*, Mart.). I found the lake nearly circular, of about a mile in diameter, and several fishing-parties were at work on it. The general sleeping apartment was a large palm-leaf shed erected on poles in the lake, at a sufficient distance from shore to secure it from the visits of carapanás. This contrivance is resorted to on all the lakes, which are abominable places for *plague* of every description.

I was disappointed not to observe a single plant in the lake, save the rank grasses around the margin; but alligators were floating on the water in almost countless numbers, resembling so many huge black stones or logs. What we had seen in the Amazon of these reptiles was as nothing compared to their abundance in the Ramos and its adjacent lakes. I can safely say that at no instant during the whole thirty days were we without one or more alligators in sight, when there was light enough to distinguish them; and we might hear their snorting or grunting all the night through. Alligators sometimes take the bait intended for pirarucú, and the line is strong enough to hold them. One morning at early dawn our men espied a young alligator about 7 feet long fast asleep in a shallow bay close by where our boat was moored. He lay with his head in the mud, and only the end of his tail sticking out of the water. They got a stout pole and drove it at him with their united force; whereon he whisked round too nimbly for his assailants to spring out of the way, spirted a shower of mud over them, and dived away.

For a description of the pirarucú I must refer to the writings of the naturalists who preceded me. It is the monarch of the fishes of the Amazon, and one of the finest fresh-water fishes in the world. When full-grown it measures 6 to 8 feet long, weighs from 60 to 100 pounds, and yields about one-third that weight of dried fish. When fresh it is capital eating, although scarcely equal to salmon, with the exception of the lower part of the belly (called the ventrexa), which being cut from the newly-caught animal and roasted on a spit over a brisk fire, is one of the choicest morsels I ever tasted; and although I call it "a morsel," it is a meal for three. It is, in fact, half luscious fat; and a notable character of most Amazon fishes is their excessive plumpness, which renders the broth made by boiling them quite as delicious as the fish itself. An Amazonian would actually as soon think of throwing away the fish as the water in which it had been boiled!

. . . . . .

During the night of the 5th our Yuma Indian gave us the slip, and took with him our montaria, the captain's cup, and the Mamaluco's cutlass, bow and arrows, hooks and lines, looking-glass and frying-pan. His tribe, like their near neighbours the Muras, are renowned for craftiness and pilfering. He had been on bad terms with the Mamaluco throughout the voyage, and took this way of revenging himself, as also of escaping to the freedom of his native forests, which were not far distant. We had now but one sailor left, the Mamaluco, who, spite of his crabbed disposition, worked well when not under the influence of cachaça—a thing which happened to him two or three times during the voyage, when he profited by my temporary absence to help himself from my demijohn; but his worn, dissipated look bore witness to habitual devotion to the fiery liquid. He was a bit of a philosopher in his way, and used to amuse me with his cynical views of life, that showed him to be as completely *désillusionné* as any man about town, or even as Ecclesiastes himself. One evening he lay on deck watching a cockroach, as it struggled to release itself from its old coat, and at length emerged—weak and tottering, but still clean, white, and new to look at; whereupon our Jacques moralised after this fashion. "How is it," he said, "that almost every animal except man renews its youth and beauty at stated seasons? Birds moult their plumage—snakes slough their skins—even this despicable little bicho, the cockroach, casts off its old covering—and all come forth bright and beautiful as in the days of their youth; but we" (casting his eyes on his brown wizened hand) "grow uglier and more discoloured every year, and the same skin in which we were born must serve unto our dying day!"

The Yuma was a lazy fellow, and we should not have missed him much, had he not taken away the montaria, which was very useful for fishing and shooting trips, and for landing at any time when it was impossible or inconvenient to take the larger boat close inshore.

Within sight of our station at the Barreiras was the mouth of a considerable river, the Mauhé, upon which, at a distance of thirty hours' journey in a montaria, there stands the town of Luzêa, anciently Aldea dos Mauhés, or Village of the Mauhé Indians. . . . Although Luzêa was not to be found on any published map in 1851, it was a place of growing importance, and boasted of a church and chapel, with a few shops and several white residents. It was founded by the Portuguese in 1800, with 243 families of Mauhé and Mundrucú Indians, the government furnishing them with iron tools and building them a church. In 1803 the population already amounted to 1627 souls, of whom 118 were whites. The progress of Luzêa has been entirely owing to its being the great centre of the cultivation of Guaraná, of which there are large plantations called guaranals near the town, as also higher up the Mauhé and on the Canomá; and near the head of those streams the plant is said to grow wild.

. . . . . .

I afterwards saw the plant cultivated on the Rio Negro, where I drew up a description and prepared specimens of it.

The Guaraná plant (*Paullinia Cupana*, Humb. and Bonpl., of the natural order Sapindaceæ) is a stout twiner, whose scandent propensities are kept down in cultivation, so as to reduce it to a compact bush with sinuous entangled branches. The leaves are pinnate, of five leaflets, each nearly half a foot long, oval, and coarsely serrated. The racemes have small white flowers set on them in clusters, and in fruit are pendulous. The fruits are about an inch and a half long, pear-shaped, with a short beak, yellow, passing to red at the point; and they enclose a single black shining seed about three-quarters of an inch in diameter, half-enveloped in a white cup-shaped aril.

The fruit is gathered when fully ripe, and the seeds are picked out of the pericarp and aril, which dye the hands of those who perform the operation a permanent yellow. The seeds are then roasted, pounded, and made up into sticks, much in the same way as chocolate, which they somewhat resemble in colour. In 1850 a stick of guaraná used to weigh from one to two pounds, and was sold at about one milreis (=2s. 4d.) the pound at Santarem; but at Cuyabá, the centre of the gold and diamond region, it was worth six or eight times as much. The usual form of the sticks was long oval or subcylindrical; but in Martius's time (1820) guaraná was "in panes ellipticos vel globosos formatum"; and Mr. Hislop had seen it made up into figures of birds, alligators, and other animals. The intense bitterness of the fresh seed is dissipated by roasting to a much greater extent than it is in coffee, and a slight aroma is acquired. The essential ingredient of guaraná, as we learn from the investigations of Von Martius and his brother Theodore, is a principle which they have called guaranine, almost identical in its elements with theine and caffeine, and possessing nearly the same properties. Guaraná is prepared for drinking by merely grating a small portion—say a tablespoonful—into cold water, and adding an equal quantity of sugar. It has a slight but peculiar and rather pleasant taste, and its properties are much the same as those of tea and coffee, being slightly astringent, and highly stimulating to the nervous system. It has had the reputation of a powerful remedy against diarrhœa, but I never found it so, although I have tried it largely, both on myself and other people. The general notion, however, is that guaraná is a preventive of every kind of sickness, and especially of epidemics, rather than an antidote against any; and Martius says of it "pro panacea peregrinantiüm habetur." Its immoderate use relaxes the stomach and causes sleeplessness—precisely the same effects as result from the abuse of tea and coffee.

. . . .

On the 15th of November we got our dried fish on board and bade farewell to the Barreiras, to my very great joy; for I began to weary of the long delay, and of the monotonous vegetation of the Ramos at that season, to say nothing of the stifling heat, the rains, and the plague of stinging insects. Most of the trees had gone out of flower, and of those that were still blooming, two species of Inga formed a continuous fringe wherever the shores were low, and we saw afterwards the same species growing in the same way along the main Amazon. On the evening of the 17th we reached a new sitio, opened a few weeks previously by a Captain Pedro Macedo from Saracá, for the purpose of fabricating Seringa or india-rubber, the tree having been found to exist in considerable quantity on the Ramos. A large space had been cleared of trees, and there the necessary huts had been erected, and a few vegetables planted, such as pumpkins, water-melons, and cabbages. We found Captain Pedro intelligent and hospitable, and were glad to accept his invitation to join him at supper and breakfast on game caught in his seringal, including wild pig or peccary, curassow, and Macaco barrigudo or Big-bellied Monkey (*Lagothrix Humboldtii*). I had hitherto rarely tasted monkey, and I thought this one rather insipid; but I learned afterwards to consider it the most savoury of its tribe, and to hold myself fortunate whenever I had one to put in the pot. After breakfast he led us into the forest, and showed us the Seringa trees, and the mode of collecting and fabricating the rubber. A track had been cut to each tree, and also to adjacent flats of Urucurí palm (*Attalea excelsa*), which, curiously enough, is almost invariably found growing near the Seringa, and whose fruit is considered essential

FIG. 7.—URUCURÍ (*Attalea excelsa*).

to the proper preparation of india-rubber. A stout liana is wound round the trunk of each Seringa tree, beginning at the base and extending upwards about as high as a man can reach, and making in this space two or three turns. It supports a narrow channel made of clay, down which the milk flows as it distils from the wounded bark, and is received into a small calabash deposited at the base. Early in the morning a man starts off into the forest, taking with him a terçado and a large calabash (called a cuyamboca) suspended by a liana handle so as to form a sort of pail, and visits in succession every Seringa tree. With his terçado he makes sundry slight gashes in the bark of each tree, and returning to the same in about an hour, he finds a quantity of milk in the calabash at its foot, which he transfers to his cuyamboca. The milk being collected and put into large shallow earthenware pans, other operators have meanwhile been filling tall, narrow-mouthed Caraipé pots with the fruits of the Urucurí palm and setting them over brisk fires. The smoke arising from the heated Urucurí is very dense and white; and as each successive coating is applied to the mould—which is done by pouring the milk over it, and not by dipping it into the milk—the operator holds it in the smoke, which hardens the milk in a few moments.

. . . . . .

Captain Pedro's own hut stood beneath the shade of an enormous Samaúma or Silk-cotton tree, which towered above all the adjacent trees. I took a sketch of the lower part of the trunk, and measured its circumference, which was 85 feet at about 3 feet from the ground; and had the tape been applied to the recesses of the sapopemas, the circumference would have been much increased. The roots embracing the trunk are those of a Fig, but there were a vast number of other twiners which the voracious mosquitoes did not allow me to sketch. I think I have

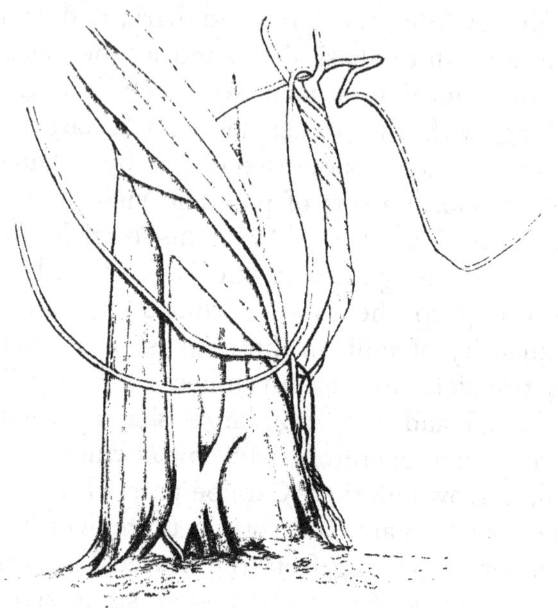

FIG. 8.—BASE OF A SILK-COTTON TREE.
Sketched in the Paraná-mirí dos Ramos, October 1850.

seen still larger trees of the same species, and I cannot doubt that it quite equals in dimensions its celebrated African relative, the Baobab, for if it be rather less corpulent it is twice as lofty. The softness and lightness of the wood render it suitable above all other trees for hollowing out the trunk into what are called cuchas or floating casks, which, being filled with turtle oil or capivi on the Upper Amazon and securely caulked, are floated down to the Barra do

Rio Negro or Pará. As we ourselves reached the goal of our voyage, one such cucha entered the port of Barra along with us, containing 1200 gallons of capivi. A merchant of that town told us he had once had a cucha made on the Solimoẽs, about 27 feet long, and so thick that in hollowing it out a man could work inside it with an adze or a short axe. It held above 300 pots of turtle oil, each pot of 12 frascos, or 6 gallons, and therefore in all nearly 2000 gallons. He had also purchased one ready-made, that had been cut down and hollowed out on the banks of the Ucayali, and into which he put 375 pots of oil, or 2250 gallons, without quite filling it. From the gauge of these enormous pipes my readers may calculate approximately the size and capacity of an entire trunk of Samaúma, 100 feet long from the base to the insertion of the first branches.

Above the mouth of the Mauhé there was no perceptible current in the Ramos. The water was very warm, and so thick with the slime of decomposed Confervæ as to be very unwholesome. We were told by parties of Indians whom we met, that the upper mouth was still closed, and that consequently we should be unable to get out into the Amazon. But on the 18th the water, although still unchanged in colour, began to run a little; and several small grass-islands and branches of trees passed us, indicating that some force was in action above. When day broke on the following morning, the water had taken a yellow tinge, and as we proceeded on our voyage several masses of scum floated by us, and the current began to run strong. There was now no doubt that the waters of the Amazon had entered the Ramú-urumuçána, as the Indians call the inlet of the Ramos; and towards night of the same day we had fuller proof of it, in occasional sudden influxes of water, converting the whole river into whirlpools. On the evening of the 19th, our one sailor was hauling by a rope, along a narrow strip of sand left bare in the middle of the river, and we on board were aiding with poles, when a sudden irruption of water flooded the sandbank, dragged the man into deep water and nearly drowned him before he could extricate himself from the rope, and whirled the canoe round I suppose a hundred times. We were drifting rapidly downwards, spite of all our exertions, and in continual danger of thumping against the side or on some sandbank, when fortunately a breath of wind sprang up, and although it did not last more than ten minutes, it sufficed to put us nearly across the river, and into comparatively still water.

The meeting of the cooler waters of the Amazon and the heated waters of the Ramos had an extraordinary effect on the fish, which floated on the surface quite benumbed and stupefied, so that we caught as many of them as we liked with our hands. On the 19th we had fresh fish in superabundance, and we salted down as many pescadas—a delicate fish, the size of a large trout—as served us for ten days afterwards. This phenomenon takes place every year, not only in the Ramos, but in many other periodically-closed channels of the Amazon; but I had not been previously informed of it, and therefore had not ascertained the temperature of the water of the Ramos before it became mixed with that of the Amazon, as I ought to have done.

On the 21st we reached a group of three houses, called "As Pedras," on account of several large blocks of volcanic rock lying on the river-bank. Here we were told that the Amazon had burst into the Ramos on the 18th, with a noise which was distinctly heard, although at a distance of nearly a day's journey; and that a montaria attempting to pass on the 20th had been split by the force of the current; which still ran so strong in the Ramú-urumuçána that there was no possibility of our getting into the Amazon, unless we were content to wait some days for the Ramos to fill, or could procure the assistance of three or four men for the dangerous pass. We chose the latter alternative, and until the men could be found I occupied myself in examining the surrounding vegetation, which I found of the same monotonous character as that of the rest of the Ramos.

. . . . . .

On the morning of the 23rd we left the Pedras, having obtained a promise of assistance on the following day to pass the mouth of the Ramos, from a brother of the half-drowned man. There was no wind to aid us, and our progress was very slow against the swift current. It was night when we reached a place where the water ran too furiously to be stemmed, about a mile distant from the mouth, which we could see very plainly. Here we anchored on the right bank, adjacent to a broad sandy delta that would soon be deep under water. After supper I started with Gustavo to explore the passage by the dim starlight; and after rounding or floundering through a good many pocinhos ("little wells," as the lagoons left in the sand are called), we reached the mouth. Here we found the waters of the Amazon entering with a force and a noise truly formidable, and ploughing through the sand in such a manner as to make a wall on each side of 15 feet high, from which the increasing torrent was every moment tearing huge masses and thus widening its bed. The grey sand and the water were so nearly of a colour, that it was with cautious steps we approached the edge of the treacherous cliff and ventured to look over. And what saw we at the foot, creeping gently along, and apparently about to ascend? A troop of tigers! Involuntarily we each seized an arm of the other and fled with no tardy steps; for not only were we unarmed, but entirely unclothed. We had run a very few paces when I stopped. "Impossible those should be Onças," said I, "in such a place; they must be waterfowl, and probably Garças Reaes (Royal Herons)." Reassured by this reflection, I again approached the bank a little lower down, and then saw clearly that the objects of our alarm were enormous masses of thick scum—now gliding smoothly along, now whirled round by some violent eddy. I called my companion to my side, and we both indulged in a hearty laugh at our late fright. We saw enough, however, of real danger to make us apprehensive about our journey of the morrow.

On the following morning, after waiting for some hours in vain for the promised aid, we resolved to attempt alone the perilous passage. It is impossible for any one to travel much on those rivers without acquiring something of the practice of navigation, and King and I had constantly taken the helm for more than half the day. We were indeed heartily sick of the protracted voyage, and glad to do anything in our power to accelerate it. On this occasion the strong cable of the anchor was secured to the foremast, and carried on shore to serve as a hauling-line, King and the Mamaluco yoking themselves to it; while I took the helm, and Gustavo stood in the prow with a pole. So long as there was water deep enough to float our vessel within five or six yards of the side we got on well enough; but when we were obliged to put out a little farther the current was too strong for our united force, and we were in great danger of being carried away. We toiled on until noon, making very little headway, and as it began to be excessively hot, we allowed the boat to take the ground, and resolved to wait until the air became cooler. In the interval we occupied ourselves in cooking our dinner, and were just about to fall on our boiled pirarucú, when a canoe came up, containing our friend of the Pedras, with two stout Indians and two boys. We made a hasty meal and by 2 o'clock were again under way. The additional hands having been placed to the hauling-line, we could now stand out more into the middle, where the stream ran fast and furious, making a deep roaring against the prow as we ploughed through it. The rope pressing on the edge of the cliff brought down, every few seconds, large masses of sand; but we stood far enough out to avoid them. My great difficulty, as steersman, was to keep the head of the vessel well out; for the force applied to the rope tended continually to draw her inshore, and had she turned in that direction, the current would have borne her violently against the bank, when she would infallibly have been swamped and buried under a mountain of sand. The exertion required was so great, that the perspiration ran off me in streams; but most happily we succeeded in getting clear out into the Amazon without once grounding, although we had rarely so much as a fathom of water. Those on shore could not have suffered less than myself, for the sun and the sand were scorching hot. It would be difficult to express what a load was taken off our minds when we found ourselves once more on the broad and breezy Amazon, and our previous silent anxiety was changed into noisy expressions of joy. The wind was blowing fair, and lasted until near sunset, sufficing to put us over to the north shore of the Amazon, along which our course now lay.

The Ramú-urumuçána and the dangers of its passage are well known to the dwellers on the Amazon. The previous year a boat, larger than ours, attempting to pass it under the same circumstances, was wrecked, from the captain's rashly scorning to seek the advice and assistance of the neighbouring settlers.

The inhabitants of the numerous sitios on the Ramos were chiefly Mestiços, of various shades of colour. The only white man we met with was Captain Macedo, and he could not be reckoned more than a visitor. Notwithstanding that the land is extremely fertile, and the lakes abound in fish and waterfowl, the people live in a state of comparative destitution; their only care being to eat up all their provisions to-day and leave nothing for the morrow. Money they rarely see, and when they have it they are unable to count it. Their sole article of commerce is pirarucú, and even that is generally sold before it is caught. At the time of my visit there was great lack even of farinha, their custom being to make it almost from day to day; and they levied frequent contributions on my biscuits, coffee, salt, etc.

At some sitios plantains were pretty extensively grown, but the fruits were all destroyed before they reached their full size by parrots, which were more numerous than ordinary that year the women told us. I suppose they were bolder, at least, from the men being away. One day I landed at a sitio in quest of plantains, and found—as was most usual—only women at home. The mistress, an elderly, grey-headed Mamaluca, had a daughter of twelve years of age, so good-looking and fair-skinned that I could not help inquiring into her parentage, and was told that her father was a Spaniard, then absent at Cametá; and further, to my great astonishment, that although so young she had been a wife a year and a half! The old woman said she had another daughter, younger and still fairer, then at school at Obidos, whom she should like to marry to me, as she had a great fancy for Englishmen; but as I had no fancy for a wife of ten years old, the negotiation went no further.

We had, on the whole, no cause to complain of lack of eatables, either in quantity or variety; for, besides our own stock of dried provisions, we could often buy fresh fish, and game we could shoot every day, often without having to leave the boat. Darters and herons we had within shot almost the day through, sitting on some overhanging branch or projecting stump, with looks intent on the water, into which they would occasionally plunge to secure a passing fish, and sometimes we could shoot them on the wing; but dry subjects like these were only eatable in default of other game. At early dawn we could sit at the entrance of the cabin, gun in hand, and (as we coasted slowly along) pot the birds as they woke up in the tree-tops; and the same in the evening, when they came to roost. In this way we would sometimes get a curassow or wild turkey, which was capital eating; and sometimes a macaw, which was tougher and less savoury, but still not to be despised. Besides these, a fat duck or a delicate quail (Inambú) would sometimes find its way into our pot; not to speak of several other kinds of fowls, whose native names would convey no idea to English ears. And game was equally as abundant on the Amazon as on the Ramos, although not quite so accessible.

On this voyage, as on other subsequent ones, I had occasion to note that the indigenous inhabitants of the Amazon valley have no idea of a habitable country, save as of land bordering a navigable river. I was often asked, "Is the river of your country large?" I once took some pains to describe the ocean to a lot of Indians, telling them of its immense extent and almost fathomless depth—how long it took to cross it, and how it had the Old World on one side of it and the New World on the other. They listened eagerly, giving vent to occasional grunts of admiration, and I thought them intelligent. When I had done, a venerable Indian turned to the rest, and said in a tone of wonder and awe, "It is the river of his land! What is *this* little river of ours" (pointing to the Amazon) "compared to *that*!" Other questions often put to me were, "Is there much open ground (campo) in your country?" "Are there extensive forests?" And they were filled with astonishment when I told them that most of our forests had been planted. "Why, here," said they, "when one wants to plant a tree, one must first cut down a dozen to make room for it!"

I have often noticed that people not born in, or not accustomed to, a mountainous country are slow to appreciate the picturesque. A Paraënse's idea of beautiful scenery supposes a land perfectly flat, with broad rivers, the stiller the better. The idea of mountains always suggests rapid rivers, with rocks and cataracts, dangerous or even impassable for canoes. If I made inquiries respecting an unvisited region, hoping to hear of "antres vast and deserts wild," they on their part would expect to give me pleasure by describing it as a "terra bonita, plaina—lá naõ ha lugares feios, nem serras nem cachoeiras," *i.e.* a nice flat country, where there are no ugly places, such as hills and waterfalls! One essential of a fine country to them, and not an object of indifference to any traveller, is that it contains "muita caça, muito peixe" ("much game, much fish").

. . . On the 29th we had fair weather, and an excellent wind lasting from 11 A.M. to 11 P.M. At 2½ P.M. we passed on the north shore the village of Serpa—almost the exact counterpart of Villa Nova, and like it seated on a small bay, where stones are rudely heaped up. The margin continued stony for a good distance, and the current was so furious that even with the strong wind we could not make head against it, so that we were obliged to creep up as close inshore as the depth of water would allow, and aid with poles. . . .

On the morning of the 2nd of December, a montaria came up with us, in which was an old man who was bound for a sugar engenho that an Englishman named M'Culloch was forming on a Paraná-mirí, separated from the main channel of the Amazon by a long island called Tamatari. I had made Mr. M'Culloch's acquaintance at Pará, so that I gladly availed myself of the opportunity to go forward in the montaria and visit him. We reached the engenho at 2 P.M., and I remained there until our boat came up, about noon the next day. There was at that epoch no manufactory of sugar on the Amazon, except near Pará, and at this distance in the interior the difficulties to be overcome in carrying out such an undertaking were immense. Mr. M'Culloch's career, indeed, as he himself sketched it to me, furnishes an instructive example of the risks and difficulties attending any enterprise on a large scale—any *empresa en grande*—in the far interior of South America. I know no better field for the skilled artisan, with steady habits, than the coast towns of both the Atlantic and Pacific sides of South America. The pay is so good that a man with a turn for saving soon accumulates capital; and if he employ it in business on his own account, and stick to the neighbourhood of the coast, he is almost certain to become wealthy; but if he be tempted to embark it in industrial, and especially in agricultural, speculations at a great distance from the seaboard, then the lack of industrious hands, the general slowness of the people, their want of good faith, and their jealousy of foreigners, nearly always cause his projects to fail.

Mr. M'Culloch was a native of Denny in Stirlingshire, and in 1850 was forty-three years of age—good-looking, muscular, and certainly an enterprising, thoughtful, clear-headed man. He had first emigrated to Canada, where he worked at his trade of carpenter and machinist; but having gone over to New York on a visit in 1832, he was met there by Mr. James Campbell of Pará, who invited him to try his fortunes on the Amazon. At Pará he continued to work at his trade, and in 1843, having by that time cleared a nice sum of money, he planned the erection of a sawmill, to be worked by water, somewhere in the interior, for the purpose of cutting up some of the immense quantity of cedar-wood floated down the Madeira and Solimoẽs every flood-time. He went, therefore, to the United States and purchased the requisite machinery. On his return he ascended the Amazon, first to Santarem, then to Villa Nova, and examined all the likely sites for a sawmill. Near Villa Nova he found an excellent fall of water at the outlet of a lake, but the people opposed his damming the outlet, on the plea that it would kill the fish in the lake. Baffled there, he next pitched on an outlet of the large lake of Saracá, and having spent several months and much money in building and in preparing his water-power, the authorities on some slight pretext refused to allow him to put up his machinery. Again driven away, he ascended to the Barra do Rio Negro, where he at last succeeded in erecting his mill on a suitable site. He even found a wealthy Brazilian to join him in the enterprise, and the two together carried on a tolerable business for two or three years. Early in 1849 his partner died, and so little protection did the laws of Brazil at that period afford to the property of foreigners, that after some litigation with the widow he was obliged to abandon everything except the machinery of the mill. Thus forced as it were to begin the world anew, he entered into partnership with an Italian merchant at the Barra, Senhor Henrique Antonij; but when they had worked the sawmill about a year it was burnt down, whether by accident or design, was never made out. I saw afterwards some of the ironwork lying at the bottom of the water at the foot of a pretty cascade, called the Cachoeira, which had been the moving power of the mill.

It was in conjunction with Senhor Henrique that M'Culloch had begun the engenho at Tamatari. He had already been nearly a year employed in clearing away forest, planting cane, arranging his water-power, etc.; and he had still much to do ere he could begin to grind cane and make spirit and sugar. The cane was magnificent—15 feet long, at the least, and as thick as the wrist—but it was so nearly ripe that he feared he should lose the first crop, from not having the machinery ready to grind it. He employed several native handicraftsmen, who worked pretty much when they listed; but the only workmen on whom he could rely were four slaves of Henrique's. He himself had to set the example in every kind of work: one day he was blacksmith; another, carpenter; another, he would be working with his spade and wheelbarrow at the embankment, harder than any of the niggers. At daybreak on the 3rd I found him occupied with a lot of wild Indians (Muras) of all sorts and sizes, who had come to work for the day. There were several small colonies of those people on the neighbouring lakes, and whenever they took it into their heads to work for M'Culloch they would come to him in the morning, as I now saw them, and he, well knowing the sort of pay they preferred, received them each with a pinga de cachaça (drop of rum). Then those who were so rich as to possess a palm-leaf hat—and, if not, they were provided with a fragment of cloth of some kind—held it out, and M'Culloch dispensed into it a cuya-full of farinha, and as much dried fish as would serve for the day.

. . . . . .

M'Culloch had fixed a gauge at the mouth of his mill-stream, by means of which he had ascertained the annual rise of the Amazon at that point to be 42 feet. Long before the water could rise to that height his dam and breakwater would be laid under water; and in effect he did not calculate on working his mill more than six months in the year, which was as much as he had been able to do at the Barra.

After leaving M'Culloch's, we had heavy rains, but very little wind in the right direction, so that we advanced barely ten miles a day; and it was not until 3 P.M. of the 6th that we reached a sitio belonging to a Captain Maquiné, where Mr. Gouzennes had given Gustavo rendezvous. He had, in fact, got there a few days before us, and we were glad to rest with him through the 7th and part of the 8th, and to compare our experiences of travel since we parted. Here, too, we gave up to him his boat, and got the loan of a smaller one and of a couple of stout Indians, and a boy to steer, for the short remainder of the voyage. We had four rocky points to pass before reaching the mouth of the Rio Negro, the first and worst being called Puraqué-coára (Electrical Eel's hole), and like the rest consisting of stratified arenaceous rock of a purplish grey colour, less granular than the Pará grit.

We entered the mouth of the Rio Negro on the morning of December the 10th. At the junction with the Amazon a bar of reddish friable rock stretches out a long way; when the rivers are full there is deep water over it, but we found it still uncovered, and had some difficulty in hauling our canoe through the furious current at its extremity. On a steep hill rising from the water's edge stood formerly the Fortaleza da Barra, built to command the entrance to the Rio Negro, but overthrown by the Cabanos in 1835. The city of the Barra, however, or Manáos, as it is now called, stands some eight miles within the Rio Negro.

The change from the yellow water of the Amazon to the black water of the Rio Negro is very perceptible, and indeed abrupt. The latter is black as ink when viewed from above, and stones or sticks at the bottom seem red; but when taken up into a glass it is of a pale amber colour, and quite free from any admixture of mud.

The Rio Negro is broader than the Solimoẽs—

as the Brazilians term the Upper Amazon—but it is less deep, and its waters are placid almost as a lake; and it looks at first sight more like the direct continuation of the Amazon than does the Solimoẽs, which starts off with an abrupt bend to southward.

We reached the Barra just after dark on the evening of the 10th, having been sixty-three days on the voyage, although the distance from Santarem is only 404 miles. I went on shore and waited on Senhor Henrique Antonij, to whom my letters of credit were addressed. He gave us a most kind and cordial reception, and at once installed us in the upper rooms of a new two-story house he had just completed, and invited us to eat at his well-furnished table.

Senhor Henrique—for by that name he was and still is known throughout Amazonland, the surname Antonij being ignored—has been the travellers' friend at the Barra for more than forty years[1]; and is spoken of in books of travel dating as far back as those of Mawe and of Smyth and Lowe. A native of Leghorn, he emigrated to Pará in 1821, being then only fifteen years of age, and in the following year ascended to the Barra, where he has ever since resided. He merits indeed the title of Father of the Barra, for when he arrived there it was going rapidly to decay, and no one did so much for its resuscitation and renovation as he, not only in building new and substantial houses, but in extending its commerce, and in opening out new channels for its industry—very profitable to the community, if not always to himself. When I knew him, in 1851-55, he was still young and fresh-looking, with

[1] [This was written about 1870.—Ed.]

a frank, good-humoured face of the genuine Tuscan type. It was his great delight to unite at his table all the foreigners who passed that way, and I recollect having once heard seven languages spoken there, by people of as many different nations. I cannot resist recording here this tribute to my old friend's hospitality and other virtues; and it was a great satisfaction to me when I was able to dedicate to him the finest new genus of plants I found on the Rio Negro, under the name of Henriquezia; one species of which (*H. verticillata*) is a noble tree of 80 to 100 feet high, having its branches and leaves in whorls, and bearing a profusion of magnificent purple foxglove-like flowers.

CHAPTER VII

AT MANÁOS: EXPLORATION OF THE VIRGIN FORESTS OF THE LOWER RIO NEGRO

(*December* 10, 1850, *to November* 14, 1851)

INTRODUCTION BY THE EDITOR

[FOR eleven months Spruce made the city of Manáos (formerly Barra do Rio Negro or "The Barra") his head-quarters, and rarely has a small tract of tropical forest in the very heart of a great continent been so well explored botanically, in so limited a time, and with the constant drawbacks of an excessively wet climate and very restricted means. During this period he appears to have kept no regular Journal, except a few notes of his more important journeys of a few days' to a few weeks' duration, with sketches of his more interesting botanical observations. He has left also a very small notebook entitled: "R. Spruce. List of Botanical Excursions, June 19, 1841—May 28, 1864." This contains a brief abstract of his early roamings in Yorkshire and other parts of Great Britain, in Ireland, in the Pyrénées, and throughout his whole South American travels. The entries are often day by day, at other times at longer intervals, occasionally giving the work of a whole month in a single short paragraph. Each movement from one place to another is regularly entered under its proper date, and the little book is thus a diary of great value in fixing his locality at any given time.

Besides these scanty materials, he carefully recorded every species of plant which he collected, its genus and natural order, and very frequently its specific name also; and always with a more or less detailed botanical description of it, made, when possible, from the freshly gathered specimens.

But what is of more use for our present purpose are the numerous letters written to Sir William Hooker, the Director of Kew Gardens; to Mr. George Bentham, the eminent botanist who had so kindly undertaken to receive his plants, and who also named them and distributed them to the various subscribers; and lastly, to his Yorkshire friend and neighbour, the late Mr. John Teasdale. These letters give us a vivid picture both of his botanical work and of his daily life, as well as of the more notable incidents and dangers of his various journeys. From these various sources I have endeavoured to construct a connected account of his travels and his work, though it is necessarily more or less imperfect, while occasionally it has been difficult to avoid partial repetitions.

An examination of the small diary shows how systematically and continuously Spruce explored the country round the city of Manáos. On the average he went out collecting every other day, the intervening day being occupied in preparing and drying, describing and cataloguing the specimens. Every road and path, every clearing, farm, or swamp, every stream or hill within reach were visited at intervals as the various trees, shrubs, or other plants came into flower. Within five or six miles east and west of the city six streams (igarapés) enter the main river, and all of these were assiduously examined either by boat or by paths overland, while several of the smaller and more accessible were followed up to their sources. Occasionally the river was crossed to examine the gapó (flooded land), and several excursions of longer duration were made to places ten or fifteen miles up the river, or into the main Amazon and some distance up the Solimoēs.

The results of this assiduous work were very gratifying from a botanical point of view. In the first year and a half of his residence in South America, he had explored the Lower Amazon at many localities and on both the north and south sides of the great river, and had collected more than 1100 species of plants. The eleven months spent at the mouth of the Rio Negro added to these no less than 750 additional species, besides a considerable number of those which had been already obtained but were of rare occurrence. Well might he say that this was the richest botanical district he had yet visited, while it produced a proportionately much larger number of new and undescribed species.

The outline map of the district around the mouth of the Rio Negro, given at p. 229, on which the various stations visited by Spruce are indicated, will enable the reader to follow more easily the extracts from letters and journals now to be given.

The term "Caatinga" being of very frequent occurrence in all Spruce's descriptions of his botanical excursions throughout the Rio Negro and Orinoco districts, it may be well to give here a note found loose in the Journal without any indication of when it was written, but almost certainly after his return to England. I will first state that Caa-tinga in the Lingoa Geral means "White Forest," applied to all woody tracts where the trees are of small height and sparse growth, so that, in comparison with the lofty virgin forests, in whose recesses there is a deep gloom, they are light and sunny. They are especially abundant on the great area of granite extending over a large portion of the Upper Rio Negro and Orinoco, where the granite rock is covered with a barren white sand, hence I think it probable that the term "white" applied to the soil rather than to the amount of light. In Central and South Brazil the same term is applied to deciduous woods, which are very common on the highlands and campos, and are due to a combination of poor soil with an arid climate. The following is Spruce's "Note":—

"Caatingas of Central Brazil have a comparatively dry climate and the trees are without leaves for some months in the cool dry season. Cacti and other succulent plants are frequent, and it is probable that Copaiferæ and other trees store up moisture to resist the drought.

"But the Caatingas of the Amazon-Orinoco region have a perpetually humid climate, and the trees are evergreen. The general character of the arborescent vegetation is to be dry and juiceless, while Cacti and similar plants scarcely exist. The effects of the moist atmosphere are seen in the mosses, Hepatics, and ferns, which form great cones at the bases of the trees, hang in festoons from the branches, and clothe even the living leaves with a fine spongy felt."

The first letter to Mr. Bentham was written three weeks after his arrival at Manáos, and the following extracts give his general impression of the vegetation, and his programme for the year.]

To Mr. George Bentham

BARRA DO RIO NEGRO, *Jan.* 1, 1851.

. . . We had a miserable voyage of 63 days (!) from Santarem; both of us were ill much of the time, and we were able to procure very few plants. Thus, what with waiting at Santarem for a passage, and what with the protracted voyage, I have lost an entire summer. The rainy season set in here some time ago, and the rain that falls far exceeds what we experienced at Santarem. However, we are in full work, and it is satisfactory to find oneself in the midst of a new vegetation—more promising, unless I am mistaken, than any I have yet met with. I have already got 10 new Melastomas, some Myrtles, Laurels, Solanums, etc.; but I have been principally occupied in securing some plants on the shore, which the river is fast overflowing.

. . . . . .

I propose making the Barra my head-quarters until the commencement of the dry season, when, if it please God, I will penetrate to the Orinoco and rifle the spoils of the Cerra Duida. . . .

It is miserable work travelling *up* these rivers.

How often do I wish it were possible to make the journeys afoot—in point of expedition I should be a great gainer. I think of buying a craft to go up the Rio Negro, but this is a thing that requires much consideration, for having the boat would not be sufficient if the crew were wanting. There is only *forced labour* here—no sum of money in the world would induce a Tapuya to work voluntarily.

. . . . . .

You cannot conceive how damp everything is here, even within the houses. Everything of iron rusts, plants mould, clothes hanging up two or three days double their weight, and the effects upon myself are a feverish cough with rheumatic pains in the limbs, etc.

I write in a rather querulous strain, but if you had seen our wan and sickly looks when we landed here (and there is not yet much improvement), you would have pitied us.

. . . . . .

To Mr. George Bentham

BARRA DO RIO NEGRO, *April* 1, 1851.

. . . . . .

I am trying to procure a boat and crew for ascending the Rio Negro, though the weather is not likely to be favourable for this until June, but, warned by past experience, I begin my preparations three or four months beforehand. I have now a collection of above 300 species [Equal to nearly 10,000 separate sheets of specimens.—ED.] made at the Barra, and would send them but for an unexpected difficulty that has arisen. In this land of forests I cannot find boards to make a packing-case! I brought a large one with me from Santarem, but how I shall get another I cannot tell. As I found no difficulty in this matter at Santarem, I did not dream of any here, but a sawmill which existed here was burnt down two years ago, and since then no planks have been prepared at the Barra.

I have just received your letter and the very welcome list of my first Santarem collection. I have no time to make any remarks upon it, but I need hardly say that it is extremely gratifying to me to find that it includes so many new species. If No. 594 be really the *Tecoma toxophora* of Martius, then was he quite mistaken in supposing it the Pao d' Arco of the inhabitants, for it is a low tree with soft wood quite unsuitable for the making of bows; the Indians call it Tauạrí do gapó, Tauarí being a general name for trees whose bark admits of being split into thin layers. There are two Bignonaceous trees called Pao d' Arco, of only one of which (148) I have yet seen the flower.

I have no doubt my Barra collection includes more variety and novelty than any previous one, but the weather has been wretched for collecting and preserving. Since our arrival on December 10 until this day, only five days have passed without rain, and these were all in February. For three weeks together I have not once stirred out without getting a thorough soaking. I have certainly not shrunk from exposing myself, and hitherto I have not felt any ill effects from it.

Two Englishmen came into the Barra a few days ago from the Rio Negro, where both had nearly died of intermittent fever. One of them is still unable to leave his hammock. Mr. Wallace, however, writes to me from the frontiers of Venezuela that he is far above the region of the ague (it commences at two days from the Barra), and that he is enjoying himself amazingly in a romantic and quite unexplored country. Were there steamboats on the Rio Negro I would not be long ere I joined him, but, alas! there are no such things; he himself was above two months in getting up, and there is nothing for me but betaking myself to the Brazilians' universal remedy, *patiencia.*

. . . . . .

The second lot I sent from Santarem, containing about 200 species, not getting away as I expected, I afterwards *arranged* (everything is aranjado here, meaning procured, collected, etc. etc.) about 100 more. These two collections I presume you would distribute together. . . . Be the number ever so small, to keep them here cased up would be to have them devoured. . . .

April 26.—The vessel which was to have taken Mr. King and this letter to Pará has been delayed by an accident not infrequent in these rivers: an igaraté (large canoe) sent to procure cargo for her in the mouth of the Solimoẽs was swamped in a storm just before reaching her destination; the cabins and masts were destroyed and others had to be prepared ere she could return. Meantime has arrived Senhor Henrique's large cutter from the Solimoẽs, nearly laden—she has now taken in all her cargo, and I profit by the opportunity for sending off all my collections to England. The dried plants are in two very large cases, and comprise between three and four hundred species. . . .

I use certain terms in speaking of localities which may require explanation. We call the virgin forest here the mato, or sometimes mato siergen; the "brush" that springs where forest has been cut down is called matinho or the little forest; deserted farms are called capociras—their vegetation is scarcely different from that of the matinho; finally, the forest bordering the rivers, which is wholly or partially under water in winter, is called gapó; and the vegetation often forms a distinct band quite different from that of the "terra firme."

I have now purchased a boat for ascending the Rio Negro; it is of 6 or 7 tons burthen, and has got a tolda da popa (poop cabin) and another da proa (at the bows), convenient for keeping my goods dry; it was built at San Carlos in Venezuela, and has made but one voyage. I have given 140 milreis for it, or £9:6:8 (at the present rate of exchange, 28d.), and I shall have to spend about another 100 milreis on it to make it suitable for my purpose. The most difficult task will now be to procure men, and I shall have to give up a few weeks to the preparation of the canoe and the hunting up of men. I can do very little just now in plants; the river is nearly full and everything has flowered on its banks that belongs to the rainy season, when the dry season commences there will be another flush in the vegetation. I propose, however, shortly going two days' journey up the Solimoẽs (the name by which the Amazon is known above the Rio Negro) to see if there is anything there different from what I get here. . . .

[A letter to Sir William Hooker of the same date as the last gives an interesting account of a week's collections made under special difficulties.]

BARRA DO RIO NEGRO, *April* 1, 1851.

Towards the end of January I crossed to the south side of the Rio Negro, to visit a campo—called Jauauarí—on which Senhor Henrique many years ago established a cattle fazenda. The grasses on this campo are of poor quality; when the winter floods are high many of the cattle perish; the neighbouring forest is much infested by onças; and, worse than all, the herdsman is of a very indolent disposition. Between the south bank of the river and the campo is an intervening gapó, or forest of low bushes and trees flooded in the rainy season, of two or three miles in width. The water had risen sufficiently to enable my boat to traverse great part of this, and it was curious work navigating among bushes. The campo is about a mile broad and three or four miles long; its southern side is skirted by the small river Jauauarí, which enters the Rio Negro near the mouth of the latter. The herdsman's house is near this stream; it is built of mud and thatched with palm-leaves, but it had fallen so much into decay that, rather than repair it, he had moved to a casa de fôrno (oven-house) which was near his mandiocca plot, and was the common property of two or three families. I had my choice between these two habitations. But the oven-house was merely a roof without any side-walls, and was so crowded with inhabitants as to leave no room for me and my work. The empty house on the campo was so surrounded by mud and water as to be inaccessible save at one corner, where a plank was laid to step on. It consisted of three rooms; there were pools of water on the floors of all these save the middle one, and in this were two opposite doorways without doors or mats, through which, during squalls, the wind swept furiously. This room I chose, preferring cold to wet, and here I remained a week, accompanied by a young fellow, a half-Indian and brother-in-law to the herdsman, who cooked my meals.

The soil of the campo is a stiff clay, while the campos I have previously visited are of loose sand; I was therefore prepared to expect something new in the vegetation, nor was I disappointed. The grasses were quite brittle in contrast with the tenacity of the soil, and I was not able to draw a single root without the aid of my knife. Both grasses and sedges were of many species, and one of the latter was an abominable "cut-grass" by walking among which my ankles were completely tattooed. As is usually the case in the tropics, these Grasses and Sedges grew in solitary tufts with bare spots of earth between them. Where the soil was rather peaty, in these bare spots grew a leafless Bladderwort with a broad three-toothed spur, and a pretty Sundew with leaves smaller than those of our *Drosera longifolia*, but with a much larger rose-coloured flower. In drier parts of the campo grew three Orchises of the genus Habenaria: one with a long raceme of greenish-white flowers; a second with shorter racemes of yellow flowers, and so abundant as to recall the Bog-Asphodel of our northern moors; the third, which had rather larger yellowish flowers, was more scarce, but it possessed the delicious odour of *Orchis conopsea*, wanting in the other two. Along with these grew a slender erect Polygala with racemes of purple and white flowers, and many other herbaceous plants, including several Rubiaceæ. . . . In drier ground grew a small species of Arum with a spathe of pure white and a hemispherical root, and a large branched herbaceous Polygala with white flowers.

Here and there large ant-hills by their decay form a sort of island in the marsh; on these grew two Liliaceæ, one with a solitary yellow flower, the other with a few terminal rather large flowers of the most delicate pale blue.

On one side of the campo the soil seems better, and Grasses and Sedges grow rank and high. This part is quite glorious with a shrubby Melastoma of 4 or 5 feet high, completely clad with large purple flowers; it is quite new to me. Stunted shrubs of a Byrsonima (Malpighiaceæ) and a Curatella (Dilleniaceæ) formed the sole woody vegetation of the central parts of the campo; but it was belted round by tall Jauarí palms, with an inner fringe of Mimosa, Myrtles, Melastomas, Malpighias, etc.; while, outside all, the dense dark forest stretched away to an immeasurable distance.

Not a day passed without rain. Sometimes there was sun enough in the morning to enable me to dry my paper before setting out to herborise. When there was not I took the paper across the river in the evening and got it dried on the fôrno, which was about a quarter of a mile up the river. This is a narrow rapid stream winding through dark forests, the climbers of which often stretch across it and are troublesome to avoid as the canoe shoots beneath them. The first time I made the passage, along with my attendant Pedro, he placed himself in the prow and I in the poop of the canoe, each of us with a paddle; but although I was well accustomed to steer by means of a rudder, I had never attempted it with a paddle, and my want of skill brought us up every now and then plump into the bushes, which I could see ruffled Pedro's equanimity no little. After we landed, I heard him say to his sister in Lingoa Geral, "This man knows nothing—I doubt if he could even shoot a bird with an arrow!" (a feat which every boy of twelve years old is supposed capable of performing). I consoled my wounded vanity with the reflection that probably the most eminent botanist in Europe would have cut no better figure than I did if placed on the stern of an Indian canoe with a paddle in his hand. Since that time, however, practice has rendered me tolerably expert at steering with a paddle.

Observing some large roots, looking like turnips but vastly larger, lying near the house, I inquired what they were, and was told that they were used in the same way as the roots of mandiocca. They showed me the grated root in a state of preparation, and gave me farinha already made from it. It is only very lately (as I learnt from these people) that the Tapuya Indians have begun to use this root, and it seems to have been first made use of by the Purupurú Indians, inhabiting the Rio dos Purus; these Indians call it Bauná. It is known also to the Mura Indians, who call it Mahaõ. The Tapuyas merely call it Maniocca-açu or the Great Mandiocca. The largest root I saw weighed 48 pounds.

On the following day I went, accompanied by an Indian, to see the Bauná plant, which grows pretty abundantly in the forest on the south of the Jauauarí. We found several plants, and I procured specimens of the stem and leaves, and dug down to the roots, but there were no flowers or fruits.

The Bauná root is still more poisonous than that of the mandiocca, though quite tasteless when fresh, and repeated washings are required to render the farinha and tapioca wholesome. A family at the mouth of the Rio Negro ate of the roasted (but unwashed) roots, and the experiment nearly cost them their lives. When properly prepared the farinha of Bauná is scarcely distinguishable from that of mandiocca; for three days I lived solely on Bauná and milk (with the exception of once eating a bit of broiled fish) and found it wholesome and nutritious.

Soon after my return from the Jauauarí, I learnt that after my departure a number of Indians residing on the river went to the herdsman's house in a body, and expostulated with his wife in the most angry manner for her thus revealing to a stranger the source of their support in times of scarcity. "The people of the Barra," said they, "will cross the river to search for this root, and will soon eradicate it. The Commandant, too, having heard of the narrow escape of this family at the mouth of the river, will send to forbid our making further use of such dangerous food." Their alarm was as great, and equally as well founded, as that of a trader up the Rio Negro, from whom Dr. Natterer procured seeds of salsaparilla. "I considered to myself," said the man afterwards to Senhor Henrique, "what a fatal blow would be struck at our trade in Salsa if this foreigner should succeed in getting the seeds to grow in his own country, where whole plantations would soon be made of it: I therefore *boiled* them before I gave them to him." I do not suppose that Dr. Natterer ever learnt how it was that his seeds had lost their vitality.

On the Jauauarí I saw a small plantation of Ipadú, a shrub of which the powdered leaves are chewed by the Indians throughout the Rio Negro. I found it to be (as I had expected) the *Erythroxylon Coca.* The leaves are roasted and then pounded in a mortar made of the trunk of the Pupunha palm, from 4 to 6 feet long, the root being left on for the bottom and the soft inside scooped out. It is made so long on account of the impalpable nature of the powder, which would otherwise fly up and choke the operator; and it is buried deep enough in the ground to be worked with ease. The pestle is made of any hard wood. When sufficiently pounded they are mixed with a little tapioca to give it consistency. With a chew of Ipadú in his cheek an Indian will go two or three days without food, and without feeling any desire to sleep. I send you the powdered Ipadú and flowering specimens of the plant. I wished to send you the mortar also, but no sum of money can purchase one. I find the greatest difficulty in inducing the Indians to part with many things of their own manufacture, the reason being that it would be a work of time to replace them, and the Indian loves ease above all things. Not long ago I saw in the hut of an Indian a fishing-line most beautifully made of the bark of some tree. All my entreaties could not induce the man to sell it. "I need it," said he, "to procure me the means of subsistence; your money will not buy me such another, and it will be the work of weeks to supply its place." Such an argument admitted of no reply, and I could only regret that he looked with such a philosophical eye on money.

There is another campo near the Barra, on the same side of the river, which differs much in every respect from the one I have described above. It is elevated about 100 feet above the river, and the soil is a loose white sand. The vegetation is chiefly shrubby, and one shrub called Umirí is so abundant that the campo is called from it the "Umirisal." It is a species of Humirium belonging to a small natural order (Humiriaceæ) peculiar to tropical America, and bears a fruit which is said to be very agreeable. Another shrub or small tree, called Yumurá-ceêm or the Sweet tree, grows in almost equal abundance, and the fruit is ripe in February. It belongs to the natural order Clusiaceæ. The other shrubs include but few species, the principal being a Myrsinea and two or three Myrtles. But what rendered the campo most interesting in my eyes was that here and there on the burning sand were large patches of four species of Claydonia, two of them exceedingly like our common Reindeer Moss, and a third with bright red fruit looking quite like our *C. coccinea.* When I add to this that everywhere among the bushes grew up a tall Fern (*Pteris caudata*) scarcely distinguishable from our Common Brake, it will easily be seen how strongly I was reminded of an English heath. There were, however, two Ferns of the curious genus Schizæa—one preferring the most exposed situations, the other nestling under bushes, and both in considerable quantity—looking so very *tropical* as at once to disperse the illusion, if it had entered my head to fancy myself at home. Besides two grasses—one very minute, the other tall and leafy—and a single grass-like sedge, the only other herbaceous plants were an Asclepiadea with narrow leaves and drooping lurid flowers, and an orchid 9 feet high, with broad fleshy distichous leaves but not in flower.

The spot where we landed in order to reach the Umirisal was rocky, and afforded me several plants quite different from those on the campo. The whole of this northern coast above the Barra, so far as I have ascended it, is rocky, and forms my most profitable herborising ground.

To Sir William Hooker

BARRA DO RIO NEGRO, *April* 18, 1851.

. . . . . .

Here, for three weeks together, I have not once gone out without returning completely soaked. Perhaps in consequence of the continued rains, the average temperature is lower, and therefore more agreeable than at Santarem. During the month of March, many days passed in which the thermometer never reached 80°, and the highest temperature I have registered for that month is only 84°. The maximum temperature for February is 88°. When the thermometer is low, that is, from 71° to 75°, in a morning before sunrise, with a tolerably clear sky, I have found it a pretty sure indication of a fine day; and the contrary when the thermometer is high, however bright the sky may be.

My Rio Negro collections include examples of nearly every natural order of plants. Leguminosæ continue to constitute a large proportion of them, but Cæsalpiniæ and Mimoseæ are more numerous than Papilionaceæ, which was not the case in the localities previously visited. I have several large-flowered Loranthi not found at Santarem, numerous Rubiaceæ, Myrtles, and Melastomas almost without end, and some curious intermediate forms between these two orders. Lecythideæ are not scarce, but many of them very difficult of access on account of their large size. The small-fruited species of Lecythis are called by the Indians Macacarecuya or the Monkey's drinking-cup, their fruit quite resembling a cup, when the lid has fallen off. Myrsineæ are far more abundant here than I have seen them on the Amazon; they are all shrubs or small trees, reminding me of the currant-berry by the aspect and often by the odour of their pendulous racemes of small flowers, which are, however, occasionally more gaily coloured. The Barra has afforded me five Myristicæ previously unnoticed, and it is worthy of remark that in every tree of this genus I have met with, the branches are arranged in whorls of five; but the secondary ramification does not follow the same law. Soon after our arrival the banks of the stream were quite gay with a small Tiliaceous tree, bearing large white star-like flowers; it agrees in most respects so well with the *Mollia speciosa* of Mart. and Zucc. (gathered also at the Barra) that I have little doubt of its being the same, although it recedes somewhat from the generic character given in Endlicher; the stamens, instead of being collected "in phalanges quinque," are arranged in ten parcels, five outer and five inner, the former having *purple anthers and green pollen*, and the latter *yellow anthers and yellow pollen*.

Grasses are less numerous here than at Santarem, but they show more novelty of form. There are three Selaginellæ in the woods, but Ferns are scarce, occurring only towards the head-waters of the streams; they include, however, a few species of Trichomanes new to me. Orchids are still not very numerous, but there are a few, both terrestrial and epiphytal, which I have not previously met with. The Palms I am much interested in; they are far more numerous than at Santarem, and I believe include several undescribed species. I expect I have amongst new species one Maximiliana, one Euterpe, one Iriartea, two Bactrides, and two or three Geonomas. I send you specimens of all these, but I should like to have time to observe them more fully before sending the descriptions. Perhaps the noblest palm in the forests of the Barra is the Patauá, of which the trunk sometimes reaches 80 feet in height and the fronds are of immense

FIG. 9.—BACÁBA (*Œnocarpus distichus*). Pará.

size. An entire spadix, laden with fruit, is a heavy load for a man. The fruits are very oily, but the only use made of them here is the preparation of a wine similar to that of the Assaí. Trunks of a few years' growth are thickly beset with slender rigid spines about 2 feet in length, pointing upwards; these are the nerves of the sheathing base of the petioles from which the parenchyma has decayed; they are called by the natives barba de Patauá. When the trunk reaches 15 or 20 feet in height, the "beard" begins to decay at the base, and the upper part being thus deprived of its support falls down in a mass. [I give here a photographic print of the allied species of Œnocarpus found at Pará, which, though common there, is not mentioned by Spruce. I was not able to figure it in my *Palm Trees of the Amazon*. The leaflets are more drooping than in the Rio Negro species, and it is not so lofty in stature, but is an elegant palm.—ED.] The Inajá (*Maximiliana regia*, Mart.) has the trunk similarly beset with the bases of the petioles, until it reaches a certain height, and an Inajá of 40 feet high looks a quite different plant from one of 20 feet. Of the barba de Patauá the Indians make the arrows for their Gravatánas or Blowing-canes. The Gravatána itself is made of the trunk of a small palm, an Iriartea, which I have met with deep in the forest at the back of the Barra. It is called Paxiúba-i or the little Paxiúba, and grows to from 10 to 18 feet high, the thickness being little more than an inch. . . .

A palm much cultivated in the Barra and the adjacent sitios, and said to grow wild up the Rio Negro, is the Pupunha, which I suppose to be the same as the Piryäô (*Guilielma speciosa*, Mart.) mentioned by Humboldt as growing on the Upper Orinoco. The fruit of this is perhaps more valuable as an edible than any other palm-fruit; the sarcocarp contains a large quantity of starch, and it is sometimes developed to such a degree that the nucleus is quite obliterated. Eaten with salt, the boiled or roasted fruit much resembles a potato, but it is also very pleasant eating with molasses. A spadix of Pupunha, laden with ripe fruit, is one of the most beautiful sights the vegetable world can show: the fruits are of the clearest scarlet in the upper half, passing below into yellow, and at the very base to green.

On a separate sheet I have written some account of a few of the edible fruits found wild in the forests adjacent to the Barra. There are several others which I have not yet obtained. Numerous Myrtaceous and Melastomaceous fruits are eaten, but few of them possess any great excellence; perhaps the best are the Guayabas, which belong to various species of Psidium. A Melastomaceous tree sent from Santarem with fruit rather resembling a Guayaba externally, but twelve-celled, is very abundant here. The fruit is called Tapiíra-guayaba or the Tapir's guayaba, but it is only insipid eating. The various species of Inga have the seeds enveloped in a sweet cottony pulp, which is very agreeable eating; the Ingá-sipó (of which I have already sent you the fruit) is the most esteemed. The Cow tree is represented on the Rio Negro by two Apocyneæ, the Cuma-i and the Cuma-açu, both species of Collophora, but only one of them known to Martius. The former is frequent near the Barra, and early in March was a great ornament to the forest, especially near the river, being profusely clad with corymbose cymes of red flowers. It grows to 30 or 35 feet high, with a diameter of about 12 inches, and the branches and leaves grow in threes. The milk flows out abundantly on a slight incision being made in the bark; it is of the consistency of new milk, of the purest white, and very sweet to the taste. The Indian mode is to apply the mouth directly to the gash and thus receive the milk as it oozes out. In this way I have many times partaken of it without experiencing any ill effects. Its extreme viscidity has suggested its employment in diarrhœa, and there is no doubt that if taken in sufficient quantity it would actually glue up the viscera. The Cuma-açu is a much larger tree, but of similar habit, and the milk is of a thicker consistence; it is said to flower towards the end of the year. The fruits of these two trees are said to be the most agreeable of any on the Rio Negro, and from their resemblance to the fruits of *Pyrus Sorbus* have been called Sorvas by the Portuguese settlers.

It is perhaps among twining plants or sipós that the greatest botanical novelties remain to be found; they are in many cases so difficult to collect that I have no doubt a great many have been passed over by travellers. I am now paying particular attention to them, and my Barra collection includes twiners of the orders Leguminosæ, Connaraceæ, Polygaleæ, Malpighiaceæ, Sapindaceæ, Convolvulaceæ, Hippocrateaceæ.

To Dr. Semann

BARRA DO RIO NEGRO, *April* 25, 1851.

. . . . . .

I wish I had time to write you a long letter, but I am over head and ears in work, packing up rubbish to send to England, and I must be brief. I hope I have now got pretty well acclimatised here, and I am beginning to enjoy myself. I cannot say that I have ever experienced that bewilderment at the multitude and variety of the forms of vegetation which some have felt on being transported to the tropics, if I except the first three or four days at Pará; but here are only trees—trees—trees! flowering, in their turn, all the year round, and never so many blooming at one time as to cause me any excess of work in preserving them, though the getting at some flowers is often a work of difficulty.

I think I am finding most novelty among the sipós or twiners, such as certain Apocyneæ, Menispermeæ, etc. Some of these climb to such inaccessible places that only the monkeys have it in their power to gather their flowers and fruit. When, however, I once see the leaves of a twiner I never lose sight of it until I find its flowers, and I generally succeed in the long run in obtaining specimens of them. . . .

To Mr. John Smith (*Curator of Kew Gardens*)

BARRA DO RIO NEGRO, *Sept.* 24, 1851.

I trouble you with a letter to ask you to compare the specimens of Palms I have sent to your museum with the Plates, etc., in Martius's great work and give me your opinion on them. I can find no one who will talk to me about Palms, and I am now coming among some that are exceedingly interesting. It is true that they are extremely difficult to collect and preserve. A prickly palm gathered in the depths of the forest at a distance from one's canoe is a load for one man, and an exceedingly unpleasant one, for one's hands are almost constantly required to cut and pull aside the twiners that obstruct the way. The Mirití which grows here in the centre of the continent is possibly distinct from the maritime species, but as a spadix is a load for two men, specimens are quite beyond the reach of a traveller like myself. However, notwithstanding all the difficulties that lie in my way, I feel that it would be quite a sin to leave so many fine things altogether unnoticed. Higher up the Rio Negro I am certain to find abundance of new palms. Mr. Wallace has just come down from the frontier and brought with him sketches of several palms, of which I have no doubt many are quite new. There are at least *two large Mauritias* quite distinct from any described by Martius. . . .

I am now describing completely every palm I find, and I hope to sketch the greater part of them, so that, with the aid of the specimens I send to England, I hope some day to be able to work them up. I am now familiar with the aspect of all the commoner palms, but I have learnt that it is very unsafe to trust to the native names for the species, these names being, in fact, in most cases *generic*; I may instance Assaí, Bacába, Marajá. The palm called Bacába at Pará and Santarem is not the *Oenocarpus Bacaba* but the *Oe. disticha.* The number of Marajás is endless.

. . . . . .

I find ferns very scarce here in the interior. I have got a few interesting species near the Barra, but they are so scarce that of some of them I have taken every individual I met with. Surely I shall find them more abundant up the Rio Negro.

To Mr. George Bentham

BARRA DO RIO NEGRO, *Nov.* 7, 1851.

Two nights ago reached me your letter of July 22, and also the Indians I had been long expecting to take me up the Rio Negro. I am now hard at work packing up my collections for you and purchasing trade goods for the voyage. It is no use taking money up the Rio Negro, and except a little copper, I am laying out my whole fortune in prints and other fabrics of cotton, axes, cutlasses, fish-hooks, beads, looking-glasses, and a host of sundries. The trafficking of these involves a serious loss of time, but there is no alternative.

We had sad news lately from Pará. Singlehurst's vessel, the *Princess Victoria*, was lost in entering the mouth of the river and nothing of her cargo was recovered. Miller went out in a boat from Pará to see the wreck and caught a severe chill, which excitement aggravated into brain-fever and speedily carried him off. . . .

Poor Miller was a very fine young man, and his loss to me is irreparable, as he was so ready to do anything I needed, even to putting himself to inconvenience. He was a schoolfellow of Gardner's,[1] and was stationed at Aracati when Gardner visited that place, where he rendered him great assistance.

. . . . . .

Since my last letter to you I have travelled about more than at any time previously, and I believe that in this collection you will find absolutely

[1] [Gardner was a botanist who collected largely in Central Brazil and published an interesting volume, *Travels in Brazil*.—ED.]

nothing common. In May, the middle of the wet season, not a tree was to be seen in flower in the forest or caapoeras, but I found that at that season precisely the twiners of the gapó began to flower, and the south shore of the river and the inundated angle between the Solimoẽs and the Rio Negro was soon quite gay with Serjanias, Asclepiadeæ, etc. The trees of the gapó do not flower until the water begins to leave them. In this month, too, I went down to the mouth of the Rio Negro (about eight English miles below the Barra), and remained there four days. I found it such an excellent station that I resolved to revisit it later in the season. I met there also an Indian carpenter whom I engaged to construct the cabin (tolda) of my canoe, and in the month of July I took her down there and remained until the cabin was completed. There is an extraordinary difference in the vegetation of the opposite shores of the Amazon at the junction of the Rio Negro and Solimoẽs. You will find in the collection some plants marked "mouth of R. Negro" and others "mouth of Solimoẽs," which the sketch plan opposite will explain.

The former plants are laved by *black* water and the latter by *white*. Any one at first sight would take the Amazon to be the continuation of the Rio Negro, from the breadth and direction of the latter; but this cannot at all compare with the Solimoẽs for depth of stream and rapidity of current. It may be long before any one exposes himself again to gather the few plants I got at the mouth of the Solimoẽs—such a place for snakes and ants in the trees I never met with. In the wet season every terricolous animal must betake itself to the trees, when thousands of miles of forest are inundated. Among plants from the forests at the mouth of the Rio Negro, none interested me more than the Cajú-açu, a tree which I had heard spoken of throughout the Amazon but could never fall in with

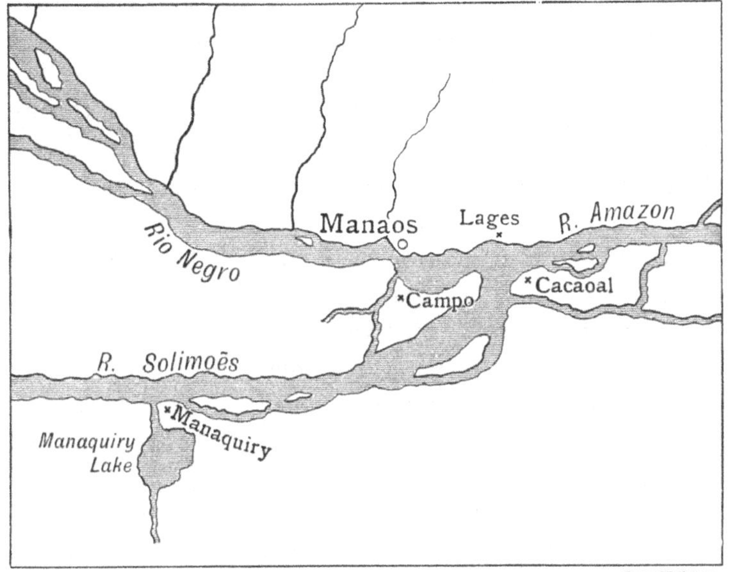

FIG. 10.—SKETCH MAP OF DISTRICT ROUND MANÁOS.
Scale about 20 miles to an inch.

previously. It is apparently a true Anacardium, but grows 90 feet high!

In the month of June I had an excursion up the Solimoēs, my destination being Manaquirý—a group of sitios on a small river and lake of the same name, lying to the south of the great river. It is accounted but three days' journey from the Barra, but it cost me a week, with four men, so strong was the current in the very height of the wet season, and so little wind was there.[1] Notwithstanding the slowness of the voyage, I found collecting very difficult. Although we crept along shore, we were rarely near enough to pluck any flowers. I sometimes stood in the prow with a long hooked pole, and when we came near enough to reach any twiner I "made a point" at it. In this way were gathered a remarkably fine Apocynum, a Mucuna, and several others; but I need not add in very small quantity. It was only two or three times that we were moored long enough during daylight to enable me to penetrate into the gapó with the montaria; yet in this way I got the few curious aquatics in my collection, a second species of your new genus Enkylista, and some other things. By the by, our little *Phyllanthus fluitans* (Euphorbiaceæ) was there in abundance. Are you sure that the embryo of this is dicotyledonous? There is a remarkable analogy (to say the least) with Hydrocharis.

I had great difficulty also in drying my paper, for, not to speak of the rain, during the whole week of the voyage *we never saw land*, and the drying had to be done on board. But when there was wind, it was difficult to secure the paper against being carried away, and when there was none I could scarcely spread it out so as not to be in the way of the rowers. I only enter into these details to show you that there may be reasons, "not dreamt of in your philosophy," why the stock of some species is not always so ample as might be desired.

At Manaquirý I paid a visit to a Senhor Zannÿ

[1] [It is about 50 miles above the mouth of the Rio Negro.—ED.]

(son of the Colonel Zanný who was deputed by the Brazilian Government to accompany Spix and Martius in the province of Pará), and passed a night with him. He told me that these naturalists passed some days at Manaquirý; it is therefore possible I may have got some of the same species as Martius gathered there. The whole region between the Madeira and the Purús is a noted country for Cacaos. In the woods behind Zanný's house I saw two species new to me and got one of them in flower.

My stay at Manaquirý, and the voyage thither and back (the latter only eighteen hours!), occupied above three weeks, but the weather was dreadful (being the fag-end of the wet season), and interfered much both with collecting and preserving. Besides, I was quite too early for the forest vegetation, and I saw multitudes of trees whose foliage was new to me, but which had not begun to show their flowers.

. . . . . .

Although I am now alone, and have to do the whole of the drying as well as the collecting, yet I think my collection is superior to that of the corresponding months of last year, notwithstanding all drawbacks. The Indians do well enough in the field when one knows how to manage them. Humboldt, from some remarks in his *Aspects of Nature*, seemed not to have attained this art. It does not do to ask them to do anything *as a task*, however much money, etc., you may offer for the performance of it. My usual invitation is "Yasso yaöatá" ("Let us go for a walk"). We get into our montaria (canoe), enter one of the igarapés (small streams), and when we reach the heart of the forest they are all alacrity to climb or cut down the trees, the gathering of the flowers being all the while represented as a mere matter of amusement. As I had no letters from Pará to the authorities here (no British Consul having been there for now more than a year and a half), I have had to send as far as Saõ Gabriel da Cachoeira for men—a month's distance, at least, above the Barra. I expected them several weeks ago, but I had news that they were ill, and I had almost given up all expectation of them when they arrived on the night of the 5th inst. There are five of them, all stout fellows, and I have "arranged" other two here (one a Peruvian Indian from Moyobamba); so that, as my canoe goes well under sail, I hope to get along merrily. I propose to make Saõ Gabriel my first resting-place. It is exactly on the Equator, in the midst of cataracts and mountains, and ought to produce something good. The Podostemons that grow on the falls are a chief article of support to the natives for one-half of the year!

. . . .

[In the MS. book containing Spruce's Journal of his voyages on the Rio Negro and Orinoco there are some notes relating to the longer excursions he took while at the Barra which are referred to in the preceding letter. The first of these excursions was from the 21st to the 24th of May, when he went down the Rio Negro to its junction with the Amazon about eight miles below the city, where was a small Indian settlement called Lages or the Ledges—from flat sandstone rocks which crop out at the river's edge. Here he spent three days, and nearly three weeks later on in July and August, to which latter period his notes on this locality apply. After his first visit he crossed to the southern bank of the Amazon, where he landed, and afterwards ascended some distance up the river, then crossing over to the angle between the Solimoẽs and Rio Negro, and ascending a few miles up the latter river before crossing over to the Barra. The following notes refer to this return journey, and will be understood by reference to the outline map at p. 229.]

On the right (south) bank of the Solimoẽs, at its mouth, or just where it takes the name of Amazon, is a flat of land which, rising a little higher than the adjacent portion, is not flooded, though it escapes by but a few inches. On this spot there was formerly a sitio and a large plantation of Cacao and other things; now all is running again to forest, but several Cacao trees remain, and there is a large flat of Breadfruit trees, which seem firmly established and even spreading, for underneath the well-grown trees appears not a plant save numerous seedlings of the same tree. Hence there appears to be some deleterious effect from decaying leaves to extraneous species.

The vegetation of the shores of the Solimoẽs is more advanced than that of inner Paraná-mirís. Yet it has a rather ragged aspect owing to the banks consisting almost wholly of terras cahidas —large portions falling away every year in the dry season and forming the great peril of the navigation.

A party was collecting turtle oil on a sandy beach of the Solimoẽs, with several canoes drawn up on land in which eggs were crushed for extraction of the oil, when an immense patch of forest on the opposite side fell with thundering noise into the river, and though there a league broad, the waves rushed far up the beach and carried away canoes, eggs, oil, and everything else laid there. Instances are not rare of canoes being swamped by the force of falling masses. Owing to this falling away, trees become exposed which had completed their growth in the crowded forest and have not the roundness of outline to be observed in the permanent forest-skirts.

. . . . . .

The banks of inland rivers should be seen early in the morning, before or after sunrise. In passing along one of these at six in the morning, when the trees had mostly acquired their new foliage, some of fine pale green, others of pink or red (here, where all is evergreen, there are no autumnal tints like those of the temperate zone), standing out from deep dark recesses, occasionally varied by the finely divided tremulous foliage of a graceful Acacia and the large white star-like leaves of a Cecropia, while here and there hang festoons of some purple-flowered Bignoniacea, white- or red-flowered climbing Polygaleas often exhaling a most delicious odour, while lower shrubs, which barely stand out of the water, are bedecked with countless flowers of various Convolvulaceæ, chiefly of a species of Batatas, mixed here and there with two or three Phaseolæ, some yellow-, others purple-flowered—in the glare of the midday all this seems comparatively tame: the eye is fatigued with looking steadfastly at anything—even green seems dazzling

—light penetrates everywhere, and the eye searches in vain for the variety of shady recesses which are at other times so pleasing. . . .

It is only in the height of the rainy season that the margins of these rivers are seen to advantage—then all is fair and pure. But let the water descend 20 feet, and there appear discoloured trunks, shaggy towards their base with black rootlets, muddy and tangled stems of shrubs which, though not normally twining, seem to have interlaced for mutual support against the crushing, sweeping water. Herbaceous twiners all dead and presenting only withered, blackened strings. Bunches of dead grass and other unsightly matter brought down by the stream hang everywhere. Yet it is in the dry season that most of the forest trees are in flower.

[The visit to Manaquirý, about fifty miles up the Solimoẽs, which occupied most of the month of June, and which is partly described in the letter to Mr. Bentham, is not specially mentioned in the Journal, except in one of the following notes, written immediately after his return.]

The leaves of the Coffee tree are often used instead of the berries in the region of the Amazon. On Lake Trombetas the leaves were strung on a stick which was stuck in a chink of the wall, and they were not used till dry. On the Rio Negro they are used both fresh and dry.

The mode of gathering rice in the lakes of Manaquirý, where it grows spontaneously, as also in many other parts of America and Solimoẽs, is very simple. When the seed is ripe, which is at the end of the wet season (June, July), a montaria is taken into the lake, and as it is rowed slowly across, the men bend the long stalks on each side of them and by a shake cause all the ripe seeds to fall into the canoe. Continuing this process, in a few hours a considerable load of rice is accumulated in the montaria.

[The following rough notes are descriptive of Lages, where Spruce went to obtain the services of a good Indian carpenter to fit up his canoe, and which afforded him a considerable number of new and interesting plants.]

The junction of the Solimoẽs and Rio Negro is now inundated, especially the angle between them; but the left bank of the Rio Negro at its mouth is high land rising far above the river.

Here the abrupt wooded hills rising above Lages are 170 feet high. From their summits are obtained fine views. Directly in front is a very large island stretching downwards towards the mouth of the Madeira; the channel at the back of this is often taken by vessels ascending in order to avoid the furious current of Lages. The river below Lages takes a wide curve to the right; the left bank is all high land, but within it appear considerable depressions constituting lakes, the first being the Lago do Alexo, perhaps a mile and a half long, and towards its extremity quite picturesque, two igarapés entering between high wooded banks. A little beyond is the smaller lake of Tapara, and again a little farther is the large lake of Puraquecóara. Looking down the Amazon, towards the right side, are dimly seen a series of islands forming the extensive archipelago at the mouths of the Antás and Madeira. From this point may be well seen the strife between the waters of the Amazon and Rio Negro, the latter maintaining its individuality far down the left bank. . . .

The deep narrow forest valley near Lages in the lower part is occupied by a grove of Mirití palms, perhaps distinct from the Miritís of Pará. The *trunk is ventricose upwards* and never reaches 30 feet in height. Mixed with Mirití is a fine grass with sub-erect leaves, 6 feet long, but with no flowers (Tripsacum).

Higher up the valley, in very marshy ground, are great quantities of tree-ferns (the same Alsophila as from Santarem), some of the trunks being 18 feet high. Growing along with the fern is an Inga (*I. versicolor*, sp. n.), the flowers with long white stamens turning vermilion after shedding their pollen, and hence giving the tree a very gay appearance.

One of the finest forest trees is the Cajú-açu (*Anacardium Spruceanum*, Bth.). The leaves, especially when young, are white above, greener beneath, and the very youngest are pink. Growing on the side of a valley and viewed from the opposite heights, they appear most beautiful, a large and densely leafy crown of white warmed with the most delicate tints of rose-colour, and spangled with scarlet fruits. The latter are exactly the same shape as the common Cajú, but are slightly smaller and the flavour intensely acid. We traced out several trees, and found them so nearly of the same size that they might all have been planted by natives' hands at the same epoch. Notwithstanding their formidable aspect, as I had determined to preserve specimens at any price, we set to work to cut one down, and after an hour's labour succeeded. This tree I measured after its fall and found it 90 feet in height by over 3 feet in diameter near the base, and perfectly straight and scarcely diminishing in thickness up to the first branch at 50 feet high. A great contrast to the common Cajú, which rarely exceeds 15 feet. The wood and bark of Cajú-açu have a resinous odour.

[I give here two long letters to Spruce's friend and neighbour in Yorkshire, Mr. John Teasdale, which serve well to complete the account of his long residence in the Barra and his more interesting excursions to Lages and Manaquirý. In these letters he writes without restraint, and exhibits his real character far better than in letters written to his botanical correspondents. We here see his interest in and sympathy with the natives, his horror of slavery, and his deep feeling for the grandeur and beauty of the broader aspects of nature around him. His few remarks (and anecdote) on the Education question I did not strike out, because it is even more to the point now than it was at the time he wrote.]

To Mr. John Teasdale

BARRA DO RIO NEGRO, *Jan.* 3, 1851.

You ask me about the temperature. The lowest I have known since arriving in Brazil was one morning at 5 o'clock on the shores of the Bay of Marajó, when the thermometer marked 70°, and everybody complained of its being dreadfully cold. I was obliged to leave my hammock some hours earlier to get additional covering. The highest temperature observed was at Santarem, where it was a little more than 90°; but I have known it higher than this in the south of France, and at Rio they have it sometimes at 110°. It is the sustained heat that we complain of here: at Santarem for many days and nights together the thermometer was never below 80°. This it is which produces the languor which preys on every one in this clime, and more on natives than on strangers. . . .

Now about turtle. Santarem lies some distance below the great turtle country, and when it appeared there it was very dear. Here we are in the very centre of the region of turtles, and we never sit down to breakfast or supper (the two daily meals of Brazilians) without turtle in various forms. We eat here at the table of an Italian merchant, Senhor Henrique Antonij, whose cuisine is excellent, and we find his turtle splendid eating. I know not how many forms it is cooked in, but we have never fewer than five dishes of turtle at table, viz. 1. Tartaruga guisada (cooked or stewed); 2. Tartaruga assada a casca (*i.e.* roasted in the shell); 3. Tartaruga picada (minced); 4. Tartaruga à la rosbif; 5. Sopa de Tartaruga. Of these the picada is the most *recherché*, but I prefer the guisada. . . .

I will now introduce you to the alligators (called here jacarés), respecting which you desire to be informed. Above Obidos we began to fall in with these elegant creatures in considerable numbers, especially when we anchored by night in the still bays. In the bright moonlight we could see them floating about in every direction, sometimes quite motionless on the surface, and only distinguishable from logs by careful inspection. Their note is a sort of grunt, such as a good-natured pig might make with his mouth shut, only rather louder. By imitating it we drew them quite near us, and 'tis little they care for a musket-ball! When, however, we got into the Paraná-mirís, and especially when we visited the Pirarucú lakes, with which the country is literally sown, we saw jacarés lying about in them like great black stones or trunks of trees. It is amusing to observe what a perfectly good understanding seems to subsist between the jacarés and the fishermen, the former waiting very patiently for their share, which is the offal. When a large fish is hooked the fishermen leap into the water, in the very midst of the jacarés, who merely sheer out of the way until their turn comes; and such a thing as a jacaré attacking a man is very rarely known. That it does, however, occur, now and then, we saw fearful evidence.

. . . .

I wish you Englishmen would agree about some general comprehensive system of Education. It is painful to read the accounts of the squabbles, and to see what narrow-mindedness exists, about a subject of such vital importance; and all this time your prisons are filling with young delinquents for whom the State has never provided any intellectual or *moral* training. Cannot this last be given apart from any *doctrinal* teaching? For really, at this distance from the scene of controversy, how insignificant to me do most doctrinal differences seem—more than half of them mere matters of opinion! Were it true, as Dogberry says, that "to be a well-favoured man is the gift of fortune; but to write and read comes by nature," then indeed we might leave nature to take its course. I remember some time ago a capital dialogue in *Punch*, between a father and son on the Educational question. The little boy asks how it is that the Queen does not educate the poor little boys and girls so that they may know better than commit the crimes which lead them to the penitentiaries and treadmills. The father answers that the Queen would only be too glad to do so if the people would let her; and this leads to some talk about the different religious opinions of various sects. The little boy asks for an illustration, and the father selects a certain dogma and attempts to explain it, but finds it rather difficult; he hums and haws, and at last says, "In short, my dear, you will know all about it some time, but now it does not make any matter to you." "Then, pray, papa," inquires the little boy (and most unanswerably, *I* think), "what matter does it make to the poor little boys and girls?"

Aug. 17, 1851.

My landlord, who lives on the opposite side of the street, a few months ago lost five slaves, who ran away from him up the Rio dos Purús, whither they were tracked by the police, and about a week ago all were returned to their owner. One of these was still so refractory that it was judged necessary to chain him by the leg to a post in the yard. At 7 o'clock the same evening his master crossed the yard to go down to the river and bathe by moonlight. In passing near the slave the latter made a spring at him with a knife which he had concealed in his bosom and stabbed him in the side; but, fortunately perceiving the movement, he sprang back and the wound was very slight. The fellow, thus balked in his murderous attempt, set the haft of the knife against the post and with desperate resolution thrust it into his own stomach. The following morning, as I went to bathe, his fellow-slaves were carrying the dead body, sewed up in a sack, down to a canoe, intending to throw it into the middle of the river. They were laughing and joking as if they carried a dead dog; nor did the event seem to produce the least impression on the neighbours. So much for the "beauties" of the slave system! . . .

Visit to As Lages

I have very lately returned from the mouth of the Rio Negro, where is a little hamlet called As Lages, about two long Portuguese leagues (or eight English miles) below the Barra, inhabited entirely by Indians and half-Indians. I visited this place (which has proved a rich botanical station) for a few days in May, and I met there with a carpenter whom I engaged to construct the cabin of my canoe. For this purpose I took her down to the Lages about the end of July and remained there about a fortnight, superintending the shipbuilding and also adding largely to my collection of plants. I much enjoy living among the Indians for a few days together, though I might tire of it were the residence compulsory and permanent. It is such a relief to get out of the town; for these Brazilians, half-savages as you undoubtedly picture them, are the greatest sticklers for etiquette and costume on the face of the earth. It is ridiculous seeing them going to Mass in the "latest Parisian costume"—toiling under the weight of black coats and hats, things which in this climate are a complete abomination. Contrasted with this, the *laissez-aller* of the Lages was delightful. Fancy me there with no other vestments than a light flannel or cotton jacket and a pair of pantaloons—no shirt (consequently no coat or waistcoat), no hat, no shoes or stockings. Even thus I was more completely clad than most of the males, who rarely wore anything beyond trousers. The dress of the women consisted of but two articles—the camisa, descending below the breasts; and the saya, from the waist downwards (corresponding to what you call a "skirt"); some-

FIG. 11.—MARIA (8 years old).

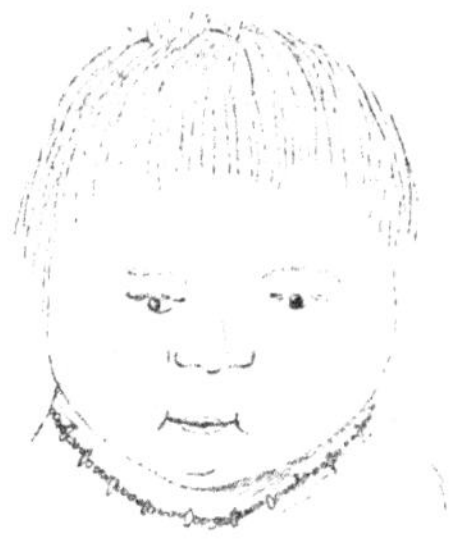

FIG. 12.—RUFINA, sister of Maria (3 years-old).

times the two meet and sometimes there is a space between them. Young girls until marriageable have rarely more than one of these garments—either upper or lower, *n'importe*; but whichever it is, when a stranger approaches one of these sitios far away up the igarapés or lakes, the bashful maiden lifts her only garment to shade her eyes from the white man's gaze; thus reminding me of what I have read of Circassian girls in the slave-market at Cairo. [While here Spruce made pencil drawings of three children of these half-breed families, which, though crude as works of art, give a very good representation of the features and expression of such children in many parts of the Lower Amazon.

FIG. 13.—ANNA, cousin of Maria and Rufina (8 years old).

They are reduced to about half the size of the drawings.] These Indians were better off than most I have met with. Each family had its roça or mandiocca field, which furnished the indispensable farinha; on the slope of the hill behind their houses each one had a little coffee-plantation; and on the summit was a tobacco-plot, which was common property. Near the houses were plantations and various fruit trees—oranges, limes, abacates, etc. etc. I should mention that at the mouth of the Rio Negro the left bank rises into a steep wooded ridge of some 200 feet high; at the foot of this and by the water's edge (which here runs over lages or beds of flat rock) stand the houses; the roças are chiefly on the shores of a picturesque lake (Lago do Aleso) a little in the interior. From the summit of the hill a fine view is obtained of the junction of the Solimoẽs and the Rio Negro, and of the downward course of the Amazon. True, nothing is to be seen but wood, water, and sky, the two former in nearly equal proportions—lakes, channels, and islands stretching away southward of the Amazon to the embouchure of the Purús on the one hand and to that of the Madeira on the other—yet the view is truly grand. It is impossible to behold such immense masses of water in the centre of a vast continent, rolling onwards to the ocean, without feeling the highest admiration; and when viewed under the setting sun (as I several times viewed this scene), and afterwards when the descending and deepening gloom blends all into an indistinguishable mass, though the tumult of the contending waters is still distinctly audible, there is excited in the mind I know not what mixture of tenderness and awe, and I have felt it difficult to tear myself from the spot. The first time I climbed this hill I carried a compass and an aneroid barometer, and took with me an Indian to carry my plant-case. I showed and explained to him the action of the two instruments. He was filled with wonder, and I heard him mutter to himself several times, "Cariúa Juruparí" ("White man is the d——l"!). Similar exclamations I have frequently heard from these people when shown anything beyond their comprehension.

At Manaquirý

In the month of June I made an excursion up the Solimoẽs. My destination was Manaquirý, a group of sitios lying on certain channels and lakes a few leagues from the south bank of that river. The journey occupies three days under favourable circumstances, but the river was at its height and we had rarely sufficient wind to enable us to stem the rapid current; the consequence was that our voyage lasted a whole week. In all this time we did not once see the real bank of the river, only islands. My host at Manaquirý was Senhor Henrique's father-in-law, a Portuguese, and not by any means a modern settler, having come out here in 1798. He is still, at above seventy, a hale, hearty man and can outwork any of his sons. I may remark of him what I have also observed in others, that those Europeans who have led the most active life in this climate, not fearing either summer's sun or winter's rain, invariably enjoy the best health; while those who give themselves up to the easy mode of life of the Brazilians (and they are the majority) become ailing, corpulent, and averse to exertion. His establishment reminded me more of an English farm than any other I have seen in this country. The house stands in a small campo (savannah) in which were to be seen horses, cows, sheep, and pigs, grazing or reposing under the trees. But these trees certainly did not look very English. They included three very fine tamarind trees of exactly my own age, having been planted in the autumn of 1817, but of a growth far surpassing my own, their girt being more than I could span; long avenues of orange trees, laden with ripe fruit, which would certainly have made their owner's fortune could he have had them in England; several large mango trees; thickets of guayabas (a sort of myrtle yielding a pleasant fruit the size of a plum). And if these had not been sufficient to give the scene a tropical character, there were to be seen groups of bananas, papaws, and, peeping here and there out of the encircling forest, various species of palms. At a little distance, on the banks of an igarapé, lay a cannavial or cane-piece, where Senhor Brandaõ had erected an engenho for the fabrication of molasses and aguardiénte, his motive power being oxen.

During my stay at Manaquirý the great annual feast took place, on the Vesper of St. John. It is a curious custom in Brazil (imitated, I believe, from an ancient usage of the mother country) to elect a governor and governess of the principal festivals of the Romish Church, who bear the expenses of the feast, being aided by alms given in the name of the patron saint. In large towns, at Santarem, for example, these "rulers of the feast" are called emperor and empress, but here they bore the more modest titles of "Juiz" and "Juiza." As might be supposed, the Juiz is chosen by the weight of his purse and the Juiza by the amount of her personal attractions. I had long been desirous to see a dance of the country, for much of the character of a people is seen in their national dances; and as I had received from the Juiz and Juiza a polite invitation to go and eat doce (sweetmeats), I resolved to profit by the opportunity. It was after six in the evening when I started, accompanied by a son of Senhor Brandaõ and a whitish young man named Estanislas—a native of Rio, but sent out here by the Government when quite a lad to aid in collecting objects of natural history. At fourteen he took to himself a wife, and now, at thirty-six, he has been some years a grandfather. As in all journeys in this country, our carriage was a canoe and our way lay on the waters. The distance was about a league, threading through the inundated forest, and had we followed the course of the river it would have been much longer. It was dark when we reached the house where the festa was held—a fazenda on the Rio Manaquirý which had been lent for the occasion, and a room in it fitted up as a temporary chapel dedicated to St. John. As we neared the place, lights innumerable sparkled on the water and on the ascent to the house, and one canoe (which bore the image of the Saint) was a perfect blaze of light, proceeding from lamps made of half an orange skin filled with turtle oil. This canoe stood in the middle of the river, and then the tiny lamps were one by one dropped into the water, forming a long line of fire which the rapid current bore swiftly away towards Old Amazon. The scene was further enlivened by the letting off of numerous rockets and muskets loaded nearly to the muzzle, and by the singing of sundry coarse voices to the music of gaitas (bamboo flutes with two holes), the hammering of a crazy drum, and several tambourines.

We landed as the Saint was brought on shore and deposited in the chapel. I was introduced to the Juiz and Juiza, who led me to the foot of the altar, where, of course, I was merely a spectator, while they and their suite arranged themselves in a semicircle, the Juiz holding the Saint, the Juiza by his side holding a long staff gaily decked with ribbons, and the rest with smaller staves similarly decorated. Vespers were then sung, proper I suppose to the occasion, the congregation assisting in the responses. In the very middle of the service, a singer a little within the door, seeing one of his companions outside, called out in a stentorian voice, "Pēther! what's th' aboot there? cum in wi' tha an' sing!" (I translate in *Yorkshire* in order to come nearer the original than I could in English). This caused not merely a smile, but a general laugh. Prayers ended, we were all invited to eat doce. A table covered with a white cloth was extended in a long verandah, on which was doce of papaw in cups, with a spoon and a tapioca biscuit to each. The brancos partook first, and afterwards the ladies and gentlemen of all colours (only two *real whites* were present—your humble servant and a Portuguese settler named Vasconcellas, for Senhor

Brandaõ's son counted Indians among his ancestors by the mother's side—the rest were Mamalucoes, the cross between a white and a Tapuya, Mulattoes, Tapuyas, and Mestiços of various shades). After doce came coffee and cachaça, the latter unfortunately in too great abundance, and meanwhile several people were occupied in lighting up around the house a number of fires, through which leaped boys and girls and several young men and women; those who made the fiery circuit a prescribed number of times being freed for the coming twelve months from all perils of plague, pestilence, and *sorcery*. A lad dressed to resemble an ox, and wearing a real ox's head and horns, was also led round the ring and made to dance and perform various pranks to the sound of the instruments and his driver's voice, the latter extemporising a song describing the past and present exploits of his ox. Other two performers were a couple of "giants" about 12 feet high, the one a lady, the other a gentleman, their faces of painted pasteboard displaying formidable Roman noses, their bodies and arms of branches and leaves of trees; within each was a Tapuya. This odd pair danced several *pas de deux* round and through the fires, which the spectators found exceedingly comic. When tired with this amusement, the verandah was cleared and a fiddle and two or three guitars put in tune for the ball. The first dances were *contradanças Inglezas*. I thought not of joining them, but the Juiz came up to me and led me to the Juiza, insisting that I should open the ball with her. I saw that it was intended to do me honour and that I should be accounted very proud if I refused. I therefore led the lady out, first casting off my coat and shoes in order to be on terms of equality with the rest of the performers. We got through the dance triumphantly, and at its close there was a general *viva* and clapping of hands for "the good white man who did not despise other people's customs!" Once "in for it" I danced all night.

We were beginning to enjoy ourselves, when about 11 o'clock I was surprised to see the dancers separate and run different ways, their looks betraying the greatest alarm. I was not long in learning the cause. A briga (quarrel) had taken place in another room between two half-breed fellows; several persons were implicated, some blows had been given, and knives were drawn. I was inclined to stay and see what a Brazilian "row" was like, but my companions seized my arm and led me away to the canoe. Not only were they afraid of being called up as witnesses should anything serious occur, but they knew that if these fellows, especially the mulattoes, once drew blood, their native ferocity would be excited and the whites would be certain to fall the first victims to it. It had been previously arranged that at midnight the Juiza should conduct as many as chose to accompany her, to eat doce at her own house, and all except the combatants were glad to anticipate the visit. The distance was about a mile, and the intervening space was soon alive with canoes. The night was pitch dark, but happily we were favoured by an interval in the rain, which throughout the night was almost incessant.

At the Juiza's we were a very canny, quiet party,

and I had there the satisfaction of seeing and taking part in several *danças de roda* or "ring dances," about which I felt most curious. These dances are chiefly of Portuguese origin, but modified by change of locality. One of the most amusing was called Picapao or the Woodpecker, of which I will try to give you a sketch. The men and women being first ranged as in our country dances, commence by dancing several times round in a ring, singing—

"Picapao para donde vai?" ("Woodpecker, where are you going?").
"Picapao para donde vem?" ("Woodpecker, whence do you come?")

They then rapidly break up the ring and fall into their places, and then follows a series of hops (intended to imitate the motions of the woodpecker)—the men and women hopping sideways but in contrary directions—at first erect, then gradually sinking down until the chin nearly touches the knees, the musician (who also leads the figure) all the while improvising a dialogue between the woodpecker and his mate. This ended, all jump up, men and women approach, singing—

"Vossé fica—adeos men bem!" ("If you stay, then adieu my love!").

with repeated clapping and snapping.

This is what may be called the *burden* of the dance, but at each repetition the musician improvises something new and varies the figure. I know not when I have laughed so much, especially at the hopping. These *danças de roda* are all eminently dramatic, and much depends on the musician; ours was excellent and contrived to make everything exceedingly comic.

Another dance was the Assaí (the name of their favourite palm-wine). After dancing and singing for some time in a ring (which must consist of an odd number of persons), at certain words in the song the ring breaks up, the dancers whirl round, and each catches in his arms some one who happens to be near him. Thus all are paired save one unlucky wight, who is forthwith shoved into the middle of the ring and condemned to sundry pains and penalties, while the rest dance and sing round him. The ladies were very fond of this dance, especially the hugging part of it, and I had often some difficulty in extricating myself from their embrace.

In the intervals between the dances we had coffee, and sometimes a genuine Indian dance in which I felt no inclination to join, though they were amusing enough to lookers-on. One of them was called Jacamîm-cunhá. Jacamîms are birds of the crane kind; there are several species on these rivers, and all have the body more or less dark with a white rump, the latter produced, as tradition says, by the birds rubbing one against the other. Cunhá is a woman. The performers dance round in a ring, and at certain phases of the tune (for all sing, and the men have nearly all some instrument —a drum, a tambourine, or a gaita) the men turn their backs to their partners and a series of bumpings follows—given with such goodwill that one of the bumpers (and as often the man as the woman) is driven to the far side of the room! Another similar dance was the Tatu or Armadillo. The songs accompanying these dances were in the Lingoa Geral of the Indians, and were of such a nature as not to admit of their being decently translated into any European language.

Among the female dancers were two very pretty Mamaluco girls, so nearly white that they might have passed for such in any part of the world; the rest were only so-so. During the course of the night I danced with every one.

Towards morning our friend Estanislas and the musician favoured us with the exhibition of a trick called "Hunting the needle," which I thought very ingenious. It is thus performed. The hunter being sent out of the room, a needle is hidden somewhere about the person of one of the party. This done, the guitar begins to "discourse" a low monotonous strain and the hunter is re-admitted. He strides into the middle of the room, crosses his arms, fixes his eyes on the ceiling, and seems lost in reverie. Then, apparently roused by the accelerated music, he commences feeling very carefully over his body, beginning at the crown of his head, as if he expected to find the needle concealed somewhere on himself. Arriving at length at the exact part where the needle is actually concealed on one of the company, he starts as if severely pricked, examines his finger, sucks it and shakes it as though in great pain. I soon found that the secret lay in the *pianos* and *fortes* of the music (though the air was never in the least changed), by means of which the two performers had previously concerted a set of signals. It now only remains to make out the person who has the needle. The hunter makes the circuit of the room, directing a scrutinising glance at each person as he passes, and the music, of course, indicates to him where to stop. He then walks up to the possessor of the needle, and at once puts his hand on the latter and draws it out.

You may well suppose that dancing in the latitude of the Equator is not a very cooling process; yet at five in the morning the dancers, though perspiring from every pore, ran out and bathed in the river. Nor was this by any means so dangerous as it would have been in your climate; for here, except in the heat of the day, the temperature of the water is generally greater than that of the air. When I was up the Trombetas and it came on to rain during the day, my Indians used to strip themselves naked and submit most stoically to the pelting of the shower; but as soon as it ceased they plunged into the water, literally to warm themselves, as you will understand when I mention that on such occasions I have found the temperature of the air to be 75° and that of the water 84°.

At daybreak preparations began also to be made for breakfast—a pig and a turtle were slaughtered and several fowls. We were strongly pressed to stay and partake of it, and my companions accepted the invitation; but as I was determined not to neglect business for pleasure, I came away at 6 o'clock with some girls who were going to a sitio near that of Senhor Brandaõ.

I very much doubt if you will find this recital a tithe as amusing as I found the actuality, but it will serve to give you an inkling of manners and customs far removed from those of Old England.

. . . . . .

[The following letter to his friend Mr. Matthew B. Slater, who was a student of British plants, gives a very vivid account of the more prominent botanical features of the great Amazonian forests, which will be more generally interesting than the details referred to in the letters to his botanical correspondents at Kew.]

To Mr. Matthew B. Slater

BARRA DO RIO NEGRO, *October* 1851.

. . . . . .

Do you now and then deign to pick up a moss or a lichen? I do not say that I have been obliged altogether to renounce Cryptogams, but in effect it comes very near it. Not only are mosses exceedingly scarce and limited in species, but I find myself in the midst of such very novel forms of higher orders of plants that it would be unpardonable to neglect them. Still, my Muscological studies have been of great use to me in giving me habits of accurate and patient analysis, and after dissecting the peristomes, etc., of mosses, I find most dissections of the parts of Phanerogamia comparatively easy. My microscope is rarely taken out now except to examine ovaries and embryos. I wish I could have you here for a week—in that time you would learn more of natural orders than in England in a year. I speak not alone of the few orders that include your European Flora, but of all those peculiar to the tropics, of which your herbariums and botanic gardens must ever give an imperfect idea. Unless Mr. Paxton's Crystal Palace could be kept at an average heat of 80°, the noble Laurels, Silk-cotton trees, etc.—the glory of South American forests—will never attain in England anything like their normal development. Nearly all vegetation here is arborescent. The largest river in the world runs through the largest forest. Fancy if you can *two millions of square miles of forest*, uninterrupted save by the streams that traverse it; for the savannahs (here called campos) that here and there occur are so insignificant, that I suppose a greater gap would be made in the largest wood in England by cutting down a single oak than any one of these campos makes in the immense Amazonian forest. You will hence be prepared to learn that nearly every natural order of plants has here *trees* among its representatives. Here are grasses (bamboos) of 40, 60, or more feet in height, sometimes growing erect, sometimes tangled in thorny thickets, through which an elephant could not penetrate. Vervains forming spreading trees with digitate leaves like the Horse-chestnut. Milkworts, stout woody twiners ascending to the tops of the highest trees, and ornamenting them with festoons of fragrant flowers not their own. Instead of your Periwinkles we have here handsome trees exuding a milk which is sometimes salutiferous, at others a most deadly poison, and bearing fruits of corresponding qualities. Violets of the size of apple trees. Daisies (or what might seem daisies) borne on trees like Alders.

The natural orders which by their frequency give a character to the vegetation of the Amazon are chiefly such as are altogether absent from the English Flora. Myrtles are exceedingly numerous, and so provokingly like each other that whoever has seen the common Myrtle of the south of Europe might swear to a Myrtle in any part of the world. They are remarkable for their simultaneous and ephemeral flowers. On a given day all the Myrtles of a certain species, scattered throughout the forest, will be clad with snowy fragrant flowers; on the following day nothing of flowers appears save withered remnants. Hence it comes that if the botanist neglect to gather his Myrtles on the very day they burst into flower, he cannot expect to number them among his "laurels." Another order, nearly allied in structure, but without anything in the European flora to which it can well be compared, is Melastomaceæ—equally abundant with Myrtles, and richer in species. Their ribbed opposite leaves afford an almost never-failing character, and there are some very pretty things among them. These two orders, with Solaneæ and Lauraceæ, form the mass of the vegetation one sees in the vicinity of the towns. But of all orders, by far the most abundant constituent of the flora of the Amazon is Leguminosæ. The species of this order constitute one-sixth of my whole collection of flowering plants and Ferns. Amongst them are some of the noblest trees of the virgin forest, some of the pleasantest fruits, and (what may surprise you) some of the strongest poisons. More than half of them have not papilionaceous flowers (the Mimoseæ and Cæsalpineæ), and would therefore be quite strange to an English botanist; some have even drupaceous fruits, and hence approach Chrysobalaneæ, an order which exists here in great abundance, resembling (I wish I could say supplying the place of) the plums and cherries of your own island. Sensitive-plants, here called Sleepy plants (Dormideiras), which you think so curious, are here so common that almost every day I scratch my fingers or my shins against some thorny member of the group.

CHAPTER VIII

JOURNAL OF A BOTANICAL VOYAGE UP THE RIO NEGRO TO SAÕ GABRIEL DA CACHOEIRA

(*November* 1851 *to January* 1852)

(CONDENSED BY THE EDITOR)

[" *Nov.* 14, 1851.—This day (Friday) I left the Barra in my canoe with six men, for the Upper Rio Negro. There was little wind, which soon failed entirely. We slept at Paricatúba, about fifteen miles from the Barra on the opposite shore, where I gathered seeds of a beautiful small tree allied to the well-known *Lagerstroemia indica* of our conservatories."

Thus begins the Journal, with entries of a very similar nature day by day. The writer notices the different characters of the soil at his various stopping-places, whether clay or sand or rock, whether sandstone or granite; and he remarks that rocky situations are at this season more prolific in flowers than sandy ones, and that everywhere he finds trees or shrubs in flower on the very margins of the river. The sudden storms alternating with calms, the various appearances of the islands and shores of the great river, the various huts or small villages passed at distant intervals, the success of his Indians in hunting and fishing, the days of good sailing or of continuous rowing, are all recorded, as well as the ever-changing character of the vegetation, the various trees and shrubs and palms which his experienced eye detected as novelties, and the many beautiful flowers he was able to gather which were not only new species but were so peculiar in structure as to constitute new genera —all this rendered the journey a continuous intellectual enjoyment to so enthusiastic a botanist.

But the daily record of such incidents during a month's journey would be monotonous and uninteresting to the general reader, and as the more important botanical discoveries are referred to in the letters to his various correspondents, I shall only give in full such portions of the Journals as describe the few incidents of more general interest that occurred at some of his stopping-places. The first of these was about the middle of the voyage, and here the Journal becomes more interesting.]

Nov. 24, 1851.—Below the mouth of the Rio Branco are the celebrated Ilhas de Pedras or Uarapanáki—granite rocks in the middle of the river on which are extensive Indian picture-writing. The figures are very numerous: some representing animals; one a number of persons joining outstretched hands, called the "Dancers"; and there is one which is plainly a rude attempt at a church, and underneath is the word *Deos*, to all appearance of the same date. The figures have been formed by scraping broad lines on the rock with some hard instrument. Sometimes the whole figure is scratched out. It does not seem necessary to suppose that all these figures are the work of the same epoch. What is certain is that for some time (possibly a hundred years) such has ceased to be executed.

I protest against the term "figure- or picture-*writing*," which supposes a hieroglyphical interpretation attached to the pictures such as I am convinced they do not possess.

A little farther on are more figures on three large contiguous granite blocks, almost paraboloidal in form, which stand on the right bank of the river. Here is the representation of a large cayman or alligator seizing a deer.

Pestana told us that the greatest number and variety of figures were on some rocks in a paraná-mirí (side-channel) whose mouth we passed a little farther on. These rocks are called Tucanaróka or the Toucan's nest. . . .

Dec. 4.—This morning at 8 we reached Cabuqüena (Moureira of the maps), standing on the summit of a range of red earth cliffs. I went on shore to try to procure farinha, of which we found great scarcity on the river, but could purchase only one basket of a man named Jacobo, a great voyager on these rivers. He had descended and reascended the Orinoco; was at Esmeralda when Schomburgk arrived there; says that this traveller should have found no difficulty in reaching the sources of the Orinoco, as he himself shortly afterwards ascended the Orinoco a month's journey above Esmeralda, till his montarias could travel no farther and the river might almost have been leaped over. He found the Guarahibo Indians quite pacific. They make great use of tururí bark for caulking their canoes. He had ascended also the Rio dos Cauaborís and the Marauiá (both of which have their rise in the same lofty serras), and there encountered with Indians from the sources of the Orinoco, who had come by a short portage. According to him, there are several cataracts on the Orinoco above Esmeralda, but they cannot be compared to those of Atures and Maypures. He knew Natterer when on the Rio Negro. Natterer ascended the rather low serras in front of Castanheiro.

Dec. 16.—Not a puff of wind to aid us to-day. After passing some trifling rapids, we arrived a little before sunset at the foot of a series so formidable that it was deemed prudent to wait for the morning before attempting their passage. They are called Jurupari-roka (the Devil's house), but possibly this appellation is derived from a large mass of granite rising with a gentle slope on the left of the falls to a height of some 40 feet, of a very sooty hue and having near the top several deep hollows. I climbed to the summit just after the sun had set and had a very fine view. Beneath me were the rapids tumbling among masses of granite with a noise which we had heard an hour before reaching them. Then spread out the glorious river, empurpled with the rays of the departed sun, shining through the interstices of five large wooded islands; while numerous shapeless blocks of granite stood out of the river here and there, some naked, others with a scanty vegetation in their clefts; the waters everywhere circling and eddying or running rapidly over some sunken ledge of rock. At my back I had dense low forest showing numerous types of foliage near to me and varied by the overtopping crown of some tall palm or the ostrich-like plume of some graceful bamboo; and all standing out in that relief and shining with those tints which only sunset can bestow. . . .

Dec. 18.—This morning opened with a gloomy but gradually clearing sky; Uanauacá appearing at back of a large Genipapa (a fruit tree, *Genipa macrophylla*) on the river's brink. At a little past noon we reached the sitio, pleasantly situated on rising ground in an artificial campo, in which are here and there trees of Tapiribá, Bacate, Oranges, Limes, etc., and including three young trees of Puxirís, which, however, are of age to bear fruit. I met with a very cordial reception from the owner, Senhor Manoel Jacinto da Souza (Tenente de Policia), who offered me a room in his house in which to arrange my collections of the voyage, and until I could procure men to take me on to Saõ Gabriel, for all those I brought with me are of Uanauacá or the neighbourhood, save one, and they wish now to work in their roças.

We remained at Uanauacá until January 6. In the interval I arranged the collections I had made on the voyage and packed the greater part of them into a case which I left with Senhor M. Jacinto to be forwarded to Pará.

I made also two excursions, one to an inundated campo on the borders of a lake on the opposite side of the river, now dry and adorned with bright blue flowers of a Lysianthus, having the aspect of *Campanula rapunculus*. The other was to an elevated campo on the same side as Uanauacá, much resembling Umirisal at Barra, and in the adjacent caapoera I gathered the Cocura-açu.

Short walks near the house afforded me several Melastomæ and other interesting plants.

[Before proceeding with the description of the perilous ascent of the cataracts between Uanauacá and Saõ Gabriel, I will insert two letters giving very picturesque descriptions of the voyage so far, the first to Mr. John Smith, at that time Curator of the Kew Gardens, giving a familiar sketch of the botanical aspects of the voyage, and the novelties he was able to collect; the second to his old friend and neighbour, Mr. John Teasdale, with a more general account of the voyage, written with much of the freedom and vivacity of familiar conversation, and constituting together a supplement to the rather formal and meagre narrative given in the Journal.]

To Mr. John Smith, Royal Gardens, Kew

SITIO DE UANAUACÁ,
BELOW THE FALLS OF SAÕ GABRIEL,
RIO NEGRO, *Dec.* 28, 1851.

"Thus far have I advanced into the bowels of the land without impediment"; and before adventuring the falls (where I may possibly get a ducking) I seize an opportunity of sending you the seeds of a beautiful Lythraceous tree which I collected on my way up. It grows on a sandy shore about 20 miles above the Barra, and I had gathered flowers of it on the 1st of October. Its habit is almost that of *Lagerstroemia indica*, but the flowers are still more showy; and as I saw no tree above 25 feet high, and all were clad with flowers almost to the ground, I have no doubt you will be able to flower it at 4 or 5 feet high. It seems to be a Physocalymma, a genus (if I may trust to Paxton) not in cultivation. My specimens give no idea of the beauty of the plant, as I was taken ill after gathering them, and they were nearly spoiled before I could get them into paper.

I left the Barra on November 14 and reached here on December 18—a good voyage considering that I worked all the way and consequently made frequent stoppages. I have dried some 3000 specimens on the voyage—a much greater number than I ever dried on any previous voyage—and I am now occupied in arranging them for packing into a case which I shall leave here to be forwarded to Pará. It was the owner of this sitio (Senhor Manoel Jacinto de Souza, a lieutenant of police) who sent me five out of the six men that composed my crew. They were under no obligation to ascend higher than Uanauacá, but they have agreed to accompany me to Saõ Gabriel, if I will only let them have a fortnight to work in their roças. It was no slight trouble to have to send 1000 miles for men, to wait three months for them, and then to have to pay them for the voyage down and for the time they were waiting for me in the Barra (for they came on me quite unexpectedly), as well as for the voyage up. Yet even on these terms I was glad to get them. So immense is the difficulty of procuring men here to do anything, that I think of removing altogether to Venezuela. . . .

I should like to ascend the Rio Negro again, because I was obliged to leave so many fine things on its banks. After passing Barcellos almost everything was new, and so many things were in flower, that I was obliged to confine myself to those which presented the greatest novelty of structure. Nothing like this has ever happened to me before. I was obliged, for instance, to shut my eyes to Myrtles, Laurels, Ingas, and several others. Between the Barra and Uanauacá I counted no fewer than fourteen species of Lecythis[1] in flower, and all but one new to me! Yet of these I got a stock of only four or five; for, to say nothing of the difficulty of preserving so many things, I found my Indians very hard to set agoing again when stopped in the middle of their work. And when you consider the time that is lost in collecting trees—for your tree is rarely on the very river's brink, but you have to cut your way to its base with cutlasses, and it has then to be climbed or cut down—you will understand why I generally contrived to make my collections when we stopped to cook our meals.

I enclose you two flowers of a Leguminous tree which was in flower all the way up the river and formed a great ornament to its banks. It is a Heterostemon (a most remarkable genus), but whether a described species I cannot say. The petals are a fine blue slightly tinged with purple, and the column of stamens is red. There are no pods ripe yet, but I will try to send you some. As it often flowers at 10 feet high, it is very suitable for cultivation. But the glory of the Rio Negro is a Bignoniaceous tree (apparently an undescribed genus) with whorled leaves and a profusion of pink flowers the size of those of the foxglove. It grows 90 feet high!

In Cryptogamia alone am I disappointed in the

[1] A genus allied to the Brazil-nut tree.

Rio Negro, though I always had my eyes open for them. The following is my Cryptogamic summary thus far: Ferns, 0; Mosses, 0; Hepaticæ, 1; Lichens, 3 or 4 epiphyllous species! Would you have expected this of the Rio Negro? I certainly hoped something better of it. In place of these tribes there are, however, plenty of Podostemons on the granite rocks which peep out of the river (and, by the by, make the navigation very dangerous), but all, *all* dead and burned up. It is here, as I remarked at Santarem, the Podostemons all flower just as the water leaves them, that is, early in the dry season; and my ascent of the Rio Negro was made towards the close of the dry season; but if I live, these little fellows shall not escape me. As their fruit is exposed to a burning sun six months or more in the year, I do not see why they should not travel safely to England in a letter, and I accordingly enclose capsules of one of the largest specimens. They ought to vegetate on stones (especially granite) barely emersed from the water of a tank; though here they never grow in *still water*—always in rapids or cataracts where the water rushes over them.

I had sad news two days ago from my friend Wallace. He is at Saõ Joaquim, at the mouth of the Uaupés, a little above Saõ Gabriel, and he writes me by another hand that he is almost at the point of death from a malignant fever, which has reduced him to such a state of weakness that he cannot rise from his hammock or even feed himself. The person who brought me the letter told me that he had taken no nourishment for some days except the juice of oranges and cashews.

Since I came to Pará the fevers of the Rio Negro have proved fatal to two of the persons mentioned in Edwards's *Voyage*—Bradley and Berchenbrinck, very fine young men both. Wallace's younger brother, who came out from Liverpool along with me, died last May. He had gone there, poor fellow, to embark for England, took the yellow fever, and died in a few days.

The Rio Negro might be called the Dead River—I never saw such a deserted region. In Sta. Isabel and Castanheiro there was not a soul as I came up, and three towns, marked on the most modern map I have, have altogether disappeared from the face of the earth. We had beautiful weather in coming up, and to this may be attributed that I and all my people arrived here in good health. . . .

Mr. Wallace came up from the Barra more than a month before me, escaped the fever on his way, but the day he set foot in Saõ Joaquim was attacked.

. . . . . .

What a beautiful little palm is the *Mauritia carinata* of Humboldt! It is remarkable for growing in *tufts*, and as I sit writing I can distinguish a cluster of perhaps fifty stems on the opposite shore of the river. It is abundant on all the Upper Rio Negro. It would fruit beautifully with you.

To Mr. John Teasdale

SAÕ GABRIEL, RIO NEGRO, *June* 24, 1852.

. . . . . .

When I wrote to you from the Barra I was on the point of starting on my voyage up the Rio

Negro, my little vessel and the Indians necessary to work it being all in readiness. I intended to have written out for you my Journal in its entirety, and I think it would have interested you, but I must content myself with a few extracts. I may premise that the voyage was on the whole a perfect contrast to that up the Amazon from Santarem, and, in short, the first agreeable voyage I have made in South America. The canoe being my own, I was master of my movements—could stop when I liked and go on when I liked. The cabin, too, was new and commodious. It was long enough to suspend my hammock within it, and I made myself besides a nice soft bed of thick layers of the bark of the Brazil-nut tree (which you will find mentioned by Humboldt under the name of Bertholletia); my large boxes ranged along the sides served for tables and the smaller ones for seats; while from the roof I suspended my gun and various things that I required to have constantly at hand. The fore-cabin or *tolda da prôa* was occupied by baskets of farinha, a few bushels of salt, and various other things which I was taking with me to barter with the Indians; it served also as a sleeping-place for the men when the weather was wet, otherwise they preferred sleeping outside. As to myself, warned by past experience of the risk of sleeping in the open air on these rivers, I constantly passed the night inside the tolda, and to this I attribute my not being attacked by the fevers which have proved fatal to so many Europeans on the Rio Negro. The cool of the evening and the early part of the night, especially when we had the moon, I was accustomed to pass seated outside the tolda, and this I could do undisturbed by the insects which are the greatest torment to the traveller on the Amazon. This is the great advantage of voyaging on black waters, that no carapaná (or zaneudo, as the Spaniards call them) interrupts one's repose. I was often reminded too of what Humboldt says in his *Aspects of Nature* (vol. i. p. 215) respecting the wonderful clearness with which the constellations are reflected in black waters. I have nowhere seen with such marvellous distinctness what might seem the "skies of a far nether world" as when anchored by night in a still bay of the Rio Negro, and looking downwards on its unruffled waters; but when moving along every stroke of the oars dashes fifty stars to shivers and thus dispels the agreeable illusion. On the Upper Rio Negro there is no lack of insect plague *by day*, in the shape of two very minute flies, called píum and maruím (the real "mosquitoes" of the Spaniards), whose bites are most annoying and cause considerable swelling and irritation. They are found wherever the river inundates granite rocks (as at Saõ Gabriel), and especially about the mouths of some affluents of the Rio Negro which have whitish water. The following extract is from my Journal of December 12, written off Sta. Isabel: "Yesterday and to-day much tormented by maruím. My hands, neck, and feet are painted with their bites. Whilst I write there is a cloud of them between my eyes and the paper, and several are feasting on my hands and face." To be exposed to such as this is no bagatelle, but I mind it little when I can look forward with tolerable certainty to a quiet night's rest. I have conversed with many people who have visited Esmeralda, on the Orinoco, and all confirm Humboldt's account of the unceasing torment of mosquitoes at that place. They tell me it is impossible to do any sort of work by day.

The crew of my canoe were all of pure Indian extraction—a great advantage, for the least streak of white blood in an Indian's veins increases tenfold his insolence and insubordination. Four of them were Barrés, one Uaupé, and one Manſoa. The last had been some years in the Barra, and took it into his head to revisit his native forests, his mother and sisters being established at Saõ Pedro, below the falls of Saõ Gabriel. On the voyage I found that he was an excellent shot, and I therefore invited him to stay with me as hunter. He accepted the offer, and has been a very great aid to me, for I am now in a country where every article of food (save farinha) must be sought for in the rivers and forests. Saõ Gabriel is a wretched place—never is there so much as an egg or a banana to be had either for love or money. This Indian, besides keeping my table supplied with game, was of great use to me in my excursions, not only for rowing my montaria, but also for climbing and cutting down trees; but though an exceedingly strong, active fellow, he was subject every now and then to attacks of acute pain in the chest and spine, resulting from a strain received in Pará in unloading a vessel; and when he had been with me about six months he had an attack so violent, attended with considerable fever, as to baffle my small skill in medicine; so that, after being confined several days to his hammock and showing no signs of improvement, I allowed him to go to his mother and stay until re-established in health. A person who came up the falls a few days ago brought me word that he was still no better, and I therefore despair of profiting further by his services, which I much regret, as I do not expect I shall again meet with one so well suited to my necessities. He was perhaps the only industrious Indian I have met with, and was never content when the "patron" had not a job for him to do. I have still with me another Indian, but he has not half the activity of the one I have lost.

Notwithstanding the greater docility of these Indians than of any others I had previously had anything to do with, they gave me no small trouble in the Barra, where they were kept waiting for me for ten days; for I was taken rather by surprise and had much to do in filling my boxes and writing my letters for England. The love of "strong waters"—inherent in these Indians as in their brethren of North America—was at the root of the matter. One old fellow made it his first business to dispose of the whole of his earthly goods (leaving himself only a pair of trousers), namely, his hammock, shirt, knife, and tinder-box, with the proceeds of which he got so gloriously drunk as to be in a state of utter helplessness for a couple of days. Yet this man, when removed beyond the scent of cachaça, proved the very best fellow in the lot—always in a good humour, always ready for work, and the first to climb any tree of which I desired the flower. The others begged money of the patron to buy a barrigada (skinful) of cachaça, and the patron had no alternative but to give it them, for if not they would have made no scruple of running away or of engaging themselves to some other patron. Throughout the Amazon and its branches the vessels are all manned by Indians, and as the latter are not sufficiently numerous for the traffic, the *negociantes* have a very bad habit of stealing Indians from one another, going themselves or sending emissaries with cachaça by night, and making the Indians dead drunk, then tumbling them into the canoe like so many logs and setting sail immediately. When the Indian wakes up from his drunken sleep he finds himself far from port and embarked on a voyage he dreamt not of undertaking; little, however, cares he for this: he is like the ass who had no fear of being taken by the enemy, knowing that it would make little difference in the weight of his burden. Temptation of this kind was not wanting to my Indians, but by exercising a little vigilance I was able to keep them all together until the hour of embarking; and once away from the Barra they were all as obedient and industrious as I could wish for.

When I left the Barra there was great difficulty in procuring provisions. Owing to the waters of the Amazon not falling as usual, no pirarucú had as yet been procured, and that is considered the staple provision for voyages in this region. As a substitute, Senhor Henrique and I bought a young bullock between us, and I had one-half of it salted down for the voyage. I bought also as many turtles as I could find in the Barra, and I bought a few more on the voyage of a man whom I encountered coming out of the mouth of a small river near

Airaó with a cargo of them.[1] I needed not, however, have put myself to this expense, for my men proved excellent fishermen, and we rarely passed a day without fresh fish. They seldom used any other weapon for killing fish than the bow and arrow; and what I more admire in this than the certitude of their aim is the acuteness of their vision. They would spy out a fish deep in the water and tell with certainty what sort it was, when I could distinguish nothing; and it was interesting to see them steal silently after a fish, in a montaria, until the fish, approaching near enough the surface, was pierced by the arrow which had been held in readiness. It was in the gapó (inundated forest) and at the mouths of the igarapés that fish were taken in this way. To give you an idea of the expertness of these men, I may mention that one morning in the space of half an hour two of them killed twenty fish in an igarapé with their bows and arrows, and the least of these was more than I could eat at a meal. My hunter also got us some excellent breakfasts and suppers with my gun. He used to enter the forest before daybreak and surprise the birds still asleep in the trees, when I could no more discern them than I could the fish in the waters; in this way he shot us several large wild-fowl, and especially mutúns (curassows). These birds are as large as a turkey, but with shorter feathers, neck, and legs, and when well cooked are excellent eating. One which we had served us all for supper, and there was enough left for my breakfast next morning. Another bird called inambú,

[1] Turtle are very rarely met with in the Rio Negro, but only on some of its lower branches. The pirarucú is a fish confined wholly to white water.

very like our partridge but larger, is to my taste the finest eating of all the game of these forests, its flesh being exceedingly white and delicate. Of this too we got a good many on the voyage. The same birds are met with in the forests of Saõ Gabriel, and various others, some good eating, and others, such as parrots and toucans, only to be eaten when there is nothing else. We get also several four-footed animals, such as cutías (agoutis), wild pigs (peccaries), antas (tapirs), etc., and I must not omit to mention various kinds of monkeys, amongst which a black monkey, called uaiapissá, is considered a first-rate delicacy.

. . . . . .

A circumstance which contributed greatly to the enjoyment of the voyage was the beautiful weather we had nearly all the way up. The season was so far advanced when I left the Barra that I was afraid of encountering naught but squalls and torrents of rain; but there is no foretelling the weather on the Rio Negro: when one looks for fair weather cometh rain, and the contrary. In order to profit as much as possible from this favourable state of things, I agreed with the men to travel chiefly by night, that is, until we reached the region of rapids, which begins a little below Sta. Isabel, after which there is no more travelling by night. Thus when there was no wind in the middle of the day, we chose out some favourable spot for spreading out my paper in the sun—such as a sandy beach, and especially a large bare rock (such as we frequently met with on the islands above Barcellos)—and there remained from 10 or 11 in the morning till 3 or 4 in the afternoon.

Whilst my men were reposing I was working—drying my plants and papers and exploring the adjacent forest for flowers. When I found any lofty tree in flower I called one of the Indians to climb it. They would then continue rowing until 10 at night, and recommence at 2 or 3 in the morning. From the Barra to some distance above Barcellos we were much aided by the trade winds, and my canoe, though anything but handsome in its cut, went excellently under sail, riding out the strongest trovoados (squalls).

It may be true, as Humboldt says, that "perils elevate the poetry of life," but I can bear witness that they have a woeful tendency to depress its prose. . . . In my own case, so long as the river was smooth and deep, my little vessel went on gallantly and my labours were uninterrupted; but when the bed of the river began to be obstructed by rocks and the current to run furiously, anxiety took the place of pleasure, and instead of working among my plants, I had to watch over the safety of my canoe and its contents. Thus from the Barra to Sta. Isabel I have much to show and little to tell, and from Sta. Isabel upwards, though I can recount plenty of perils of waters, I can produce but few plants gathered by their margins. . . . In many places the river spreads out to an enormous width, nothing being known with certainty of great part of the northern shore. Frequently it is sprinkled with islands, and sometimes opens out a lake-like expanse, so wide that were it not for the lofty skirting forest the opposite coast would be invisible. The idea of a river studded with islands no doubt suggests to you a variety of pleasant views; but when you are told that the islands are all a dead level, clad with an unbroken forest, and many of them as large as Castle Howard Park, while the channels between them are sometimes no wider than the Thames at London Bridge, you will justly conclude that they offer only a monotonous aspect to the voyager. When the river begins to be narrower, and its waters to run with a perceptible current, then the islands are smaller, and, when rocky, often picturesque.

. . . .

[Leaving Uanauacá to ascend the falls, Spruce stayed the first night at the village of Saõ José, on the left bank, where there was a half-breed Inspector of the District, who had been a traveller as far as Guiana by way of the Rio Branco. At his door was an old blunderbuss fixed on a block of wood, used to frighten the Macú Indians, who were sometimes troublesome. Here a nocturnal disturbance occurred which is described as follows. —Ed.]

In the middle of the night, lying awake in the tolda, I was startled by hearing a long scream from a woman, followed by a report of a musket, and, shortly after, the explosion of the Inspector's blunderbuss and of several other firearms. This continuing for some minutes and being accompanied by wild shouts, I very naturally fancied it to be caused by an attack of Macús, and called to my pilot, who was lying near the cabin door, to ask what he thought of it. He was quite nonplussed. The shouts, he said, were not those of people engaged in combat; still, the Macús might have shown themselves in the adjacent forest, and the people might be trying to scare them away. All at once turning his face to the sky, he burst into a fit of laughter, which it was some time ere he could repress so as to speak intelligibly. At length says he: "It is the moon, patron—it is the moon; come out and look!" "Lord save us," thought I, "but this is a novel form of lunacy, which affects simultaneously a whole township!" and I bolted out of the tolda to interrogate Diana thereupon. But though the sky was clear save a few fleecy clouds, and the moon ought to have been in mid-heaven, nowhere was she to be seen! I at once perceived she was totally eclipsed; and in about a minute she showed her obscured face from behind a small cloud which was passing at the time I first turned my eyes towards her.

I learnt from the pilot, and from the people themselves this morning, that they were afraid the moon was about to leave them altogether, and that the firing and shouting were to frighten her back again! They asked me if we did the same in my country when the moon showed signs of absconding, and heard with surprise that we did not. The noisy demonstrations were kept up, at first briskly, then at lengthening intervals, until the eclipse had nearly passed off.[1]

. . . At about 10 we reached the most formidable rapid below the great caxoeira, where the river is divided into two narrowish channels by a long island and across both stretches a broken ridge of rock which gives rise to the rapids. The

[1] I have since learnt that on occasions of an eclipse the Indians are accustomed to shoot a number of arrows towards the moon and in the morning to pick them up again, believing that with them their aim will be unerring in the chase.

difficult spot was the turning of a point where the granite shore juts into the current, and all our efforts to pass it were unavailing. A sitio stands close by, and we invited the owner to help us, which he very readily did. I took the helm, though very ill-disposed for the task, the pilot leaped into the water with two or three more, applying their shoulders to the canoe, whilst the rest on board lugged at a rope made fast on shore beyond the point. In our course lay a sunken rock, which it was thought the canoe might pass; but, instead, she struck on it and immediately fell over on one side. The boat swung round, forcing the rope out of the hands of the men, who instantly leaped into the water, not showing much consideration for the safety of my goods, and I was then left alone. I stuck pertinaciously to the helm. The canoe again swung round and fell over on the contrary side, and all thought this time she would have gone clean over; but she did not. Another revolution and she swung fairly off the rock, righting at the same time. I set her head to the fall and she shot down like an arrow. In a few instants she reached an eddy of the current, and I was able to take advantage of a slight reflux to set her head to shore and bring her up in a small bay, where my men speedily rejoined me. All this took up scarcely more than a minute; whilst it lasted I felt nothing like fear, but when it was over, I fully realised the peril I had been in, and made a mental resolution to have no more to do with the helm in rapids.

A council was now held, and I determined to send across the river to a sitio where aid might possibly be procured. After waiting two hours, my montaria returned with one man, and we again made the attempt; but even then we should have failed to surmount the rapid had it not been for the aid of a brisk wind which sprang up. As it was, we could only advance by inches, and it took us half an hour to ascend what we might have descended in half a minute.

Jan. 10.—This morning at 8, Senhor Pailhête took me across the river to view the Serras de Curicuriarí, which lie directly at the back of his sitio (a day's journey, but there is no path), and on the east side of the river Curicuriarí. . . . From our point of view they might have been clearly seen had there not been much vapour in the air. The highest has much steep rock, mottled with brown and white, and quite inaccessible on the south side, but its summit might possibly be reached by taking a col between it and the flat-backed wooded mountain to the right.

This afternoon I had a walk in the virgin forest, where I saw much that was new to me, though few things were in flower.

Jan. 11.—Dull morning with slight rain. My pilot and one of Senhor Pailhête's men went a-hunting early this morning, and returned at 10 with three mutúns (curassows). At midday we embarked, my crew being augmented by a Tapuya lent by Senhor Pailhête, who was a good proeiro, and another of the Tochana's men, so that I had now seven oars. Still, the rapids were so frequent that we got on but slowly. This afternoon we reached the mouth of the river Curicuriarí at about sunset, and made fast for the night. . . .

Jan. 12.—This afternoon at 5.30 we reached

the foot of the great rapids of Camanáos, considered the commencement of the caxoeira of Saõ Gabriel, and I immediately sent off my pilot in search of the "pratico das cachoeiras," a half-Indian named Dyonisio; but his sitio was some distance up on the left bank (to which we had just crossed with considerable difficulty and risk), and I had miscalculated the time necessary for reaching it against the rapids. It was dark when my messenger arrived there, and he found the pilot laid up with a wound in his leg caused by falling on the stump of a tree. In the morning he procured a substitute—a Tapuya named Quintiliano, who I suppose to be much inferior to Dyonisio.

Jan. 13.—This morning Quintiliano presented himself at the canoe about 9, and at 10 we got under way. We were aided nearly throughout the day by some people who were working in a roça near, so that I had constantly eleven persons employed, and sometimes more. From the shallowness of the water and the depth of my canoe, we had great difficulty in passing many of the falls and rapids, and often scraped the rough granite rocks. I had taken the precaution to fasten my heaviest boxes to the sides of the cabin, and it was well I did so, or when the canoe fell over on her side (which was not infrequent) they would have fallen upon one another and might have caused considerable destruction.

Opposite the pilot's house is a fall considered one of the most dangerous. Here there are two channels separated by a ridge of granite, and we passed along the wider of the two, that adjacent to the right bank of the river, without much difficulty;

but in the rainy season it is necessary to take the narrower channel, and the fall is so great that the canoe has to be unloaded and the cargo passed over the rocks to above the fall.

Our mode of progression was as follows. I will suppose it necessary to turn some point of granite rock round which water rushes furiously, or perhaps falls at once a few feet. Our five-inch cable was made fast to some rock beyond the point, the Indians carrying it thither partly through the water and partly across the granite blocks that stood out of the river, a very laborious and perilous task; the end which remained on board was then passed round the mast, the stout oars laid across the tolda in pairs and secured so that the men might rest their feet against them whilst tugging at the rope in a sitting position. A shorter three-inch cable was also fastened to the prow, and two or more men yoked themselves to it, pulling rather inshore, their object being to prevent the canoe from falling outwards with the force of the current. As many men as could find room to work having taken their places at the five-inch cable on board the canoe, the pilot stood out into the rapid as far as was considered necessary in order to clear the rocks, and the men commenced tugging with all their force. If the water was deep enough we got through without accident, the only risk being, firstly, in the men not being able to draw in the rope fast enough, when the canoe was brought up violently against the rocks; but as I had always men stationed there prepared for such a contingency, and the pilot and two or three of the men always leaped into the water and assisted in holding the canoe off the rocks, we sustained no damage in this way; secondly, in the breaking of the rope—a very possible occurrence, for this piassaba is a very brittle material, and as it strains and crackles one watches with intense interest every successive inch that is passed round the mast (especially when the canoe is one's own): from this casualty also we happily escaped. But by far the greatest danger is when some sunken rock lies in the way, over which the prow of the canoe passes without touching, but on which the poop strikes. The current having now a *point d'appui*, becomes irresistible, for our course against it is always more or less oblique. The men at the shorter rope are dragged under water, and did they not leave go would be dashed to pieces, and those on board may try as they like they cannot prevent the catastrophe; the canoe whirls half round and falls over on her side; the men hold on as best they may, and then leap into the water to prop up the canoe from going over altogether, and to right her again if possible. This happened to us several times, and once (on the second day) I thought it was a gone case, so completely and apparently irrecoverably did the canoe fall over. My cooking apparatus was a large superannuated pitch-cauldron (of Welsh manufacture, by the by) given me by Senhor Henrique; this, half filled with earth on which three large stones were placed, made an excellent stove. It was placed in the poop, and when the accident happened, notwithstanding its great weight, it pitched over the tiller and fell splash into the water. Fortunately the pilot had already leaped overboard on the contrary side, or it would have demolished him.

I bade adieu to it; but when the canoe had passed the fall, my Indians fished it up again, without any directions of mine. I should mention that it cost us an hour to get the canoe off this rock, for even after she had been righted she several times again fell over, and I feared she would have to remain there. Some idea of the force of the current may be formed from this circumstance. Once, when ascending a rapid with cables, which a man had carried in a montaria and made fast to a rock ahead, the montaria returning with all the velocity of the current and the man aboard her incautiously approaching too near the canoe, the montaria was immediately sucked underneath it. He had presence of mind to seize hold of the canoe with one hand, still retaining his paddle in the other; in an instant he leaped across the canoe, but the montaria had already passed beneath and was floating bottom upwards at several yards' distance. He did not hesitate to plunge into the water, reached the montaria, seated himself astride, and having guided it into stiller water, turned it over, put the water out with his paddle, and made the best of his way up the stream again.

My position was usually close to the mast, and my occupation was confined to a general vigilance over the canoe and its contents, to cheering on the men, and occasionally lending a hand when there was room for me.

. . . . . .

Rain came on at 5 P.M., and it rained afterwards nearly throughout the night in drizzling showers. Though we gave up early, the men were very much fatigued. Instead of fishing or skipping about as on previous nights, they lighted up their large fire and at once betook themselves to their hammocks.

Jan. 14.—This day passed like the last. We ascended one high fall, called Cojubí, where it was necessary to carry the heavy cargo overland. In the wet season there is another formidable fall round some picturesque rocks called the Fôrno

FIG. 14.—ROCKS BELOW THE FÔRNO IN THE CATARACTS OF SAÕ GABRIEL. (R. S.)

(there being on one of the highest a large flat stone supported on two erect ones bearing some resemblance to a mandiocca oven); but we were able to pass it without unloading.

. . . . . .

Jan. 15.—Rose this morning with a sensation of weariness and disgust scarcely conceivable. The idea of having still another day to pass through like the two last was most depressing. The excitement had had time to evaporate and a mental reaction was taking place. However, Saõ Gabriel was in sight, and the sun rose beautifully clear, dispelling the mists from the serras and tinging them with gold. To a mind alive to the beauties of nature such a scene has always a soothing and enlivening effect; and this being further aided by the stimulus of a fragrant cup of coffee, "Richard was himself again." We had one considerable fall to ascend just after starting, but after this we had only rapids easily passed until reaching the worst of all the falls, at the foot of the hill on which Saõ Gabriel is built. It is commonly called the "cachoeira da praya granda" from a wide sandy beach stretching below it, on the left bank of the river. Here we had again to pass the heavy cargo overland. A broad path has been made from below the fall up into the town, but the distance is much greater than from above the fall. I walked up, however, to have an interview with the Commandant, and found the path sufficiently fatiguing—up and down hills of granite, heated by an unclouded sun. Thanks to Senhor Manoel Jacinto's recommendation, he had procured me a house, the best in the place. Having ascertained this, I returned to see the canoe dragged up the fall. There was now no want of hands, for several soldiers of the garrison came to lend their assistance, attracted probably by the expectation of a pinga of aguardiénte. Still, it took an hour and a half to surmount the fall, though fifteen men were yoked to the ropes.

I sat down under a cliff of granite, watching with anxious eyes the passage of my little vessel; and when at last she had plainly cleared the perilous spot, a load was, as it were, removed from my heart, and I mentally returned thanks to a kind providence who had thus brought me safely through all the dangers of the voyage, and had permitted me to reach its termination without losing either my vessel or a single article of her cargo, the latter to me invaluable. For my life I had never any fears. Throughout the ascent of the caxoeiras I kept as lightly clad as possible, in order not to be incommoded in swimming should it ever be necessary to abandon the canoe, which it happily was not, and I think I could have swum out of any place we passed. My Uaupé Indians did not hesitate to swim down the most furious of the falls; they even seemed to delight in doing it, using only their legs in swimming and stretching out their arms under water in front of their head and chest, which they thereby saved from any blow of a sunken rock.

It was past 4 o'clock ere I got the canoe unladen and the goods stowed in my new residence, and the Tochaua and his men were not paid and sent off until nearly dark. I found looking-glasses most in request with them, and one little fellow took a couple. Next to these were terçados (cutlasses). The Tochaua had done but little, yet, as he had furnished me the men, I gave him a gay handkerchief. They all seemed highly contented, and went their way rejoicing. They were really a set of fine fellows, always in good humour, and when the patron wished for anything it was which could get it for him first. One of them, called Ignacio, had during the voyage offered to stay with me in Saõ Gabriel, and I have accepted his offer. He is a tall, stout, handsome fellow, and appears remarkably good-natured.

Nov. 16.—At 6 P.M. the barometer on the threshold of my house (which is opposite the church) stood at 30.470, and in the part below the Commandant's house at 30.570, indicating a difference of altitude of 85.5 feet.

. . . . . .

CHAPTER IX

THE CATARACTS AND MOUNTAIN-FORESTS AROUND SAÕ GABRIEL

(*January* 15 *to August* 20, 1852)

[THIS chapter is made up from parts of two letters to Mr. Bentham, mostly devoted to a description of the botanical features of the district, the difficulties of travelling and of procuring food, and other matters of interest as illustrating the obstructions in the way of a working naturalist in these remote regions. The remainder consists of such portions of the Journal as deal with subjects of general interest. These comprise a rather lengthy account of the ascent of one of the isolated rocky serras, which is given in full for two purposes. It gives a very interesting and readable account of the curious caatinga forests of the great granitic region, so strikingly different from the usual virgin forests of the Amazonian plains; and, secondly, it shows clearly the great labour and loss of time, as well as expense, of making such ascents, and the extreme poverty of the results. Here, as in other cases, almost all the plants of novelty or special interest were found on the level ground at the foot of the mountain, hardly anything on the mountain itself, though its summit was forest-clad. This will serve to explain why he afterwards rarely ascended such mountains, and even made no attempt to ascend the great mountain of Duida on the Upper Orinoco, though when he left the Lower Amazon he had spoken of making an attempt to "rifle its botanical treasures."

I print also a detailed description of a native Indian festival, because a large number of readers are interested in the customs and folklore of savage peoples.]

To Mr. George Bentham

SAÕ GABRIEL, RIO NEGRO, *April* 15, 1852.

. . . . . .

I found it a great advantage travelling in my own canoe. I had it fitted up so that I could work comfortably and stow away my plants when dried, besides being able to dry my paper on the top of the cabins when it was inconvenient to stop in the middle of the day. I was also master of my own movements; could stop where and when I liked, save that it was necessary to keep the Indians in good humour. When the weather was cool they did not like to be interrupted in pulling, but when they were toiling under a hot sun they rather liked a stoppage now and then. Towards the end of the voyage they got into the habit of peering into the trees as we went along in the hot afternoons, and would call out to me—busy among my papers in the cabin—"O patraõ! aikué potéra poranga" ("Patron! here's a pretty flower"). I of course turned out to see if it was anything new, as it often proved to be.

Lecythis were very numerous, and I had not time either to gather or preserve all I saw. I hoped to get some of them here in fruit, but I cannot see a single Lecythis in the gapó of the falls.

The Leguminosæ (*Diplotropis nitida* and other varieties) were frequent nearly all the way up. . . .

The *Dicorynia Spruceana* (a tree 80 feet high) was frequent and very ornamental from a little below Barcellos nearly to the base of the falls. About the falls its place is supplied by another Cæsalpineous tree (*Aldina latifolia*), which I gathered in flower and hope to get also with ripe fruit.

Shortly after I reached here my montaria broke from its moorings one night and went over the falls. I sent my two men in quest of it. They were out all night, and returned next day with the montaria, which an honest Indian had found almost uninjured wedged between two rocks. They brought me also a branch of a tree in flower which proved to be a small-leaved Dicorynia. Three or four days afterwards I went down the falls to get more of it; but the flowers were nearly all gone, and, strange to say, we could find only that one tree from which the men had plucked the branch.

Gustavias were tolerably frequent, but it was scarcely possible to preserve their flowers on account of the number of caterpillars bred in them.

It would surprise most people to be told that Proteaceæ are so numerous on the shores of the Rio Negro (in individuals, not in species) as to give a marked character to the vegetation. I am acquainted with three or four Proteaceæ (Andriapetala) of the terra firme, but I have never been able to find them in flower or fruit. All that I have hitherto gathered (including the one from Santarem) are of the gapó. All are remarkable for the leaves of the young plants being polymorphous—pinnate, pinnatifid, or laciniated, though this is not noted by Endlicher under Andriapetalum.

The finest tree on the Rio Negro is an apparently undescribed Bignoniacea. If the genus be new, I hope you will allow me to call it Henriquezia, in honour of Senhor Henrique Antonij, a native of Leghorn, but for more than thirty years settled at the Barra do Rio Negro, where he has constantly rendered every assistance to scientific and other travellers during that period, as you may see by referring to all the works that have been lately written respecting these rivers.

. . . . . .

Above Uanauacá all was rapids; indeed, there had been little else from Sta. Isabel.

. . . . . .

It is not very pleasant work here to be always among cataracts in my excursions. I have been once the whole length of the falls and up again. I was out four days, but two of them were lost time. I made my station at the house of the pilot of the falls, at the foot of the latter, and arrived just in time to see the commencement of one of their great festas. Much against my will, I was compelled also to see the end of it, for no one would stir until after two days of drinking and two nights of dancing. I was interested to hear the legend of the discovery of the mandiocca-root sung in the Barré language, but this was poor consolation for such a loss of time; and you may imagine how I fretted in my imprisonment on a small rocky island, begirt with foaming waters, where I could not find a single flower that I had not already gathered. In returning, with four men, we passed all the falls without accidents until reaching the great fall above-mentioned; here, in dragging the boat up the rocks, it filled with water, and a large parcel of plants in paper, about 3 feet high, was so completely soaked that two men could scarcely carry it. Two large vasculæ full of fresh specimens floated out, but we secured them, and I lost only a few plants that were loose in a basket. I was much fatigued, having been on the water from 6 in the morning till 5 in the afternoon, yet I had now the soaked parcel to open out and the plants to transfer to dry paper, which occupied me until midnight. To some of them the mischief was already done—the leaves had begun to disarticulate—but you must take the specimens as they are, as I shall probably not find the same again. Whatever advantages Saõ Gabriel may have as a station, on account of its interesting vegetation, it has disadvantages so great that if I had commenced my South American collections here I daresay I should have given them up in despair. The house I am in is very old; the thatch is stocked with rats, vampires, scorpions, cockroaches, and other pests to society; the floor (being simply mother earth) is undermined by saúba ants, with whom I have had some terrible contests. In one night they carried off as much farinha as I could eat in a month; then they found out my dried plants and began to cut them up and carry them off. I have burnt them, smoked them, drowned them, trod on them, and, in short, retaliated in every possible way, so that at this moment I believe not a saúba dares show its face inside the house; but they demand my constant vigilance. Then the termites, which are more insidious in their approaches, have covered ways along every post and beam. They have already eaten me up a towel and made their way into a deal packing-case, where fortunately they found nothing to eat. But the greatest nuisance at Saõ Gabriel is one I had not foreseen. Almost the sole inhabitants are the soldiers of the garrison, and do you know how the armies of Brazil are recruited? When a man commits a crime which entitles him to transportation, he is enlisted and marched off to one of the frontier posts. Thus, of the fourteen men composing the garrison of Saõ Gabriel, there is not one who has not committed some serious crime, and at least half of them are murderers. Judge with what security I can leave my house for a few days. It has already been twice entered during my absence, and about two gallons of spirits, a quantity of molasses and vinegar, and some other things stolen from it.

I have in the house with me two Indians—a hunter and a fisherman. One at least is an absolute necessity to prevent my dying of hunger, for here nothing is to be bought, not even an egg or a banana. For farinha I have had to send to the Rio Uaupés. The hunter I brought with me from the Barra. He is an excellent shot, and keeps me mostly well supplied with game. He is also useful to me for climbing trees and rowing, at both of which he cannot be excelled. But he is a terrible fellow for cachaça, like most of his race. I induced one of the Uaupé Indians who came with me from Uanauacá to become my fisherman. He was with me about two months when the Commandant of the fort seized him for the service of the corréo (post) to the Barra. Indians to row the courier's canoe are obtained in this way. A detachment of soldiers is sent by night to enter the sitios and seize as many men as are wanted, who are forthwith clapped into prison and there kept until the day of sailing—in irons if they make any resistance. The voyage averages fifty days, and these poor fellows receive neither pay nor even food for the whole of this time. The Indian, however, never dies of hunger when his brother Indian has food, and these men call at the nearest sitio to replenish their supply of farinha from time to time. But such treatment is a great disgrace to the Government, and it is not to be wondered at that the Indians hide themselves in the forests when they get wind that the courier is about to be dis-

Fig. 15.—Saõ Gabriel, with Serras do Curicuriarí
As seen from a few yards east of the church, looking down the river. (R. Spruce, July 1852.)

patched. Within these few days I have been fortunate enough to engage another fisherman. It is worth my while to keep these two men solely for the sake of accompanying me in my excursions, for it is not safe to venture among the falls with fewer than two oars.

. . . . . .

The serras around Saõ Gabriel were a great attraction to my establishing myself here. I began with the lowest, which rises at the back of Saõ Gabriel. In streams about its base I got several Ferns, but on the serra itself nothing. I then undertook to ascend a serra which appears in front, on the right bank, when one goes half a day's journey up the river. On Schomburgk's map it is marked Mount Wanarimapan, but no one knows it by

FIG. 16.—SAÕ GABRIEL DO RIO NEGRO, LOOKING UP THE RIVER. The Equator passes through the high peak on the left. (R. S.)

this name. The Indian name is Urucú-initéra (or the hill of Anatto), but it is more generally known by its Portuguese name, Serra do Gama. . . . I succeeded in reaching the very highest point of the serra, but it cost me above a week, and here also the serra itself proved barren of novelty, being clad with lofty forest to its summit and destitute of water save near its base. It is 1600 feet high above the river at Saõ Gabriel. All these serras are huge masses of granite rising abruptly out of the plain. You have no idea what work it is climbing them: towards the base they are strewed with blocks as big as churches, all enveloped in forest and netted over with twiners. In a caatinga at the base of the Serra do Gama I made an interesting collection. There are also other caatingas or "white forests" in the neighbourhood: the soil a thin covering of white sand over granite, the trees low, twiners scarcely any, trunks hung with Ferns and Orchises, branches with Hepaticæ. The Ferns are very interesting, the Orchises numerous but insignificant, the Hepaticæ few in species. Scarcely any of the trees are now in flower, but they seem all peculiar.

. . . .

I am now entering another great Guaraná country. I have

seen a few plants in the sitios, but it is across the frontier that it is cultivated and used in the greatest quantity. The Barré Indians of Venezuela drink it in immense quantities, especially the first thing in a morning, in place of coffee, and they use only the fresh berry, grated, without sugar. Their name for it is cupána. . . .

To Mr. George Bentham

SAÕ GABRIEL, RIO NEGRO, *Aug.* 18, 1852.

Since last writing to you I have been able to add scarcely anything to my collections. My hunter some three months ago was taken seriously ill, and perhaps he will never be able to bear any exertion more. With my Indian disabled came the festa of Saõ Gabriel, commencing on the eve of the Ascension and lasting above a month. During this time no one would either hunt or fish; fishing, indeed, was scarcely possible with rod and line, from the rising of the waters. Never was I so near dying of hunger. I was reduced to take the gun on my shoulder and go out early in the morning into the caapoeras in quest of parrots and japus. Unless the rain came on very furious I always succeeded in procuring my dinner, but I once passed three days solely on xibé (farinha mixed with water), which the Indians drink, and sometimes take no other food for several days; but to a person unused to it, it causes great flatulency and does not allay hunger. When the streams began to swell, the larger kinds of game retired deep into the forest, and it was necessary to go by water to some distance, pass the night in the forest, and with the dawn of morning proceed on the chase. But it is almost useless a person hunting here who has not been used from his infancy to threading the forest and to spying out the game in and among the trees, which it requires an Indian's eye to do.

The day was so far broken into by my morning's shooting that I could rarely get more than a short walk in the afternoon. The falls, too, became so dangerous that I could not venture into them with fewer than three Indians in my montaria, and rarely were so many to be had. Throughout the months of June and July there were really scarcely any flowers to be had; not a tree was in flower in the great forest, and scarcely any in the gapó. There is scarcely any breadth of gapó here, consequently the herbaceous and woody twiners which I used to gather near the Barra by rowing about among the tree-tops are all but absent here. The trees of the gapó are just beginning to flower, and I think I am going up at a good time.

. . . .

My canoe gives signs of not holding together long. As I did not understand things of this kind at all, I relied entirely on Henrique in the purchasing of it; but I afterwards found that the man who had it to sell was a much older friend of Henrique's than myself, and that I had been taken in considerably. Vessels built up here in Venezuela (as mine was) are not expected to last more than from three to five years; mine is already three years old and will hardly last another year.

. . . .

My last dates from England are a year old. Neither newspapers nor anything else ever reach me now. I seem to have taken my last leave of civilisation. . . .

JOURNAL (*continued*)—*January to August* 1852

THE BLOOD-SUCKING BATS

Saõ Gabriel is terribly infested with vampires, and my house, which has an old, decayed roof, has more than its share of them. When I entered it there were large patches of dried-up blood on the floor which had been drawn from my predecessors by those midnight blood-letters, and my two men were attacked the first night, one of them having wounds on the ends of four toes, three on one foot and one on the other. The same has happened every night since, and the bats do not stop at the toes, but bite occasionally on the legs, fingers' ends, nose and chin and forehead, especially of children. . . .

A curious circumstance occurred to the family of my next neighbour since I arrived here. The children were much tormented by vampires, being bitten in various parts night by night. A cat was observed to be very expert at killing bats in the doorway at nightfall. One night, by accident, the cat was allowed to remain in the house, and whenever a bat alighted on the children's hammocks she pounced upon it. When morning came they had not once been bitten, and now the cat is their constant nocturnal guard. She also evidently knows her office, for as regularly as the children lie down at night to sleep she takes up her station by their hammocks. Poor Pussy! the good deeds of those who call thee "ungrateful" and "perfidious" seldom shine with such lustre on a naughty world! From my youth up I have been a lover of cats, and sagacious dames have at divers times foretold of me that for that reason I should die a bachelor, which, if I live not to get married, is likely enough to come true.

On the granite rocks near my house the sheep belonging to the inhabitants often pass the night, and in the morning regularly leave behind them pools of blood from the bites of the vampires.

This vampire is a small species with the membrane connecting the ears very narrow. A leaf-nosed bat in my house at the Barra was nearly thrice the size, the ears very large and the connecting membrane very broad.

As I wear stockings of a night, wrap myself well in my blanket, and often cover my face with a handkerchief, I have hitherto escaped being bitten, but they often come to my hammock in search of a vulnerable point. The best preventive against them is to keep a lamp burning all night, but oil is unfortunately a very scarce article here.

Surgeons boast of their painless operations nowadays, but the vampire beats them all. I have never yet met with a person who was awakened by a vampire biting him, but several have had the vampire fasten on them when awake, and these confirm the account of the animal fanning with his wings whilst sucking. The wound is a round piece of the skin (often the whole thickness and with some flesh besides, as once happened to myself) taken completely out as if cut out with a knife. The quantity of blood lost is generally trifling unless the vampire happens to light on the small veins. It prefers the toe-ends, and next to them the finger-ends or nose-end.

They cause great destruction at times among fowls, which are allowed to roost in the open air, sucking them in the head and drawing so much blood as sometimes to cause their death in three or four blood-lettings. I had a hen which I was obliged to kill on this account.

In the fort they were exceedingly abundant. A soldier called at my house at 6 one morning and showed me his feet, so completely covered with wounds and fresh blood that at first I thought he must have fallen into a bed of prickly palms. The wounds were all bites of vampires, and in one great toe there were no fewer than eight holes. The toes, heels, and ankles had been the worst used.

My Uaupé Indian was quite naked with the exception of the tanga when he entered my service. I gave him cloth to make a shirt and trousers. His companion was the tailor, and when the trousers were completed I was present at the ceremony of trying them on. You have seen a child in England don his first buttoned clothes, what mixture of uneasiness and self-satisfaction he displays, and how awkwardly he steps out, and how he twists his neck in the vain attempt to obtain a view of the remotest back-settlements (reminding one more of a turkey cock than of anything else). Fancy all these movements exaggerated in a stout young man of twenty, with an ingenuous countenance, and you will have an idea of the figure Ignacio cut on this occasion. I was highly amused, but forbore laughing for fear of hurting the poor fellow's feelings.

One of the commonest weeds in Saõ Gabriel

is a shrubby Solanum (*S. Jamaicense*) 4 to 6 feet high, which furnished food to thousands of black Hemiptera in the dusk of the evening and about sunrise of a morning in the month of January. At their feeding-times they hover over the plants like swarms of bees and the bushes are almost black with them. Standing at my door one evening after sunset, a flock of these settled down on a Solanum bush close by. I fetched a small pint bottle and commenced filling it with the insects; but though I frightened away twice as many as I put into my bottle, in ten minutes scarcely anything was left of the leaves save the midribs. It is from 1¼ inch to 1⅜ inch long, and is remarkable for the very diminutive thorax and for the tumid abdomen protruding much beyond the elytra.

Expedition to the Serra do Gama

Soon after reaching Saõ Gabriel I formed a plan for ascending the serras which lie half a day's journey up the river on the right bank. The sitio nearest their base is occupied by an old man (nearly seventy) named Gama, and his father occupied it before him. Hence these serras are now known by no other name than "Serra do Gama." . . .

On Friday, March 5, I removed with my apparatus to Gama's sitio, whence I sent one of my men on to Saõ Joaquin to purchase an ubá, my little montaria being ill-fitted for buffeting with the caxoeiras. During his absence I employed myself with exploring the environs of Gama's house.

The caapoera is of loftier trees than usual, but slender. . . . Adjacent to the caapoera was a caatinga-soil, a thin covering of white sand over granite. There are no Selaginellæ on the ground and few twiners in the trees, hence the forest is easily traversable. The mass of the vegetation is a Cæsalpinieous tree (gathered also in the caatinga at Uanauacá) which does not exceed 50-60 feet in height. There are also scattered loftier, thicker trees. In interspaces are smaller trees with few, weak branches (including two Melastomea, other two which from their habit may be Olacacea, and a few others). The most frequent is an Amyridea. All are remarkable for slender stems not exceeding 10-15 feet, and scanty, long, irregular, weak branches. The few twiners are mostly herbaceous.

An Acanthea whose succulent stems crawl by means of rootlets and occasionally twine is frequent on the smaller trees, rarely reaching up them farther than 3-4 feet. But frequent above all is an Orontiacea, whose slender green woody stems are branched and closely clasp the supporting tree by means of ring-like roots; it sometimes ascends the highest trees, but prefers the Amyridea, which it not infrequently kills, while from the summit of the dead tree it sends out a pendulous crown of distantly leafy branches. It is one of the sipós called Timbó-titica so useful for cordage, but there is a better kind than this, with larger leaves and very tough stems. The stems of this are rather brittle.

Beyond the caatinga lies the caá-uaçú. Here much of the undergrowth consists of a slender Myrsinea, 10-18 feet high, with pendulous panicles of small pale pink flowers, followed by black shining drupes the size of a wild cherry. The same is abundant all the way up the serra. There is also a Rutacea (apparently a species of Galipea) tolerably frequent, remarkable for its simple stem 6 to 30 feet high with a corona of large digitate leaves and racemes of cream-coloured flowers at its summit. It is one of the plants used under the name of Timbó for killing fish. A large twiner of the same genus as the Flor do Espirito Sando from the Barra is also frequent.

In caatingas near the base of the serras the trees are still lower and they are mostly clad with Mosses and Jungermannia to their slenderest twigs, the same tribes forming often a conical sheath at their bases. Amongst the Mosses are perched Ferns (several species of Acrostichums), Bromeliaceæ and Orchids, the last chiefly small-flowered species. Mosses also grow on the ground in some places and on fallen trunks.

In caatingas at Uanauacá, which were very moist and apparently with water standing on them in winter (though not inundated from the river), the rootlets of the trees project from the soil in a dense netted mass, called by the Indians Samambáya (the same name they give to Ferns).

On Wednesday, March 10, I sent Gama and

my two men to clear the path to the serra, which they accomplished and returned at night. But this making a road through the forest is not the heavy task which might be supposed. The great point is to know in which direction to steer, and at this the Indians are remarkably sagacious. The road consists merely of twigs broken half through and bent to the off side on each hand in passing, and occasionally of a sipó cut through when it obstructs the way. Sometimes advantage is taken of the sandy bed of a stream—when the water is not over knee-deep—to walk along it for some distance, and in doing this twigs on each side are in a like manner broken down. Such a track is very difficult to follow, to an unaccustomed eye, and I when alone am obliged to trace it with slow and cautious steps; but an Indian trips along as securely as if he were on one of the Queen of England's highways, and so securely fenced in on each side as to render it impossible to stray.

The expedition was fixed for next day, but whilst the men were tracing out the path in the forest comes a trader from Pará with a boat-load of dry and wet goods, whose house in Saõ Gabriel having been burnt down, sought a residence in Gama's sitio, his pilot, by the way, being Gama's eldest son. As usual on arriving off a long voyage, the trader "stood treat," and there was great firing off of rockets, drinking, and dancing for the space of two days, after which a third day was necessary to recover from the effects of the debauch.

On Friday afternoon, having heard that the corréo had arrived in Saõ Gabriel, I went to see if he had brought anything for me, accompanied by Senhor Gama and the newly-arrived trader. In returning we were overtaken by a tremendous tornado, which in two minutes left us without a dry thread about us. Rain beat into the men's eyes so that they could scarcely see how they were rowing. The roar of the thunder was scarcely distinguishable from that of the rapids. Night was coming on, but the lightning every few seconds illumined every object and lit up our faces with a spectral red glare. I was sitting in the canoe with my head resting on my hands and my hands on my knees—the usual position in these small craft—and when we reached our destination my clothes were so surcharged with wet that I could scarcely step on shore, and the rain ran off my trousers in streams. The Saturday, too, was gloomy and showery.

On Sunday morning, March 14, at 7 I started for the serra, accompanied by Senhor Gama and four Indians (my Uaupé Indian hid himself in a neighbouring sitio in order not to go on the dreadful enterprise, and whilst assisting some women to crush cane came a detachment of soldiers and seized him and two others to row in a canoe about to be dispatched to the Barra with the post). We carried farinha for three days, roast fish for one, a bottle of rum, and as much salt and capsicum as we were likely to need. Our arms were three muskets, two cutlasses, and four carving-knives. We had gone but a little way when I found it necessary to walk barefoot on account of the number of streams to be crossed, and the having in many places to walk for some distance along them. We crossed streams above twenty times; the last we encountered had to be forded four or five times. It is the largest we met, being 4 or 5 yards wide in its upper part, the depth now rarely more than to the knees, but in the flood averaging about 4 feet. It is called Uíwa-igarapé (or the river Arrow), and does not run directly into the Rio Negro, but into the Curicuriarí, proving the latter to deviate much in its upper part from its direction near the mouth. Hence also the Serra do Gama may be considered a continuation of the Serras do Curicuriarí, though there is apparently a great gap between them. The Uíwa has a sandy bottom and clear (not black) water. On its banks we chose a place to pitch our tents, having arrived as near to the serra as we judged convenient. In its sands and on rocks standing out of it, I got some interesting Ferns. Close by our resting-place (which we reached at 1 P.M.) was a large Loureira, at least 100 feet high and very straight. This we tapped and a good draught from it twice a day was my allowance whilst we stayed. The milk was thinner than I had before met with it, and the Indians say that the milk of all milky trees is more copious and flows more freely in the wet season than in the dry. There were also some very tall Assaís and Paxiúba barriguda palms, perhaps over 100 feet high.

My men (three of them) set to work to erect a couple of huts, one thatched with Assaí, the other with Paxiúba. For each two trees were selected at a convenient distance for hanging the hammocks below, and to support the roof short sticks were tied across the trees, forming triangles. The huts were but just finished when the rain, which had been growling for some time in the distance, came on, and continued till past midnight. The other two men had gone to hunt, and they also returned just before the rain, bringing each of them a mutún (curassow). We had also killed a mutún on the way, so we were amply supplied with provisions. A fire was lit between the two huts and a stage erected over it on which to roast the mutúns, but unfortunately a sufficient stock of fuel had not been got together before the rain came, so that we passed more than half the night without fire. The position was sufficiently dismal. The rain rendered the air so cold that I found it impossible to sleep till near morning. To make it worse, I had no covering, having left my blanket in order that my men might have as small loads as possible. We were in the most utter darkness, for even at midday the place was illumined only by "a dim religious light," like that in an old cathedral, and now there was neither moon nor stars to pierce the thick gloom. We were serenaded by the lugubrious croaking of frogs until near midnight, to which the raindrops pattering on the leaves and plashing in the stream formed an appropriate accompaniment. Other sounds I could distinguish none, though at times I listened attentively.

. . . At daybreak we heard a tiger (jaguar), but it was at a great distance; but in the evening following after we came down from the serra, the two hunters plunged into the forest with their guns and fell on the track of a cutía (agouti), and whilst following this they unexpectedly came up with a tiger, who also seemed in chase of the cutía. The foremost hunter presented his piece, but it missed fire, and the tiger, instead of retreating, advanced upon him. He was preparing to attack it with the butt-end of his musket when his companion came up and fired, wounding the tiger severely, yet not preventing him from making off at such a pace that they were unable to come up with him again.

In one of the huts there was room for two hammocks, in the other but for one; those who could not extend their hammocks slept on palm-leaves laid on the ground of the huts; but on the following night, which was fine and dry, they hung their hammocks on trees outside and kept up a roaring fire all night.

March 15.—This morning before break of day the three hunters started off in quest of game, and as they did not return Senhor Gama and I breakfasted and set out alone to climb the serra. We followed up the stream until the ground began to ascend on our right, when we left it and commenced climbing.

. . . . . .

We went on continually aiming for the highest ground, as well as the blocks of granite and net-work of sipós would allow us. We struggled on, sometimes climbing steep inclined planes of slippery stones by the aid of the sipós and roots on them, until we both began to feel rest needful. We sat down, and opening the barometer I found we had already climbed 1000 feet. I felt sure, therefore, that we had already reached above half-way up, and I bade my guide take courage. Our companions here joined us and we resumed our march. In a short time we emerged on a narrow ridge which sloped rapidly down on the opposite side, and we correctly judged it to be a shoulder of the mountain connected with the terminal peak. We therefore followed it very much at our ease, for the ascent was very slight and the ground was comparatively clear of twiners. The chief vegetation was an Ubim-rana, with a few plants of Bactris. Senhor Gama thought he saw proofs of habitations having formerly existed on this ridge (as there is a tradition) in the absence of any trees of larger size; though up to this ridge the forest had been lofty, as it also was above it to the summit of the highest peak. The peak soon began to show itself quite near, looming through the mist, and it shot up so abruptly that we had some fear of not being able to surmount it. We proceeded, however, with much difficulty till we came to a perpendicular wall of above 40 feet high, on which were a few scattered shrubs and sipós, by the aid of which we contrived to climb it with greater ease than I had expected. A few minutes of gentle ascent and then came another similar wall, which we also climbed in safety, though not without some apprehensions of finding it much more difficult on our downward passage. After this, though the ascent was abrupt, we had no more escarpments until we reached the very summit—a slightly convex platform of about 20 yards in diameter, thickly clad with tall trees and bushes, mostly of the very same species as occurred in the plain below. There were, for instance, some Inajá palms—one about 40 feet high. I was about to place my barometer on what seemed the highest point, when I found that a strong colony of wasps had already taken possession of it, and I was obliged to stand at a respectful distance and hold it at the altitude of the culminating point. During this ascent of the peak we were in the midst of a thick cloud and were soaked by the wet dripping from the trees. Though we cut a way at the summit to the side from which we should have had a good view of the rest of the serra and of the river, and waited some time, the clouds only now and then partially rolled away so as to show the first lower ridge round the base of which we had skirted in order to reach the foot of the highest peak. It seemed to be continuous with the latter, being joined by the shoulder before mentioned and forming with it a kind of cirque. We were on the top exactly at noon.

In descending it was not a very pleasant look downwards from the top of the perpendicular walls, but the actual descent of them was accomplished without accident. My long legs and arms stood me in good stead in reaching from one branch of sipó to another, and I retained my vasculum slung across my shoulders all the way. These rocks were adorned by pendulous masses of a large Selaginella, silvery on the underside. The rock throughout was granite. We were just descending from the shoulder spoken of in the ascent, when the sun broke forth and the clouds rolled rapidly away; but it was not worth while to reclimb 500 feet merely for the sake of the view, even had the sky been certain to continue clear.

On returning the following day I spent nearly two hours in the caatinga, where I gathered a good many Ferns and Hepaticæ.

[By careful barometer observations at the top and the foot of the serra, and the average at corresponding hours of the whole month at Saõ Gabriel, the height was found to be 1635 feet; adding to which the height of Saõ Gabriel above the sea, by other observations, the total height may be estimated at about 1800 feet, with a probable error of 50 feet. Spruce gives his full calculations.]

Mode of obtaining Salsaparilla

March 23.—There is a small plantation of Salsa in a tabocal (bamboo grove) a little way down the falls, whither I went this day with the owner to witness the mode of taking up the roots. The plant selected had five stems from the crown, and the numerous radiating roots extended about 3 yards on every side. The roots were first bared, and had the Salsa been the only plant occupying the ground, the task would have been easy, but they are often difficult to trace among the intricate mass of roots of other plants, which require to be cut through with a knife or small cutlass. The earth, which is only a thin covering, is scraped away by the hand or by a pointed stick. The roots being at length all laid bare (in this case it was the work of half a day, but of large virgin plants it sometimes takes up a whole day), they are cut asunder near their base, a few of the more slender being left in order to stay the plant in its place. A well-grown plant will yield at the first cutting from one to even two arróbas. In a couple of years it may be cut again, but the yield is much less—the roots slenderer and yielding (say the Indians) less starch.

An Indian Festival

April 17 and 18 I was present at a Dabocurí (or festa of Barré Indians) on an island near the base of the falls, a little above the ancient village of Camanáos. The house was pleasantly situated on rising ground, the walk up to it fringed with Coffee trees laden with berries, amongst which were three or four clusters of Pupunha palms and here and there a Cocura tree. A flat, semi-circular space of hard sand in front of the house had been clean swept to prepare it for the dancers. This space was skirted by spreading Ingas, under shade of which benches had been put up with backs to them, the seats of strips of Paxiúba palm laid close together.

There had been prepared beforehand a quantity of cauim (distilled from sugar-cane); two flageolets of Paxiúba about 6 feet long (made by Indians on the river Içanna) and three or four smaller ones; a number of gaitas of a single internode of the slenderer branches of some Cecropia, with a wind-hole cut on one side through which the pith had been extracted and into which the performer blew; a quantity of carajurú in powder for painting their bodies; and an immense quantity of ipadú (coca).

The performances commenced early by the blowing of flageolets and gaitas, and the company kept arriving till past 9 o'clock. A dance was then commenced by the men and boys inside the house, by forming themselves into a ring, each holding the flute to his mouth with the right hand and placing his left hand on the right shoulder of the person in advance of him, and then moving round to the slow, almost monotonous, cadence of the gaitas. The steps were merely a succession of dactyls—one long step followed by two short ones—the body being bowed forward with the long step, and again elevated with the short ones. After dancing in this way a few minutes, they turned out upon the terrace, where they were joined by the women and girls. Each man now passed his left arm round his partner's neck, and she her right round his waist, and the dance continued to the same tune and step, but gradually increasing in quickness until it almost reached a run. When the flutists were completely out of breath, the ring broke up and the dancers, giving a general shout, retired to repose themselves on the benches or inside the house. (There were also benches and boards, apparently permanent, placed along the front of the house under the projecting eaves.) After a short repose the men started up to renew the dance, and so kept on till about 3 in the afternoon, when news was brought up from the port that the ruler of the feast and his attendants had arrived. These formed a party about equal to that already assembled at the house, and they brought with them a number of aturas· (baskets) filled, some with roots of mandiocca, others with baked fish, besides several shallow baskets of beijú and two or three alqueires (bushels) of farinha. Each person was furnished with an ambaúba or drum made of the trunk of *Cecropia pettata*; those of the men were about 3 feet long and 5 inches in thickness, the diameter of the bore being about 4 inches; those of the boys were smaller. They had been bored by means of firebrands and the lower end closed with leaves beat down with a pestle. Two rectilinear oblong holes were cut near each other adjacent to the upper end of the tube, by which it was held, the thumb being inserted into one hole and the fingers into the other. The lower end for the breadth of a few inches was painted black, and about the space of a foot near the middle was painted with fantastic devices according to the taste of the fabricator.

Several formal messages now passed between the giver and the ruler of the feast; the latter was invited up to the house along with his party and the gifts he had brought. They did not, however, make their appearance for nearly an hour, and meantime the company assembled at the house occupied the time in painting themselves with carajurú, the men, being naked above the waist, on their bodies, faces, and arms, and the women on their faces and arms. At length they appeared ascending the hill in a file, beating the ground with their drums, and, arrived on the terrace, formed themselves into a ring, still drumming away. Cauim (rum) was now brought out in bottles and large cuyas, from which it was decanted into small cups, and cups of caraipé (pottery) (the latter were two together at each end of a short rod of the same material, the whole gaudily painted). The flageolet-players now headed the procession, followed by little boys bearing the cups of cauim, and the whole made the circuit of the ring, each Ganymede in succession offering his cup to every drummer, who was obliged by the etiquette of these assemblies to sip of every one. Attendant men and women replenished the cups as they were emptied, and after the drummers had partaken, cauim was in like manner handed round to the rest of the company.

All the women were now sent down to the port to bring up the gifts (being the contribution of all the ruler's party).

The fish, farinha, and beijú (Port. Cassiwa cake) were deposited in the house, and the roots of mandiocca piled in a heap in front. The women immediately set to work to make caribé of the beijú, and filled several gassabas with it. The drummers now began to dance round the heap of mandiocca, the step being a sort of skip which finally quickened to a gallop, and singing to the beating of their drums. Their songs seemed to have been divided into short stanzas, each ending in a sort of refrain. The first song was the legend of the discovery of the Mandiocca in the Barré language, and this is the substance as translated to me. Like the Tree of Life in the Garden of Eden, the Mandiocca tree stood solitary in the midst of the forest. It was an immense tree, as large as the Samaúma nowadays, and every mortal shunned it, knowing its deadly properties. At length the bird called japú showed an Indian how the roots might be divested of their poison and converted into a nourishing food. Every one flocked to supply himself with the wonderful root, until the tree had no more to yield. They then set to work to cut off the branches. Each branch was the size of the stem of the Mandiocca plant as it now exists, and being stuck into the ground, produced tubers like those of the parent plant. Each main branch gave a variety distinct from the rest, hence all the Mandiocca and all the varieties of it now cultivated; and it may now with truth be called the Tree of Life to the dwellers of the Amazon and its tributaries.

Afterwards came another song recounting the offerings they had brought, and praying the giver of the feast to accept them. Part of it was in substance as follows: "Receive, we pray thee, these products of the earth and the waters which thy brethren offer thee. We bring them not to thee expecting of thee payment for the same, but because in days of old thy grandfather gave to our grandfathers to eat of his fish and farinha and to drink of his caribé,[1] as thy father also gave to our fathers, thou to us, and as hereafter thy son shall give to our sons." There was much more in the same strain, but my interpreter spoke Portuguese so imperfectly, and his ideas were becoming so mystified by the cauim he had drunk, that he could explain no more intelligibly. The songs being ended, the heap of mandiocca was cleared away, and the singers retired into the house to refresh themselves with caribé, which was handed to them in large cuyas. The ruler of the feast also dispensed the fish to such as chose to eat, but these were few, and during the two days and nights the feast lasted there were some who ate not a morsel, supporting themselves solely on cauim and ipadú. And here it may be mentioned that throughout this time ipadú was every few hours handed about in large cuyas, along with a broken tablespoon, with which each one helped himself, the customary allowance being a couple of spoonfuls. After taking a dose of ipadú, they generally pass a few minutes without opening their mouths, adjusting the ipadú carefully in the recesses of their cheeks and inhaling its delightful influences. I could scarcely resist laughing at their swollen cheeks and grave looks during these intervals of silence. I tried two or three times a spoonful, but it had scarcely any perceptible effect on me, and assuredly did not render me insensible to the calls of hunger, though it did in some measure those of sleep. Probably I took too small a dose.

The ipadú is not sucked, but allowed to find its way insensibly to the stomach along with the saliva. I am told that no ill consequences result from its use even in very large quantities.

As night closed in, fires were lighted up at the corners of the terrace, sufficing to light the dancers in their movements. We had now two rings, one of the drummers and the other of the flutists; the former being more noisy and their step more lively, were decidedly the favourites with the ladies, and a little after midnight the latter resigned the field altogether, contenting themselves for the rest of the night with dispensing the cauim and ipadú. How I wished for the pencil of a Teniers to delineate the scene before me! The dancers in their picturesque costume, their heads adorned with tiaras of the feathers of the toucan, their bodies fantastically streaked with carajurú (chica), and the long drums, whose beats kept time with the movements of their feet, also gaily painted, occupied the open space on the terrace, whilst grouped around the fires or on the benches sat the old people discussing cauim and ipadú; and the glare from the fires, and the strong shadows deepening till blended with the impenetrable

[1] Fish, farinha, and caribé are to a Barré Indian precisely what beef, bread, and ale are to an English peasant.

gloom of the encircling forest, gave to the whole an aspect scarcely earthly. The wild and somewhat mournful sounds of their music added to the effect, and if heard from a distance, in the dead of night, were well calculated to inspire with terror a person ignorant of their origin.

The dances did not cease till after sunrise. In the afternoon an attempt was made to renew them with the intention of continuing them through another night; but the cauim had done its work so successfully on most of the performers that they were not to be roused to further exertion.

The head-quarters of the Barré nation above referred to is now at San Carlos del Rio Negro, and people of that nation are scattered throughout the whole of the Casiquiarian region even to Maypures on the Orinoco. They seem originally to have inhabited much lower down the river, and to have gradually extended northward, and even at this day as far south as Castanheiro and Camanáos, below the falls of Saõ Gabriel, the old Indians are still Barrés. The portrait here given of a Barré girl eight years old named Maria was made during my short stay at Castanheiro on my voyage up the Rio Negro.

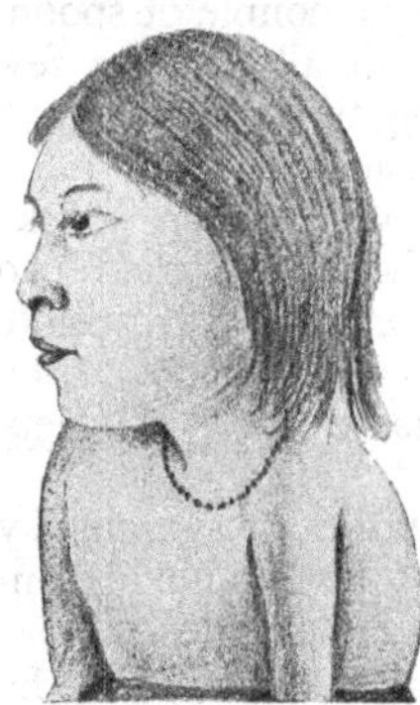

FIG. 17.—MARIA, a Barré Indian (8 years old).

[In the original Journals there is no record of Spruce having stopped at Castanheiro either on his upward or downward voyage. He probably made the drawing at a place near, called Mazarubi, where he stopped to buy farinha on his voyage up to Saõ Gabriel.]

CHAPTER X

AN EXPEDITION TO THE CATARACTS AND UNEXPLORED FORESTS OF THE UAUPÉS

(*August* 21, 1852, *to March* 7, 1853)

[THIS chapter has had to be made up of very fragmentary materials. Partly because *I* had made two excursions up the river before him, partly also because the very rich and novel flora he found there occupied every moment of his time in its collection and preservation, Spruce kept no regular Journal, except during the few days occupied in short excursions up the river. I have therefore had to utilise so far as possible his letters to his botanical friends at Kew, and one which he wrote to myself while he was residing here. This is the more unfortunate as he made no less than ten very careful pencil portraits of Uaupé Indians of different sexes and various ages, and though my friend was no artist he was a very painstaking and accurate draughtsman, and from my own knowledge of these people (and of several of the very individuals represented), I can certify that they give a faithful idea of the features and expression of this fine Indian type.]

On Saturday, August 21, 1852, I left Saõ Gabriel

for Panuré (or Saõ Jeronymo) on the Rio Uaupés. The river had not reached its flood-mark of last year (which was rather high) by about 3 feet, and it had now descended about 4 feet; still, it ran with a swift current. I had nine Indians (eight Uaupés and my own Tapuya), yet when we reached the foot of the caxoeira of Saõ Miguel at 1 P.M., I found them insufficient to pass it. I was therefore obliged to cross the river in my ubá (small canoe) to two sitios in search of aid. It was night when I reached the more distant of these, and I remained there till break of day, but I succeeded in obtaining seven men additional for the day's work. With the aid of these, and with immense labour, we passed Saõ Miguel, and I persuaded four of them to go on with me to Saõ Joaquin, at the mouth of the Uaupés. With this large crew it took us five days to surmount all the caxoeiras, and we passed our fifth night a little below Saõ Joaquin, where we arrived by daybreak the following morning. We had to completely unload the canoe in order to get up two of the caxoeiras. . . .

We did not reach Panuré until nearly midnight of Tuesday, the 7th of September, making the whole voyage consist of eighteen days, whereas in a montaria it can be accomplished in seven. I, however, profited well of the last thirteen days, and made a very fine collection. The weather was tolerable for the Rio Negro, though we had some tremendous thunderstorms.

[There is here a gap in the Journal for more than six weeks, to the time when Spruce went on a short journey up the river to the Jauarité caxoeira; but his time was too fully occupied in exploring the new and very rich locality he had found to allow time for anything but his regular botanical work. He was fortunate in finding at Saõ Jeronymo three white traders (Brâzilians or Portuguese), who were very serviceable to him, and whose presence alone rendered his stay there for four months at all possible, as will be seen by the extracts from his correspondence which I now proceed to give.

In a letter to myself (who had just returned home, having left Saõ Jeronymo six months before his arrival) he says:—]

"Saõ Jeronymo is now very lively. There are two brancos constructing large canoes—Chagas and Amansio. It is pleasant to have their society, but they occupy nearly all the male population in cutting timber, etc., so that there is no one left to fish, and the land is not very farta (well supplied) just now. The people complain of having passed a dismal winter—'não se-achen nada para se comer' ('nothing could be found to eat'). I ought to have told you that I am inhabiting a quarto (room) in Agostinho's house;[1] I have, in fact, had the house to myself till three days ago, when he returned from the Barra. I have three Indians in my service, but they are vadios (vagabonds), and I really think I should be better off in the way of comeres (eatables) if I were alone. . . .

My first excursions round Saõ Jeronymo were by water to the caxoeiras, all of which I have explored for carurús (Podostemas). The estrada grande is

[1] [Agostinho was a young Brazilian trader who, with his young wife (also white), was at Saõ Jeronymo when I was there, and with whom I stayed a few days. See my *Travels on the Amazon*.—Ed.]

singularly barren, but a caatinga lying north of it and another on the south side of the river have afforded me much novelty. The weather has been for some days very sunny, and butterflies are everywhere abundant. How long I may stay here is uncertain. I ought to stay twelve or even fifteen months, but in that time I should have to go to Marabitanas or somewhere to seek planks for making more boxes. I am now arranging with Agostinho to accompany him as far as Jauarité caxoeira in about a fortnight, and I do not propose staying there more than two or three weeks. If I go to the Juripari (devil-caxoeira)—from which the Lord deliver *you*—it will be in January with Jesuino."

At the end of the letter he says: "Don't forget to tell me how yourself and your collections reached England, and especially, what progress you are making in the English tongue, and whether you can by this time make yourself understood by the natives.—Your faithful friend and quondam companion through this wilderness,

RICHARD SPRUCE."

[The last paragraph of this letter refers to the circumstance that when we met at Saõ Gabriel on my way home, we found that we could neither of us talk English together without so frequently introducing Portuguese words and sentences as to form about one-third of our speech. Even when we said: "Now, let us speak English for a little while," we could only do so for a few minutes by much watchfulness, and the moment we got interested, or had to tell some anecdote, in came the Portuguese again!

I will here give the description in the Journal of the various falls he visited during his excursion to the Jauarité caxoeira with Agostinho, above referred to.]

The Falls and an Excursion up the River

The first fall on the Rio Uaupés is that of Panuré, less than a mile above the village of Saõ Jeronymo. Panuré is the ancient Indian name of the village, and has been lately restored to it. The river is here divided into two narrow channels, in each of which there is a dangerous fall. The height is apparently not great, but from the narrowness of the channels, and the rocks obstructing them, the waters are very tumultuous, and even Indians who fall in here mostly perish. The only person known to have gone down the falls of Panuré alive was an Indian boy who was in a canoe that was carried away by the current, filled with water, overset, and was sucked down by the whirlpools at the base of the fall. After going down several times and coming up again as often, to be rapidly whirled round and round, it at last floated out, and the boy, who had never released his hold on it, had sustained only a few bruises. The marvel is that both boy and canoe were not dashed to pieces, for large trunks of trees caught in that whirlpool go down root foremost and either stick at the bottom or come up again torn to shivers.

There is a portage of some half a mile ascending a rather steep path into the forest and again descending to the river at a point well above the fall; this is on the left bank. Less than an hour's rowing above this brings us to the Pinô-pinô caxoeira, where there are four falls, separated by islands. These are really cascades, especially the one on the right, by which is the customary route, and where the water falls at once nearly perpendicularly some 10 or 12 feet. At the very margin the water is shallow—indeed, the stones are said to be completely dry in the height of summer, so that canoes are dragged up and let down without much risk as far as the waters are concerned. The peril is in approaching the fall from above, where there are violent currents and eddies, and also sunken rocks, among which it demands a practised hand to steer so as to shoot into a small bay at the very edge of the falls.

The falls of Pinô-pinô, with the intervening islands, are really picturesque when viewed from below. Here was anciently placed the village, that of Saõ Jeronymo being on a modern site.

A little above and in sight of Pinô-pinô falls is a bay on the right bank, in the recess of which is the residence of one of the most powerful Indians on the river—a Tariana named Bernardo. His house, called Urubú-coará (the Turkey-buzzard's nest), is one of those very large church-like fabrics which would seem anciently to have been the normal habitations of these Indians; and it contains, besides the families of his sons and daughters, those also of numerous dependents.

From above Pinô-pinô the river is again wide, and in many places there is not the least current perceptible; there is also a wider gapó than is usual below Panuré.

We entered the mouths of two igarapés which we found led speedily to caatingas. Throughout the Uaupés the greater portion of the forest is said to be caatinga, and as far up as I have seen it the report is correct. At from two to three days above Pinô-pinô (according to the size of canoe) are the next cataracts, those of Jauarité, at and below the junction of the river Paapurís, which enters from the south. The falls of Jauarité are about equal

FIG. 18.—URUBÚ-COARÁ, ABOVE THE PINÔ-PINÔ CATARACTS IN THE RIO UAUPÉS. (R. S.)

in length to those of Panuré, but less difficult to pass. The Rio Paapurís is full of cataracts from the mouth to a distance of three or four hours within, where is a very formidable line of cascades across the river called Aracapa caxoeira. The river is here precipitated through narrow channels, between two islands and the mainland, a height of perhaps 15 feet, which in one fall is nearly perpendicular. Canoes are dragged across one of the islands—a distance of perhaps 30 yards—by a narrow path which has been partially smoothed over the rocks among low forest. The scenery is really beautiful, and there are small Indian sitios near. There is also on the rocks some of the clearest and best executed picture-writing I have met with, and it is the only instance I have found of a distinct tradition connected with its execution.

[This is described at the end of Chapter XXVII.]

. . . . . .

A few hours from the mouth of the Paapurís is a mallóca of Pira-Tapuya Indians, and a little above this another mallóca of Tucáno Indians. Near the head-waters (from which there is a short portage by land to a tributary of the Japená) is the country of the Carapaná Indians.

Below the Jauarité caxoeira (*i.e.* just at its base) is the village which goes by the same name —or sometimes Povoacaõ de Callistro—the name of the existing chief, Tushaua of all the Tariana Indians. This contains some twenty houses, ranged chiefly along the brow of a steeply rising bank, which is of reddish sand in the upper part and rocky at the base by the river. The number of Pupunha palms[1] standing in clusters on the hill-side and among the houses gives a very pretty appearance to the village. At the back of the other houses stands the large house of the Tushaua, at present considerably decayed and partly fallen away at one end, so that I could not ascertain its original length, but its breadth inside is 76 feet.[2] Stretching away from this house towards the forest is a very broad sandy path, somewhat exceeding the width of the house,

[1] Humboldt in his *Aspects of Nature* describes the Pupunha palm (Piriguao or Pijiguao, as it is called in Venezuela) with a smooth and polished trunk between 60 and 70 feet high; but in his *Personal Narration* he correctly says it has a thorny trunk more than 64 feet high. Again, in specifying what he considers requisites of beauty in Palms, he speaks of the *heaven-aspiring* fronds of the Pijiguao, whereas they are remarkably *earth-pointing*, and the pendulous plume-like fronds of this palm are one of its most striking features.

[2] [I give the length as 115 feet (*Travels on the Amazon*, p. 198), and I made a sketch of it, reproduced in my *Palms of the Amazon*, Pl. xxxvi.—ED.]

which is kept constantly cleared and levelled for the dances of the Dabocurís. By its sides are several Umirí trees, of considerable size; and

FIG. 19.—CALLISTRO (Uiáca), Tushaua (Chief) of the Tariana Indians, Uaupés River (above 60 years old).

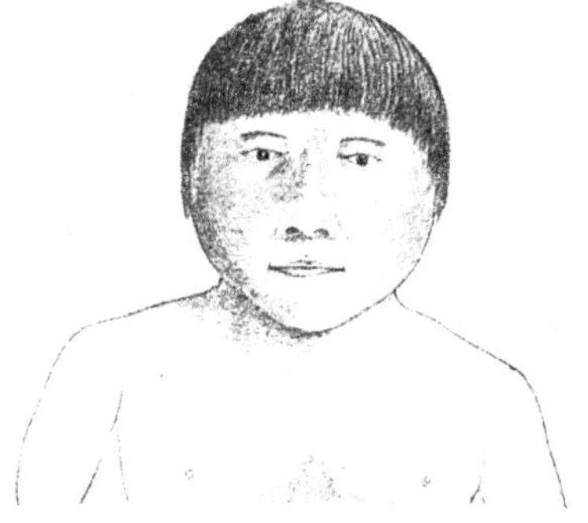

FIG. 20.—CÁALI (=Mandiocca), youngest son of Callistro (20 years old).

here and there twines the indispensable Caapí (a powerful intoxicant), which like the Umirí has been planted.

Callistro (or Uiáca), the Tushaua of the Tarianas,

FIG. 21.—ANÁSSADO, grand-daughter of Callistro (6 years old).

was a fine old man, with a head-piece (speaking both literally and figuratively) which would not have disgraced a European. The Brazilian traders did not much like him, but I could see no other reason for it except that he would not allow himself to be duped or outraged by them. He allowed me to take his portrait (which is here reproduced), and the Indians were so delighted with the likeness of their chief, that I verily believe every one of the tribe came to have a look at it.

FIG. 22.—CUMÁNTIARA (=Duck's down), daughter of Bernardo of Urubú-coará (17 years old).

[The example of their chief rendered others willing to sit for their portraits. Cáali, the youngest son of Callistro; and Anássado, Callistro's granddaughter, a little girl of about six years, taken sitting in a hammock, also well show the characteristic features of Indians at different ages. Two other females — Cumántiara, a daughter of Bernardo, the headman of Urubú-coará, mentioned above, and Paramháada, a girl of fifteen, both of the same tribe as Callistro — are fair examples of the better types of young Indian women; but Spruce states that some of the younger and prettier ones were too shy and frightened to allow themselves to be delineated by the white stranger.

FIG. 23. — PARAMHÁADA (baptized Itelvina), Tariana Indian, Jauarité Caxoeira (15 years old).

The above are all Tariana Indians, the most extensive tribe on the Uaupés, but there are also portraits of individuals of three other tribes — the Pira-Tapuyas (fish Indians), the Tucános, and the Carapanás.

Of the Pira-Tapuyas, the young man, Icanturú, is about twenty-five years old, the woman Tschéno is

FIG. 24.—ICANTURÚ, Pira-Tapuya Indian, Jauarité Caxoeira (25 years old).

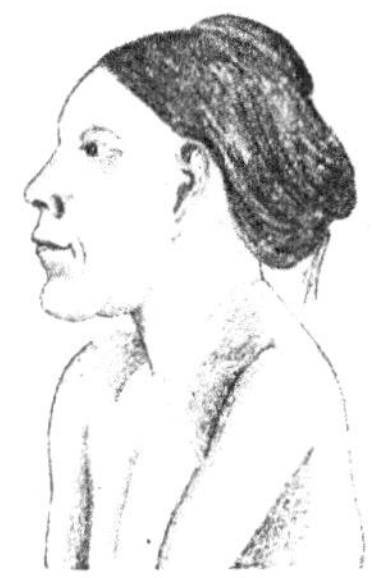

FIG. 25.—TSCHÉNO (baptized Anna), Pira-Tapuya Indian (40 years old).

about forty. The Carapaná is a young man named Cuiauí, probably under twenty. He has the comb commonly worn by the younger men. The Tucáno

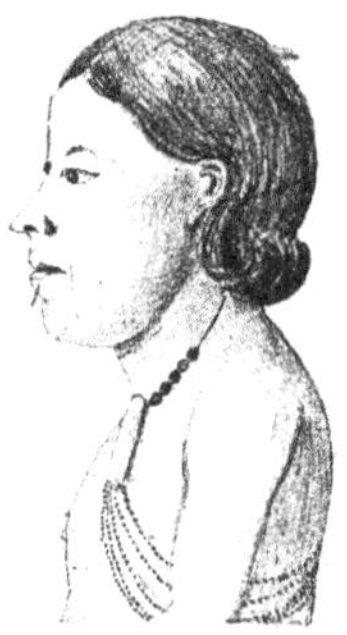

FIG. 26. — CUIAUÍ (baptized Salvador), Carapaná Indian (a young man).

FIG. 27. — KUMÁNO (= Summer), Tucáno Indian (50-60 years old).

is an old man named Kumáno, probably between fifty and sixty years of age; while Yepádia (Cáali's wife) is also a Tucáno. These four tribes all intermarry, and possess little of physical differences, although they have each a distinct language or dialect. Owing to the complete absence of beard and the custom of wearing the hair long, the younger men look remarkably like women; but it will be seen that in the case of three of the younger men and women the features are exceedingly well-formed, and except for the slight obliquity of the eyes, there is little to distinguish them in this respect from many Europeans. The Uaupé Indians are, however, among the finest of the South American tribes.

FIG. 28. — YEPÁDIA, Tucáno Indian, wife of Cáali (20 years old).

The following extract from a letter to Mr. Bentham explains why Spruce could neither go farther up the river Uaupés nor remain at Saõ Jeronymo as long as he would have done under more favourable conditions.]

To Mr. George Bentham

SAN CARLOS DEL RIO NEGRO,
June 28, 1853.

. . . I found it impracticable to ascend high on the Uaupés for several reasons, the first being the impossibility of getting up a large stock of paper and goods (in place of money) in a single small canoe, for only such can ascend a river full of cataracts; and here I felt especially, what indeed has been a great lack ever since I left Pará, the need of a trusty companion accustomed to navigation and to manage the Indians. I had, besides, no place of security in which to leave the bulk of my goods at Panuré, where are only Indian houses, with doors frequently of straw and, of course, without locks. I was fortunate in finding three whites (two of them with their families) established at Panuré for the summer, for the purpose of building large canoes; had it not been for these I do not see how I could have stayed there at all, as my house could at any time be entered when I was away from it. I used to leave it in charge of the wife of one of the whites,[1] and but for her it would once have been entered by some Indians, who had begun to make a hole at the back when she came on them. I would gladly have stayed to complete the twelvemonth at Panuré, for I have occupied no station so rich in respect of plants, and not at all to be complained of in respect of eatables, but I found it impracticable to remain there alone, and we (the whites) all left on the same day.

. . . . . .

[A few lines from a later letter to myself (dated "San Carlos, July 2, 1853"), and referring to his life at Saõ Jeronymo, will serve to wind up the account of his visit to the almost unknown and extremely interesting river Uaupés.]

"Besides myself, there were three brancos in the place, Agostinho, Chagas, and Amansio, all three building large canoes. We generally all supped together, and passed the evening very agreeably, "a rir et a nos divertimos" ("laughing and amusing ourselves"). You, who go of nights to Geographical Societies' meetings and other long-faced reunions, will perhaps despise our mode of passing

[1] [Agostinho, referred to in his letter to myself.—Ed.]

the time, and yet I daresay you would have liked now and then to listen to tales of frades and moças (friars and girls), and of men who could turn themselves into bútas and cobras grandes. We all left Saõ Jeronymo together. You know Chagas—a "homen muito serviçal" ("a very useful man") and a great scoundrel—with a face exactly like the back of a Surinam toad. He rendered me much assistance in my passeios (excursions), etc., and also took a special delight in cheating me in our little negocios (dealings). He sent another expedition up the Paapurís to steal curuminis and cunhã-tãs (boys and girls), your friend Bernardo being at the head of it. Even I was in some sort an accomplice, having lent a gun to Tushaua Joãn (Bernardo), though without knowing for what purpose it was intended. For this, and for other of his good deeds, our friend Chagas is now in prison at the Barra, but I know not yet what is likely to be the result."[1]

[Two short essays appertaining to the Uaupés appear in the Journal, and will appropriately come at the end of this chapter.]

Customs at Death and Burial

1853.—On January 2 died an old woman in Panuré. The decease took place about noon, and the relatives who were on the spot immediately commenced their lamentations, keeping up a regular song and often pointing to the dead body as it lay in a hammock. The burden of their song I under-

[1] [I have myself given some account of this man's former evil deeds and of his escape from punishment.—Ed.]

stood to be "My mother! why did you die, my mother!" varied by an occasional angry exclamation directed against the pajé (wizard) who was supposed to have caused her death. For among all the Indians on the Uaupés there seems to be a belief that death is always caused by some evil wish, or witchcraft, or putting secretly some poison into the food which should sooner or later prove mortal. In this case another old woman—very old —was pitched on as the pajé, and had not her relatives been the most powerful family on the river (for she is aunt of Bernardo of Urubú-coará), there seems little doubt that they would have killed her.

In the afternoon a grave was dug inside the house, and the body put into it, the lamenting going on all the while without intermission, and by sunset the whole population had assembled on the spot. As many as there was room for seated themselves round the grave, some of them with pieces of wood in their hands, beating the earth hard down over the corpse, in order, said they, that the pajé who had caused the woman's death should not carry off her body. When night closed in a large fire was lighted on the grave, that its occupant might not suffer cold deep down in the ground. Into the fire was thrown everything that had belonged to the deceased—her hammock, saya, baskets, tinder-box, etc. The fire was kept up and the people sang and wept round it all through the night.

I inquired if they considered that nothing remained of the deceased save the body under their feet, and was told that her anga (soul) was now in the place of her birth (she was a native of the Paapurís), where it would reappear probably in the form of some animal.

In reference to this superstition it may be mentioned that these Indians have a great repugnance to kill the larger quadrupeds, such as the deer, the tapir, etc., believing the bodies of these animals to be the resting-places of the souls of their ancestors. "How should we kill the stag," they say, "he is our grandfather?" They are, however, ready enough to kill fish, and when the white man kills and cooks a tapir they rarely refuse to eat of it.

Some days afterwards a quantity of caxirí and caapí was prepared, and a very large company assembled from the village and adjacent sitios to repeat their lamentations for the deceased. At intervals parties of eight or ten of the men partook of caapí and sallied forth from the house with lances and arrows, bestowing mortal blows on the ground, as they said they would do on the pajé were he or she in the same place.

Rise and Fall of Rio Uaupés

Like the Rio Negro and Solimoẽs, the Uaupés is said to be at its height near June 24, but does not fall perceptibly till the beginning of August. When I reached Saõ Jeronymo on September 7, it was gradually lowering, and so continued, only occasionally filling again a few inches with a heavy rain. But on the 20th of November it began to refill, and by midnight had risen 20 inches. Afterwards it rose very slowly until December 5, when it again began to fall, the whole rise not having exceeded 3 or 4 feet.

It went on falling a few inches each day, or on some days neither rising nor falling, till December 19, when it began to rise again, and by the 23rd had reached the height of its former rise. Thus it continued (save that on one day, the 28th, it fell a little) until midnight on the 31st, when it began to subside.

Jan. 9. River had fallen 2 ft. 10 in.
(On 11th rose slightly, but on 12th again fell.)
Jan. 16. River had fallen 3 ft. 3 in.
Feb. 1. " " 4 ft. 8 in.
" 5. " " 5 ft.

Then about midnight it began to rise rapidly, and continued rising until February 15, having reached within a foot of its former rise. Then it fell so rapidly that on the 25th it had fallen a foot lower than in last fall. *This is the lowest point the river attained during this season*, and it was so far from attaining the ebb of former years that the Indians said the summer had passed without any vasante (ebb) properly so called. In other years the river is said to have dried so much that scattered rocks peeped out all the way across in front of Saõ Jeronymo. This year a group of rocks appeared only about the middle, and a praya (sandy beach) on the opposite side was considerably exposed. The high water seems also to vary considerably, and I could not ascertain the point it reached most usually. One line which was shown me was about 10 feet above the highest rise above recorded; supposing the river reached this at the last flood, then the whole ebb was but 15½ feet.

Customarily after the river begins to fall in July,

there is but one partial flood, which the people call the Boia-assú, and are accustomed to look for in November or December; but this season, as is seen above, there were three partial floods, viz.—

From Nov. 20 to Dec. 5.
„ Dec. 19 „ Dec. 31.
„ Feb. 5 „ Feb. 15.

So that no one could say to which of these the term Boia-assú should be applied. . . .

[The following notes from the Journal and letters of the characteristic features of the vegetation observed on the shores of the Uaupés will be of interest to botanists.]

To Mr. George Bentham

SAN CARLOS, VENEZUELA, *June* 25, 1853.

. . . .

You will find a great many interesting things in the collection, especially among Triurideæ, Burmanniaceæ, and Voyrieæ. I hope you will be satisfied with Triurideæ, which sometimes exceed 4 feet in height! I do not know whether any one has seen an affinity of these plants with figs. To me it seems striking, but I have not time to specify the reasons on which my opinion is based. I got at least five distinct species on the Uaupés. The Burmanniaceæ are more numerous, but the individuals (like those of Triurideæ) grew widely and rarely dispersed through the forest, and some of them were gathered by only two or three specimens at a time: only two species could be called common. When growing they are easily distinguished, notwithstanding their minuteness, by their colour and by the notchings of the perianth, and on examination good characters are found in the structure of the anthers and stigmas, especially in the tail-like appendages of the latter (wanting in some species). The species of one-flowered Voyrias seem almost endless, but some of them were very scarce.

By means of fish-hooks, Jew's harps, and beads I was able to enlist a troop of little Indians in the search for these plants, and they were a great help to me, especially when I could go into the forest with them and point out what I wanted. They were also expert at hunting out fungi, which were tolerably numerous near Panuré; I got indeed so many species that they seem to me worth distributing, and I enclose along with the flowers two parcels of fungi (and two or three large species wrapped up separately) which I will thank you to forward to Mr. Berkeley, to whom I am writing at this time.

I was rather disappointed with the Podostemeæ, though perhaps I ought not to have been. Very few species grow together in any one place, and the time they last is so short that it is impossible for one person to get many species. There are plenty of raudalitos (rapids) on the Casiquiari, Orinoco, and Cunacunúma, so that I may hope to get a good many species yet.

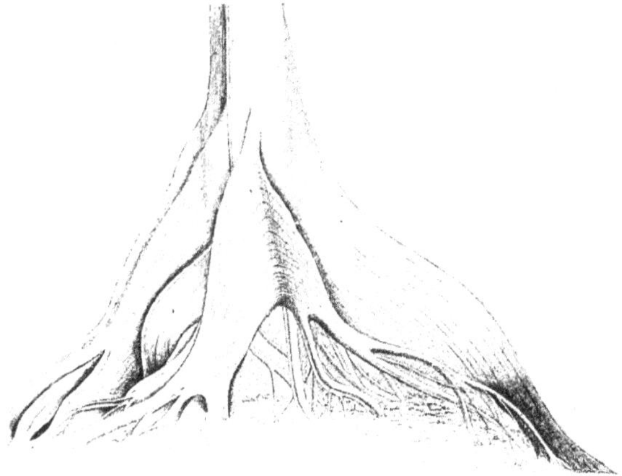

FIG. 29.—SAPOPEMAS OF A LEGUMINOUS TREE (*Monopteryx angustifolia*), PANURÉ, RIO UAUPÉS. (R. S.)

[The trunk of this tree is 4 feet thick and 80 feet high. It has racemes of rose-coloured papilionaceous flowers. It grew on the rocky banks of the cataracts.]

I was at Panuré at the best time of the year for everything but Ferns, and these seemed mostly the same as I had already got at Saõ Gabriel. Among trees of the forest and caatinga I made a splendid harvest, and the species of Vochysiaceæ and Cæsalpinieæ seem to me peculiarly interesting. You will be pleased to see two additional species of Aptandra, Miers, the first gathered in the gapó in flower and fruit, and the second in the caapoeras in fruit only. . . .

[A letter to Sir William Hooker, of the same date, gives further details of his work on the

Uaupés, with a general account of the results of his expedition which will be of much interest to botanists, and also, I believe, to all who are interested in natural history and in the difficulties of the collector in such remote and savage regions.]

To Sir William Hooker

SAN CARLOS DEL RIO NEGRO,
June 27, 1853.

. . . .

I had a very interesting excursion on the Uaupés, lasting from the end of August (if I include the voyage from Saõ Gabriel) to early in March of the present year. My collection contains a greater number than any preceding one of the tallest forest trees, among which are several undescribed Vochysiaceæ and Cæsalpinieæ. There are also a great many new things among the minutest tribes of flowering plants, such as Podostemeæ, Triurideæ, Burmanniaceæ, and the leafless Gentianeæ (Voyrieæ). I suppose that of the whole collection, numbering some 500 species, about four-fifths are entirely undescribed. I unfortunately made myself ill by working too hard both in and out of doors in the heat of the day, and was visited by some distressing attacks of vertigo from which I am yet scarcely free.

The mechanical labour of drying plants is so great here that I have little time for making geographical and other observations, and as Mr. Wallace had preceded me on the Uaupés, and his occupations leave him much more spare time than mine do, I scarcely attended to anything but botany there. I determined the latitude of Panuré, or Saõ Jeronymo, an Indian village at the foot of the first falls, which I made my principal station, to be 0° 13′ N. My watch has proved almost useless in determining longitudes, and I much regret I did not bring with me a telescope. I purchased indeed a telescope in the Barra of a Franciscan friar, who had bought it at Rio Janeiro; and it has proved of the greatest use to me in my herborisations, enabling me to distinguish green flowers on a tree at the distance of a mile, and when sailing near the bank of a river to ascertain the form of the leaves of the adjacent trees; but it barely shows the satellites of Jupiter, and is not sufficiently powerful to take an observation of them with accuracy.

. . . .

Characteristics of the Vegetation of the Shores of the Rio Uaupés as far as Panuré

(*From the Journal*)

The shores are flat, yet as they consisted almost entirely of terra firme, I could often step off my canoe into the virgin forest. Sometimes the banks were 15 to 20 feet above present height of water.

Very little rock was exposed. In one place there was a round convex granite island where the water ran rather swiftly. Here I got a Podostema. About half-way up were steep white banks, on the right bank of the river, consisting of numerous strata of alluvium (apparently clay and sand). The water was scarcely so black as that of the Rio Negro—perhaps owing to its now running off rapidly, and therefore in its most turbid state.

Where any gapó exists it is mostly indicated by the presence of Jauarí palms; these constitute the mass of the vegetation of some inundated islands in the upper part, with a fringe of low laurels and an Inga. Two other Ingas were frequent—one (*I. micradeniæ*) near going out of flower when the other (*I. rutilans*, Spruce) was just opening.

The plant most remarkable for its abundance and the delicious odour of its small cream-coloured flowers is the *Strychnos rondeletioides*, sp. n., which in some places hung in masses of many feet in breadth from the tops of the trees to the water's edge, and, especially in the evening and the early morning, perfumed the whole gapó.

Another great ornament to the banks is a small Apocyneous tree with odoriferous white flowers, which I was assured is the true Mulongo of which corks, etc., are made. It proves to be the same as a species I had gathered near Saõ Gabriel (*Hancornia laxa*, A. D. C.).

Campsiandra laurifolia (Legum) was tolerably frequent, and I gathered a narrow-leaved form, or perhaps distinct species (*C. angustifolia*). Nothing can be more abundant throughout the Rio Negro (as also the Tapajoz, as far as I have seen it) than this tree, which often occurs in continuous beds where the river, retiring in the dry season, leaves a wide sandy beach. The first trees met with in crossing one of these beaches when the river is at its lowest are these Campsiandras, two or three small Myrtaceous trees (such as are called Araçás), and many small Chrysobalaneæ. . . . With the large flat seeds of the Campsiandra the Indian mariners amuse themselves with making "ducks and drakes," and on the Orinoco, where it is said to be equally abundant, the seeds grated and treated like the root of mandiocca yield a large proportion of nearly pure starch, of which cassave-bread is made.

The Lingoa Geral name is Cumandú-assú or Great Kidney bean, cumandú being a general name for seeds of every species of bean.

In ascending the falls of Saõ Gabriel I saw, but was unable to gather, two very interesting twiners occurring in some abundance. They are represented but sparingly on the Uaupés. One is an Apocynea with large bright yellow flowers, and the other is a Menispermea (*Anomospermum Schomburgkii*, Miers), interesting from the structure of its flowers and from their having a strong odour of mellow ribstons or golden pippins. Alas, that it should be "odor et praeterea nulla"!

Two Clusiacea were very frequent, both with odoriferous flowers, and looking very much alike though distinct. The larger one has flowers of four yellowish petals opposite the sepals, and the smaller one five white petals with four sepals, along with somewhat less coriaceous leaves.

. . . .

A fine Cæsalpineous tree seems an undescribed Tachigalia, with silky leaves and very dense terminal racemes of yellow odoriferous flowers, the calyx being tinged externally with purple. The most remarkable feature is that at the apex of the petiole is a trigono-fusiform sac which is constantly inhabited by a colony of ants which pour out of a small hole bored underneath the sac to attack the hands of the too-eager botanist.

Very frequent was a Humirium, attaining sometimes 40 feet or more, with a diameter of 4 feet and a very bushy mode of growth. . . .

Melastomaceæ at present in flower included only two or three Tococas and a pretty Memecylea with very small shining leaves and yellow odoriferous flowers. Among the Myrtaceæ I saw nothing that looked very new, and these, like Lauraceæ, I was obliged mostly to forswear.

A very ornamental tree, an Anonacea (*Xylopia Spruceana*, Bth.), grew some 25 feet high, and its pinnate branches and small dark-green, crowded, distichous leaves gave it a very pretty cedar-like appearance. It grows also on the Casiquiari and Guainia.

The caatingas around Jauarité caxoeira are of the loftier sort. Their vegetation has much general similarity to that of Panuré, but it is less rich. There is a large Byttneriaceous tree (*Myrodia brevifolia*, sp. n.) frequent, which I have not seen at Panuré.

In the forest, especially near the Paapurís, grows a very lofty Vochysiaceous tree with pale yellow flowers. Another tree of the same family (*Qualea acuminata*, S., sp. n.), with large white odoriferous flowers, is frequent in the gapó. The shores of the Paapurís are peculiarly rich. . . .

Plants which the Indians on the Uaupés are accustomed to plant in their Roças or near their Houses

Edible Fruits

Cocura	=	Pourouma sp.,
Ingá-sipó	=	*Inga spuria*,
Ingá-chichí	=	*Inga Spruceana*, Bth.,
Ingá-péna	=	Inga sp.,
Pupunha	=	*Guilielma speciosa*,
Umari	=	Poraqueiba sp.,
Quiinha	=	Capsicum (many species),
Guayaba	=	Psidium (two sp.),
Namaõ	=	*Carica Papaya*,

and four others.

Edible Roots

Uaramá Uarcá	=	Marantaceæ,
Paacua-rana	=	Urania sp., long springy root,

and five others.

[The following very curious incident, which occurred at Panuré, forms the conclusion of a lengthy article on the author's experiences of venomous snakes, insects, etc., a considerable portion of which is given in the chapter on Spruce's residence at Tarapoto, where the events described occurred. Other portions appear in the Journal, especially the experiences of ant-stings and snake-bites at San Carlos as described in the next chapter. The anecdote of the trumpeter and the snake, with the reflections that follow, serve as a pleasant conclusion to this somewhat meagre chapter.]

What is it in the constitution of certain animals—notably of some birds—that renders them invulnerable, or nearly so, to the bites of venomous snakes? The agamí or trumpeter (*Psophia crepitans*) is said to be quite unaffected by a snake-bite. I can testify that it is a fearless and indiscriminate snake-hunter, and that it seizes a snake by any part of the body, so that the snake might easily seize it in return, and perhaps does so sometimes. When at Panuré, on the river Uaupés, we had a tame agamí which so attached itself to me that it would follow me about like a dog, and never failed to kill any snake that came in our way. One day I was alone with the agamí in a caatinga about four miles from the village, where I lingered about a good while in a spot comparatively clear of underwood, but abounding in certain minute plants (Burmanniaceæ) which I was much interested to gather. Whilst I hunted for plants the agamí hunted for snakes, and had already caught three or four, which it brought and laid before me as it caught them. I suppose I had not noticed and praised its prowess as I usually did, for at length—apparently determined to attract my attention—it laid a newly-caught snake on my naked feet, when I was standing erect, absorbed in the examination of a little Burmannia with my lens. The snake was scarcely injured, and immediately twined up my leg. To snatch it off and jerk it away into the bush was the work of a moment; but ever afterwards I took care to leave the agamí at home when I started for the forest. A professional snake-hunter, however, could hardly do better than enlist a pair of agamís in his service. The Brazilian Government might promote the keeping of these birds in large numbers, for the express purpose of reducing the pest of snakes in the neighbourhood of towns, instead of the few that are now kept in private houses merely as pets. They require no training to hunt snakes, but only to be encouraged to seek them out. The agamí might even be introduced into British India with very great advantage. It would find there a congenial climate; it is harmless and affectionate, and it likes the society and the protection of man. One does not see, indeed, why the native mongoose has not been more utilised in that country for the same purpose, but the agamí would prove (I think) a far superior snake-hunter to the mongoose.

It is amusing to watch a lot of hens pounce on a snake, tear it up and devour it, and to contrast that with their terror at the sight of a scorpion, and especially with the horrified note of warning of a hen to one of her brood which she sees about to peck at a scorpion; for the latter, although stabbed through by the chicken's beak, would curl up its long scaly tail and sting it in the head, no doubt fatally.

Swine are great enemies to snakes, and eat them greedily. A person who kept large herds on the savannahs of Guayaquil told me he had never known any but very young porkers bitten by a snake. The pig, he said, was a very quick-sighted animal, and when a snake darted on it, it immediately erected all its bristles, so that the snake's fangs never reached its skin.

Man is not invulnerable to snake-bites like the agamí, nor has he the quickness of sight and movement which some other animals possess, certainly in a far greater degree than fowls and swine. All he can do, in traversing the forest, is *to note where he places his hand and his foot.* That is the rule I mentally proposed to myself, but I did not always succeed in acting up to it, and I have at times incurred very great risk through neglecting it.

CHAPTER XI

SAN CARLOS AND THE CERROS OF UPPER RIO NEGRO

(*March* 8, 1853, *to November* 27, 1853)

[THIS chapter consists of extracts from a rather detailed Journal, and from letters to his friends in England.—ED.]

JOURNAL

1853.—On March 8 left Panuré, and after a voyage of thirteen days reached Marabitanas at 9 A.M. on the 21st. There was scarcely anything in flower, but a few plants were in fruit; and the vegetation of the Upper Rio Negro appeared very similar to that of the Uaupés. Here and there in the forest appeared a Japurá, its large round head completely red with fruits. The water has considerable admixture of mud, owing no doubt to the river filling rapidly, but possibly in part to nearing the mouth of the Casiquiari. We here began to be visited about sunset, and whilst the moon showed any light, by a small black mosquito, nearly silent but biting virulently; fortunately it was not in great numbers.

MARABITANAS, FRONTIER TOWN OF BRAZIL

At Marabitanas I saw a tree of Retama (*Thevetia neriifolia*, Juss.) planted by the Commandant's house. It is an Apocynea, very milky, low and widely spreading (20 feet high), the trunk about

1 foot thick, branched almost from base. It bears flowers and fruits all the year round. The flowers are the size of those of *Campanula latifolia*, smelling like primroses, borne few together near apex of the shoots, very fugacious, falling off in a few hours after expansion. The tree is said to be abundant on Rio Apuri, but always near houses.

The obtusely trigonous drupes, about the size of garden cherries, are greenish or yellowish when ripe. I observed that the Commandant's fowls picked up the fruits as they fell and greedily devoured the fleshy covering; so, thinking that what was food for them could not be poison for me, I ate three or four of them, and found them to have little taste—very slightly sweet—nor did I experience any ill effects. Yet the milky juice of the tree is a deadly poison.

The bony endocarps, of the same shape as the drupes, are perforated and strung a great many together on long strings by the Indians, who wear them wound round their ankles so as to keep up a continual rattling in their dances.

At Marabitanas, as elsewhere, I saw fowls watch all day under the male trees of the Papaya and pick up the flowers which fell, almost in a shower, especially in the after part of the day.

[While at Marabitanas Spruce made the sketches here given of two Indian girls of the Macú tribe, of which he gives the following account:—

"The Macús are one of the few wandering tribes, with no fixed residence, who exist in the forests of the Amazon and are met with through nearly the whole length of the Rio Negro, but principally to westward of it. Two Macú girls taken in a marauding expedition at the head of the Içanna had been recently purchased by the Commandant of Marabitanas when I visited him in July 1853. The few men I had seen of that nation were mostly such miserable specimens of humanity that I was greatly surprised to find in the elder girl one of the finest faces I had seen; and, notwithstanding her brown skin, I think it very probable she may have had white blood in her veins. The poor creatures were downcast, as might be expected of captives, and could converse with none of those around them, for they were ignorant of both Portuguese and Lingoa Geral; but with the aid of signs I obtained from them a good many words of their language. Unfortunately, the note-book containing them was lost."

Spruce appears to have stayed at Marabitanas twelve days (March 20 to April 1), probably to

FIG. 30.—MACÚ INDIAN (age about 9).

FIG. 31.—MACÚ INDIAN (age about 16).

obtain stores of food or to make some needful repairs to his boat; but there is no other record of this period in his Journal than the preceding note; nor does he appear to have done any botanising in the neighbourhood, since in the descriptive register of his plants (carefully bound in volumes) there is no entry between the last on the Uaupés river and the first at San Carlos more than a month later. This almost implies that he was ill, though he does not mention it.—ED.]

While staying here I met a young man from Uruana on the Orinoco, where Humboldt observed Otomacs eating earth; this person, however, has never seen them do the like and does not believe the custom exists.

But on the Alto Rio Negro, for example at Marabitanas, Indians occasionally eat the white clay (called tabatinga) exposed in some places on the banks of the rivers when the water is low. The clay is kneaded in the hand into a small ball and roasted by fire until it begins to turn red, when

FIG. 32.—PIEDRA DEL COCUÍ, FROM THE OPPOSITE SIDE OF THE RIO NEGRO. (R. S.)

it is eaten without being again melted in water. This is only an occasional practice, and the Indians do not consider clay sufficient to sustain life.

Children on the Rio Negro are much addicted to eating earth, and numbers die from that cause. To cure them of this practice some are hung up to the roof in a basket and only let down to meals, etc.

April 1 (*Friday*).—Left Marabitanas for San Carlos. On the afternoon of 3rd we reached the frontier, where is a detachment of three soldiers on the right bank exactly opposite Piedra del Cocuí or Hawk's Rock, which rises directly out of the plain at a short distance from the left bank. Its height is perhaps 1000 feet. . . .

April 6.—In the afternoon I took the montaria and crossed to the left bank of the river to visit some sitios in search of eatables. At the first I found some orange trees, and we filled a basket with the fruit, now a great luxury to me as none exists on the Uaupés. But at this and at a second sitio there was not a pig or a fowl, or even a morsel of baked fish. I was directed to a third within a caño (igarapé), which we entered and sought about for the sitio. The caño was about as wide as the Derwent at Kirkham, and there was a wide low gapó on each side appearing to pass into caatinga on the dry land. The sitio was at last discovered, well hidden in the forest, and we found in the house an elderly Indian woman with some boys. She was so rich as to possess three ducks, two of which were the parents of a lot of ducklings nestling under a basket in the middle of the floor. I immediately proceeded to bargain with the woman for the odd duck, which she showed no anxiety to part with. With some difficulty I induced her to set a price, which she did at the moderate sum of three dollars! I offered her an ell of strong calico (worth a dollar here). "No," said she, "if you give no more than one ell I'll keep my duck," and she pressed the favourite affectionately to her breast. At length she noticed a small cutlass I had in my hand and asked if I would give it for the duck. The offer was gladly accepted, and we bore off the duck in triumph. The cutlass was not worth more than two dollars, yet it was still a high price for the duck.

[There follows here a gap of more than three months in the Journal, but the chief events are rather fully described in several letters which must be here introduced.

In a letter to Mr. Bentham (June 25, 1853), Spruce writes: "I left Saõ Jeronymo on the Rio Uaupés in March, and for above a month afterwards literally found no rest for the sole of my foot. Since April 11, when I reached San Carlos, until the present date my time has been taken up in procuring materials for a miserable existence. I write now under most unpleasant circumstances, and God only knows whether I shall live to close this letter." The circumstances here referred to are much more fully described in a letter written a week later to his friend Teasdale, which takes up the story as follows:—]

To Mr. John Teasdale

SAN CARLOS, VENEZUELA, *July* 1, 1853.

. . . The only other foreigners in San Carlos besides myself are two Portuguese young men, but established here for some years and having families. The Venezuelans, like the Brazilians, have a great dislike to Europeans settling among them, knowing the greater industry of the latter, and that consequently they get the better part of the trade into their hands. The native racionales (whites) have for a long time back, as it would seem, given occasional pretty broad hints to the Indians as to the desirableness of getting rid of the Portuguese, the said racionales (Spaniards) being, by the by, not any of them popular with the Indians, and by no means secure of their own skins should the Indians once draw the blood of any white. For some time previous to the feast of San Juan (June 24) there were obscure rumours that a general massacre of the whites had been planned for that occasion, and as the Portuguese passed along the streets the Indians called out from their houses that the Feast of St. John was coming, when old scores would be paid off. Some said that they had submitted long enough to the whites, and that on the Orinoco it was quite a common thing to kill a white man and throw his body into the river, and there was no more heard of it. A fortnight before the festival the Comisario took himself off, as it now appears, to be out of the way should any novidade occur, leaving another white to supply his place; but this man also disappeared before daylight on the morning of the 23rd. I should mention too that the Comisario before he left displaced a respectable Indian who had held the office of captain for many years, and appointed in his stead one of the most drunken and ruffianly Indians in the place.

Early in the morning of the 23rd a number of women arrived in the port from their cunúcos (mandiocca fields), bringing with them great quantities of bureche (rum), which they had been employed some weeks in distilling. The proceedings of the feast were forthwith commenced by the firing of muskets and blowing of carizos (musical instruments made of bamboos and used in a peculiar dance which also bears the same name, carizo), and the demijohns of bureche were broached. Shortly after daylight the two Portuguese came to talk to me of the posture of affairs, and to tell me that we were deserted not only by the Comisario but by his Suplente. The housekeeper of one of the Portuguese is the daughter of one of the principal Indians, and she had heard during the night her relatives talking over their plans, from which it seemed certain that a general massacre of the whites was resolved on, either for the night that was coming or for the night of the 24th. I had been here a very short time and had had no quarrel with any one of my colour; but I was accused of the crime of having a white skin and of being a foreigner, and as with my little stock of merchandise I found myself the richest merchant in San Carlos, pretty pickings were calculated on in the sacking of my house. Such being the case, I declared my readiness to join in any plan of defence that could be devised, and we agreed that the best way would be for all three to unite in a house which should be fortified as well as we were able and defended with all the arms we could raise. I had three guns, one of them double-barrelled, which I at once proceeded to put in order and load with ball. Unfortunately, one of the Portuguese a few days before had lent a double-barrelled gun, and a formidable blunderbuss had been disabled by giving salvos to the Comisario General on his recent visit; still, we mustered altogether seven firearms, two swords, and cutlasses without end, and we were well supplied with ammunition. The day passed over without our being molested, save by parties of Indians coming occasionally to ask for rum—visits unpleasant enough, for any man when drunk is disagreeable company, but a drunken Indian is the most annoying animal under the sun. I did not leave my house all day, but at the hour of

Ave Maria, when all were praying in the church, I betook myself to the place of rendezvous, where I found my companions already assembled with their families. Our dispositions were speedily completed, and we set ourselves to await the event, our arms being so placed as to be seized at a moment's warning. But though throughout the night parties of drunken Indians paraded the streets with tambourines and carizos, it passed over without our being attacked. You may imagine our state of anxiety, which must have been greater on the part of my comrades than on mine, surrounded as they were by their trembling families. Whenever a drunken party was heard approaching the house with shouts, beating of drums, and occasional firing of muskets, our conversation was suspended, and with our hands on our weapons we awaited what for aught we knew might be the commencement of the attack.

Towards 4 o'clock in the afternoon of the following day, though there still remained a considerable quantity of bureche, and indeed fresh supplies had come in, every one had left off drinking. At sunset not a person was to be seen in the streets and all was still as death. The Portuguese, who had lived in San Carlos many years, and had never seen the night of St. John's Day passed otherwise than in drinking, dancing, and quarrelling, were filled with apprehension that this unwonted silence was the prelude to an attack, and that the Indians were merely keeping themselves sober for the sake of making it with more effect. We have reasons to conclude that such was really their intention, one of the principal being that in the morning the drinking, etc., were resumed and kept up for several days afterwards. When night closed in we remarked that two men were walking up and down the street in front of the house; these were a sort of scouts or sentinels, and were changed at short intervals throughout the night. The Indians, however, never screwed up their courage so far as to venture to attack us. They knew of our warlike preparations, and, as it would seem, calculated that a good many of the foremost in the assault must necessarily forfeit their lives. Of their ultimate success against us there can be little doubt, for they were 150 against three. My firm resolve, in case of being attacked, was not to allow myself to be taken alive, and so suffer a hundred deaths in one.

On the following day the Indians removed with their bureche to the other side of the river, where they remained revelling until their stock of the precious liquor was exhausted. We knew too that their powder was exhausted in the firing of salvos, so that we were relieved from further apprehension for the present. . . .

[While the events just recorded were in progress, Spruce wrote a very long letter to Sir William Hooker, in which, among much other matter, he gave a full account of the results of numerous inquiries he had been making through traders and Indians as to the sources of the Orinoco and the mountains in which it rises, with the object, if possible, of reaching these mountains, which had hitherto been unvisited and even unapproached by any European. These inquiries may be of value to any future traveller who attempts this great journey, which Spruce was unable to carry out, and they will be also useful to geographers as supplementing what Spruce was able to perform in the exploration of this almost unknown region. I therefore think it right to give it here; but as it is mainly of interest to the geographer, it is printed in smaller type.]

To Sir William Hooker

SAN CARLOS, *June* 27, 1853.

I mark every day the maximum and minimum of the barometer, and it is interesting to observe with what regularity the atmospheric tides recur on the Equator, being apparently totally uninfluenced by changes in the weather. During the space of nearly two years, it has only twice occurred that the minimum has been considerably retarded beyond its usual hour, which is from 3 to 4 o'clock, while the maximum is attained between 9 and 10.

Ever since I have been on the Rio Negro I have made inquiries respecting the position and possible means of reaching the sources of the Orinoco, without any expectation, however, on my part of being able to solve this interesting geographical problem. Quite unexpectedly, the means of doing it seem about placed within my grasp. We were lately visited at San Carlos by the Comisario General of the Canton of the Rio Negro, Don Gregorio Diaz, who resides at San Fernando de Atabapo; and on my mentioning to him how much I should like to reach the head-waters of the Orinoco, he at once entered ardently into the project, saying that it was what he had all his life been longing to do, and that if I would promise to accompany him he would arrange as many men well-armed as he could, to start on the expedition early in the year 1854. Nearly all the whites in the Canton seem eager to join us, being possessed with the idea that there is certainly an El Dorado at the source of the Orinoco. Don Gregorio is at present making a progress through his dominions, having come to San Carlos by the Atabapo and Guainia, and returning by the Casiquiari and Orinoco. He proposes to ascend above Esmeralda as far as the mouth of the Manaca, and to enter three days' journey within this river, where is a pueblo established a few years ago. He engages to make everywhere inquiries as to the best route for reaching the sources of the Orinoco, and the facilities or hindrances we may expect to encounter. I heard from him the other day, from about midway along the Casiquiari, and he promises to write me again, should there be opportunity, from Esmeralda.

As to the modes of reaching the sources of the Orinoco, besides that of following the river itself, there seem to be several. When I was at the Barra, the most direct route seemed to be by the Rio Padauirí, whose mouth is a little eastward of the 64th meridian. This large river has its sources in the Serra de Tapiíra-pecú or Ox's-tongue, and the Orinoco is considered to rise on the north-eastern slopes of the same serra. Persons who have ascended high up the Padauirí, in quest of salsaparilla, assure me they have met Indians from the sources of the Orinoco. The river Padauirí, however, gives dysentery and the ague to every one who enters it, and it was here my countryman Mr. Bradley caught the illness which proved fatal to him, while cutting piassaba with a party of Indians. The Marania is the next large river entering the Rio Negro on the same side, but its course is ascertained to be much shorter than that of the Padauirí. The Rio Cauaborís, which enters the Rio Negro on the 66th meridian, probably extends nearly to the Orinoco. In its lower part it makes a large curve to westward, nearly parallel to that of the Rio Negro, and I have been assured by Indians at Saõ Gabriel that it ran not much to the eastward of that place. From Marabitanas, the frontier town of Brazil, I could distinctly see, though at a great distance, the serrania called Pirá-pukú or The Long Fish, whose base is laved by the Cauaborís. This lofty ridge seems to run westward, trending slightly northward, and the portion of it seen from Marabitanas extends through an angle of about 90° (from E. nearly to N.), its prolongation westward being hid from view by the forest on the opposite side of the river. With my telescope I could distinguish steep escarpments, bare of forest, but in no part could I distinguish the trees, the forest-clad portion being only recognisable from its colour. I suppose that in their highest part—an abrupt truncate peak about midway—they may be nearly 4000 feet above the plain. Those who have ascended the river Cauaborís describe it as very picturesque and possessing a peculiar vegetation. Certain curious plants, said to resemble both palms and ferns, from the description given me can only be Cycades. I was delighted to meet with a Cycas in the Uaupés, though it never showed signs of flowering; it is the only species of this tribe I have seen in South America.

The Rio Cauaborís is easily reached from San Carlos by proceeding up the Pacimoni, a tributary of the Casiquiari, and up its southern branch the Bariá, from which there is a short portage to the Cauaborís; but nothing of bulk could be taken this way, and I have reason to believe that the Cauaborís does not reach the Cerro de Tapiíra-pecú.

A more likely route for us is by the Siapa, the longest tributary of the Casiquiari, called in its upper part the Rio Castanha, and certainly having its sources in the above-named cerro. The only objection to it is that several steep rapids have to be passed; but these may be avoided by making a circuit through the upper mouth of the Casiquiari and going up the Manaca, from which there is a short passage by land to the Castanha.

We have discussed these and other routes principally with the view of avoiding the hostile Guaharibos, the more especially as it is believed that these Indians do not extend to the actual sources of the Orinoco, but that tribes inhabit there with whom friendly communication has been held by the Castanha and Padauirí. On the whole, I think we incline to first risk a battle with the Guaharibos, and I have little doubt that with fifty men well armed we should be able to force our way.

Shortly after the separation of Venezuela from the mother country, and whilst there was still an armed police in the Canton del Rio Negro — there is none of any kind now — the Commandant of San Fernando was sent with a considerable body of armed men to endeavour to open amicable relations with the Guaharibos. He reached the Raudal de los Guaharibos with his little fleet of fifteen piragoas, and as the river was full, the whole of them might have passed the raudal, but it was not considered necessary, and his own piragoa alone was dragged up, the rest being left below to await their return. A very little way above they encountered a large encampment of Guaharibos, by whom they were received amicably, in return for which they rose on the Indians by night, killed as many of the men as they could, and carried off the children. One of these captives is still living near the upper mouth of the Casiquiari, where I hope to see and converse with him. Treatment such as this of course is calculated to confirm, and perhaps it was the original cause, of the hostility of these Indians to the whites. The same sort of thing seems to have been practised anciently near the head-waters of all these rivers. On the Rio Negro, where the Portuguese had formerly large fazendas reaes (royal farms), in which were cultivated great quantities of coffee, indigo, etc., it was the custom to recruit from time to time the hands required for working them by sending armed men up the various rivers debouching into the Rio Negro and Japurá to make pegas (raids) among the indigenous inhabitants. The fazendas reaes have disappeared, and the Brazilian Government has promulgated edicts against the seizing of the native inhabitants and reducing them to slavery, yet the practice still exists and is carried out. I speak of this with certainty, because since I came up the Rio Negro two such expeditions have been sent up a tributary of the Uaupés, called the Rio Paapurís, to make pegas among the Carapaná

Indians. . . . I have also seen and conversed with two female children stolen from the Carapanás in these expeditions.

. . . .

To return to the Orinoco. I have met at San Carlos several people who have been as far as the Raudal de los Guaharibos. The most intelligent of these, and the person who perhaps of all others knows most of the country between the Casiquiari and the sources of the Orinoco, is an old gentleman called Don Diego Pina, residing now in Solano (a little within the Casiquiari), but when Schomburgk passed this way residing in San Carlos and acting as Comisario. He is unfortunately quite blind, and cannot therefore point out anything on my maps, but his memory seems perfect for distances and bearings. According to him, it takes a month to reach the raudal from Esmeralda, travelling as traders are accustomed to do here, that is, stopping at all the caños, within which the Indians usually fix their habitations. The Orinoco above the raudal is still a large river, which in the force of the rainy season might be navigated by piragoas[1] of considerable size. He is of opinion that the real sources of the Orinoco are very much to the eastward of what is supposed by Humboldt in his *Aspects of Nature*; and it seems to be clearly made out that they are at least considerably to the east of the sources of the Rio Branco, or, in other words, that the system of the Rio Branco *overlaps* (if I may so say) that of the Orinoco—a circumstance not without parallel in other river systems.

Don Diego is perhaps the only white now living in the Canton del Rio Negro who recollects Humboldt in Venezuela. He was making turtle oil on the Orinoco, on a playa near the mouth of the Apuré, when that distinguished traveller passed on his way towards the cataracts. A person died in San Fernando two or three years ago who had seen Humboldt and Bonpland at Esmeralda, and remembered the difficulty they had in procuring the flowers of the Juvia (*Bertholletia excelsa*), for which, said he, they offered an ounce of gold. At the season of fruit of this tree the Guaharibos descend much below the raudal in order to collect it for food, and at that time the Indians of the Casiquiari, in parties of not more than five or six, lie in wait for them and carry off such as they can lay hold on, making of them slaves for cultivating their cunúcos. Many Indians on the Casiquiari can show lance-wounds received from the Guaharibos in these expeditions.

I should mention that Don Gregorio Diaz has also travelled much in the rivers eastward of the Casiquiari, and in his voyages about the head-waters of the Siapa must have come very near the sources of the Orinoco.

---

[1] The piragoa of Venezuela is the same as the igaraté of Brazil, and has for its foundation a hollowed tree-trunk, above which are fastened three or more planks on each side.

I have been twice to the junction of the Guainia and the Casiquiari. The water of the latter is not very white, which is explained by its having received during its course from the Orinoco two considerable rivers of black water, the Pacimoni and Siapa. The Guainia and Casiquiari seem of nearly equal bulk, but neither can compare with the Uaupés. It should be noted that the name "Guainia" does not extend below the mouth of the Casiquiari, the junction of the two constituting the Rio Negro. "Quiare" is the ancient name of the Rio Negro,[1] and "Casiquiare" has evidently some connection with it, but what I am not prepared to say. *Possibly* the prefix "casi" is pure Spanish (Lat. "quasi"); for the Rio Negro is here considered the continuation of the Casiquiari ("as it were the Quiarë"), and not of the Guainia.

I am now preparing a boat to ascend the Casiquiari and, if possible, explore the mountains at the back of the Duida of Esmeralda, for which purpose the preferable course seems to be to enter the Rio Cunucunúma, whose mouth is half a day's journey on the Orinoco, below the Casiquiari. The summit of the Duida is said to be inaccessible on account of the perpendicular walls of rock on every side of it; yet everybody seems to know perfectly well that there is a round lake on the very top, inhabited by a large turtle, the "genius" of the mountain. Whether I shall proceed direct from the Cunucunúma towards the sources of the Orinoco, or first return to San Carlos, will depend on the intelligence I receive from Don Gregorio on his reaching San Fernando.

The gratification I naturally feel at finding myself fairly *in terra Humboldtiana* is considerably lessened by various untoward circumstances, not the least of which is the very great difficulty experienced here in procuring the necessaries of life, so great indeed that it occupies nearly all a person's time, especially when the river is filling, and we think ourselves well off at San Carlos when we can eat once a day. Anciently when there were missions in most of the pueblos on the Orinoco and Rio Negro, travellers had in them a ready resource; but for some twenty years past there has not been a padre resident in the Canton del Rio Negro, and scarcely one on the Orinoco out of Angostura. A country without priests, lawyers, doctors, police, and soldiers is not quite so happy as Rousseau dreamt it ought to be; and this in which I now am has been in a state of gradual decadence ever since the separation from Spain, at which period (or shortly after) the inhabitants rid themselves of these functionaries in the most unscrupulous manner. . . .

[I will now give, in the order of their occurrence, the record in the Journal of the chief excursions

[1] See Baena, *Ensaio Corografico sobre a Provincia do Pará*, p. 530.

Spruce made from San Carlos before leaving for his great journey up the Casiquiari and Orinoco to Esmeralda, and up their most important tributaries the Cunucunúma and the Pacimoni. It will be seen, by a note at the end of the next chapter, that the sources of the Orinoco have been reached by a French traveller in 1877, but nothing is yet gained but the bare fact of their being accessible.]

Ascent of Piedra de Cocuí

Júly 19, 1853.—I started at 6 A.M. from the second Brazilian sitio on the left bank of the Rio Negro, below the mouth of a widish igarapé. It took us two hours of ascent to reach the mouth of the narrow igarapé which leads to the serra. It was difficult work pushing the canoe along this because of overhanging and entangled branches of trees. Two hours more, herborising by the way, brought us to the base of the serra. The vegetation was much as that around the serra of Saõ Gabriel: the same common ferns, and, on large blocks strewn at the base of the serra, the same delicate Selaginella grew on their steep faces. . . .

Much of the forest is of the loftier caatinga. The Cerro de Cocuí is not less than 1000 feet above the river. The side fronting the river is destitute of vegetation almost to the apex, the rock actually overhanging its base and being destitute of furrow or fissure. In the concave part it is streaked with white, yellow, and pink, perhaps from decomposition of its surface, but there seems also to be there an unusual proportion of mica in the granite. As very little water runs over this part it is not clad, as is most of the rest of the rock, with the same blackish Conferva which invests all the exposed granite in this region.

At the top it is clad with forest and two bare rocks stand out like paps; these are seen to be the apices of the two parts into which the mountain is deeply cleft when it is viewed from a little farther up the river. Going higher up still the left-hand peak resolves itself into two, and the whole mountain then presents the form of a truncated pyramid with a three-toothed apex.

At the base the view of the immense overhanging mass is very imposing. There is one very grand scene when, looking down into a ravine, the infant river is seen emerging from beneath a mass of rocks, of which the uppermost, spanning across all the rest, is an immense parallelopipedon perhaps equalling the Royal Exchange in magnitude; and the frowning mountain rises at the back of a thin strip of forest and shuts in the picture.

We skirted along the base of the mountain until we reached the opposite side where the rock sinks down to the plain by more gradual undulations. The forest has here straggled up the sloping rock to nearly one-third the whole height of the mountain, and the rock presents a singular aspect from being clad with roots of trees which are closely applied to its surface and extend in almost parallel lines to apparently interminable lengths. Their direction is that of descending floods which must come down from the upper part of the mountain with every heavy rain, and whose tendency is to wash them away; this effect being prevented by the roots presenting the least possible surface to the onslaught of the flood.

When we reach the upper margin of vegetation, where the inclination of the rock is at least 45°, we find that the plant which by its resistance to the full force of the flood helps to sustain all those below it is a Bromeliacea with leaves somewhat like those of the pineapple but less rigid, and its dead flower-stems 6 feet high. Below this are beds of an Orchis (*Sobralia dichotoma*), whose tufted, distichous, leafy stems rise 5 or 6 feet high and bear at the summit a few large, handsome, aromatic flowers, of which all the parts are white save the lip, which is yellow within with vermilion streaks. About the roots of this orchis were the tufts of a moss (a Calymperes) in fruit, apparently the same species as is frequent in the low sandy forest throughout the Alto Rio Negro. The only other herbaceous plants seen here were a Cyperacea (Scleria sp.), a Scrophulacea with red flowers about the size of those of *Antirrhinum majus*, a minute Utricularia with greenish white flowers in places where water constantly trickled down, and a slender Dioscorea. Twining along with the last over the Orchis and the Bromeliacea was a suffruticose Echites, remarkable for its large white bracts with roseate tips. The arboreous vegetation comprises an Ivy, a Cordia (*C. graveolens*) with aromatic leaves, a Melastoma with large white flowers, and a Malpighaceous shrub with erect compound racemes of yellow flowers; the same species grows in large patches at the mouth of Guainia, and at a short distance bears a great resemblance to the common broom; and some others which, not being in flower or fruit, I could not refer with certainty to their genus or order. None of these rose above 12 feet high, and they constituted a band of some twenty or thirty yards broad, skirting forest of a loftier description though still very low compared with that of the plain.

Above this there is a large extent of bare swelling rock, quite dry except in two or three places where water trickles down shallow furrows. At first sight it seems impossible to ascend it, but on trial I found that the asperities of the surface sufficed to prevent my naked feet from sliding. I climbed only so far as to have a clear view of the country, for the descent is fearful and can only be safely accomplished on both hands and feet with the face towards the sky; but my Indians, agile as monkeys, climbed up to the next belt of vegetation, at about midway of the mountain, and brought me down a quantity of the Orchis above mentioned, along with two other species of the same order, one with large delicate red flowers quite withered by heat when they reached me, the other an Epidendrum, with small pink flowers and roundish leaves almost as fleshy as those of a Mesembryanthemum. They told me that it would be possible to ascend the mountain still higher though the rock was very steep, but it must be early in the morning ere the dew had passed off, for the heat made it slippery, besides scorching their feet.

From the point I attained, where a slight dimple in the rock allowed me to sit down, I had a view of a range of mountains called Pirá-pukú, extending from S.S.E. to E., and its prolongation hidden by forest at the base of Cocuí. . .

By far the most striking object, however, was the mountain at my back, and when I stood up and turned to view it, it seemed the finest object for a painter's pencil I had seen in South America. It is impossible to do justice to the scene in words. The two peaks stood out in all their distinctness, that on the right (east) being slightly higher than the other, of an exact sugar-loaf form, quite destitute of vegetation save a scrap at the summit, and I suppose absolutely impossible to ascend. The peak on the left has a broader top, and bears a good deal of forest, among which I thought I could distinguish two palms, probably Inajás, for my Indians found an Inajá palm growing at the highest point they attained, and I have previously seen this palm ascending to greater heights.

Effects of Ant Stings

(*Journal*)

Aug. 15, 1853.—Yesterday I had the pleasure for the first time of experiencing the sting of the large black ant called tucandéra in Lingoa Geral. . . .

I had gone after breakfast to herborise in the caapoera north of San Carlos, where there were a good many decayed trunks and stumps. I stooped down to cut off a patch of a moss (Fissidens) on a stump, and remarked that by so doing I exposed a large hollow in the rotten wood; but when I turned me to put the moss into my vasculum I did not notice that a string of angry tucandéras poured out of the opening I had made. I was speedily made aware of it by a prick in the thigh, which I supposed to be caused by a snake, until springing up I saw that my feet and legs were being covered by the dreaded tucandéra. There was nothing but flight for it, and I accordingly ran off as quickly as I could among the entangling branches, and finally succeeded in beating off the ants, but not before I had been dreadfully stung about the feet, for I wore only slippers without heels and these came off in the struggle. I was little more than five minutes' walk from my house (for I was returning when the circumstance occurred), and I wished to walk rapidly but could not. I was in agonies, and had much to do to keep from throwing myself on the ground and rolling about as I had seen the Indians do when suffering from the stings of this ant. I had in my way to cross a strip of burning sand and then to wade through a lagoon, partly dried up and not more than two feet deep. Both these increased the torture: I thought the contact with the water would have alleviated it, but it was not so.

When I reached my house I immediately had recourse to hartshorn. No one was near but an Indian woman (my cook), and she, without my telling her (though I was about to do it), bound a ligature tightly above each ankle. After rubbing for some time with the hartshorn, and experiencing no relief, I caused her to rub with oil, and then with oil and hartshorn mixed. None of these seemed to have any effect; when the oil was made hot it relieved me a little, but very little indeed, and the wounds which were least rubbed, ceased to pain me the soonest, one that had not been touched being the first cured.

It was about 2 P.M. when I was stung, and I experienced no alleviation of the pain till 5. During all this time my sufferings were indescribable—I can only liken the pain to that of a hundred thousand nettle-stings. My feet and sometimes my hands trembled as though I had the palsy, and for some time the perspiration ran down my face from the pain. With difficulty I repressed a strong inclination to vomit. I took a dose of laudanum at 4, and I think this did more than anything to lull the pain. I had been stung on the two big toes and on the soles of my feet, but the stings that caused me most suffering were four close together among the fine veins below the left ankle. When the pain of all the others had subsided, this continued to torment me, and pains shot from it all over the forefoot and some way up the leg, notwithstanding the bandages.

After the pain had become more bearable, it returned with great force on two occasions, at 9 o'clock and at midnight, when I stepped out of my hammock on my left foot, and each time caused me an hour of acute suffering. Towards morning I slept, and when I woke up I felt no inconvenience beyond a slight numbness in the feet, but the inflammation continued unabated for thirty hours. It is curious that nothing was visible externally more than would be caused by the stinging of an ordinary nettle. Possibly swelling was prevented by the application of hartshorn and oil, for I have heard of cases where the swelling was considerable. Rubbing in the ingredients served to increase the pain both at the time and afterwards.

My vasculum and one slipper were left on the

field of battle. To obtain the former, which is to me a priceless article, I ventured to-day to revisit the spot, and cautiously picking my steps, I succeeded in drawing away with a long hooked stick both shoe and vasculum, nor did I disturb a single tucandéra.

I came worse out of this encounter than any other in which I have been engaged since entering South America. Many times have I been stung by ants and wasps, but never so badly. Once, near Saõ Gabriel, in my visit to the falls of Camanáos, I was making my way to a small campo; a branch hung inconveniently across my way and I made a cut at it with my cutlass, not noticing that a wasps' nest was suspended from it; but I was not left a moment longer in ignorance, for a cloud of the vile insects "buzzed out wi' angry fyke," and attacked me tooth and tail. I ran back, beating away the wasps; my hat fell off and a good many of them remained with it, but not a few still followed me, got into my hair, and stung me all over my head and neck. When I fairly got free from them I sat down on the ground, for I was dizzy and stupefied, and it seemed as if my head were bursting, for I suppose I had not fewer than twenty stings in the head and face alone. It came on to rain smartly, and I allowed the rain to beat on my head and neck, which in a few minutes seemed to relieve me much. After a while I was able to recommence my journey, though still in great pain, and I cut myself a track through the bushes so as to give a wide offing to the wasps' nest. The pain grew gradually less acute, though it did not fairly pass off all day. An Indian whom I was taking with me, and who had lagged a great way behind, had the luck to thrust his head into the same wasps' nest, and also got considerably stung. . . . I have been twice stung by the common house scorpion, but the pain was not greater than that produced by an English wasp. There is a larger kind whose sting is said to be far worse. The bite of the common Scolopendra (centipede) is about equal to that of the scorpion, but I have never been bitten by the immense Scolopendra that is seen in heaps of timber or among rubbish in deserted houses.[1]

. . . . . .

Up to the present date (August 1853) I have through God's mercy been preserved from the bites of venomous snakes, nor have I yet seen any one under the actual influence of a snake-bite, though people have been bitten in my very near neighbourhood. The venomous jararáca is frequent in caapoeras and in rubbishy places near houses through all the Rio Negro. At Panuré (on the Uaupés) I was one afternoon putting dry paper to my plants, and I had a quantity spread out drying in front of my house, when chancing to look through the open door, I saw what seemed to be a large greenish beetle bobbing about among the sheets, and I bolted out to seize on it; but fortunately ere doing this I discovered what I took for a beetle was the head of a jararáca. Like most venomous snakes, this is fortunately very sluggish in its movements, and I had no difficulty in killing it with a stick which was at hand. A few days afterwards while similarly engaged I

[1] Among some planks piled up by a sawpit at Tomo I found a Scolopendra 11 inches long, 1 inch broad.

heard a slight shuffling noise near me, and on looking up saw a poor toad crossing the floor with all speed and a jararáca in close pursuit. I sprang up and the jararáca faced about and retreated to a fôrno (mandiocca oven) which was near; beneath this it succeeded in hiding itself ere I could lay hold of anything with which to attack it.

. . . . . .

[The following cases of snake-bite ending fatally which Spruce heard from the relatives of the sufferers, together with one case treated differently in which he witnessed the recovery, gave him an amount of experience which enabled him, nearly two years later, undoubtedly to save the life of an Indian of Peru, and by so doing not improbably his own.]

Oct. 11, 1853.—Two days ago a boy of about twelve years old was bitten by a rattlesnake while hunting peccári with his father and mother in the forest at some distance from their cunúco, which is nearly a day below San Carlos. He was standing at the time on a spot remarkably clear of bush, and his mother had only just left his side when (as it would seem) the snake sallied out of a thicket near by and bit him in the back of the leg just below the calf. He was taken home with all speed and gunpowder was applied to the wound and given him to drink. The grated skin of the buta (tonnina) was also given him, great faith being put by the Indians (apparently without any reason) in this remedy. Notwithstanding these applications, the wound speedily proved fatal. He was bitten at 1 P.M., and by three of the following morning he was a corpse. The body was brought to the village to bury. . . .

The juice of lemons rubbed over the wound and taken internally has a great reputation against snake-bites at San Carlos. Most people, however, die who are bitten. Not long ago a neighbour of mine lost a daughter, a fine young woman, from the bite of a jararáca, and the woman who cooks for me lost her father a few years ago from the bite of the same snake. When I was at Saõ Gabriel, a little before I arrived, a young woman, daughter of the pilot of Camanáos, died from the bite of a jararáca.

March 1854.—When I returned from my expedition up the Casiquiari and Orinoco, I found recently established at San Carlos a mulatto trader who had married the daughter of a wealthy man of colour at the Barra, and with her fortune had embarked in the trade in the products of the Rio Negro. Both he and his young wife were fond of shooting, and one day were in the forest together in quest of game when he was bitten in the foot by a jararáca. The reptile reared itself for a second stroke when it was shot by his wife. They hastened to their home, which was near, and washed the wound freely with vinegar; but finding no relief from that, and feeling assured he should die, he determined to drown the pain and the fear of death, as far as possible, by copious draughts of strong rum. With such a will had he applied the opiate that when his wife fetched the Comisario and myself half an hour afterwards, we found him completely stupefied. I observed that the wound had bled copiously, and with the blood some portion of the venom had doubtless escaped, so that the danger of a fatal result was all the less. He dozed a good while, and although his frame was now and then shaken by a convulsive shudder, his pulse was gaining strength. When he woke up he was quite out of danger, although he still complained of occasional shooting pains and fancied he must inevitably die; but a cup of strong coffee completed the cure and the next day he was going about as usual.

[About this time, having been over six months in San Carlos and the neighbourhood, Spruce collected together in his Journal his various notes as to numerous insect plagues which abound in the district and seriously interfere with personal comfort and power of work. It must be noted that since his arrival in the Spanish-speaking districts belonging to Venezuela, and thenceforth throughout his travels, he uses the word "mosquito" (as do all the inhabitants) for various small biting flies ranging in size from what we term sand-flies up to horse-flies, while for the gnat-like insects which we call mosquitoes the local term "zancudos" (long-legged flies) is used. It will be seen that in this region of the Upper Rio Negro the former group are much more numerous and a much greater "plague" than the latter. I give here these notes as being interesting in themselves and as serving to elucidate what must have been the still greater "plague" he encountered on the Orinoco.]

Insect Plagues on the Rio Negro

Since the Casiquiari began to fall (July 28) there has been no lack of mosquitoes at San Carlos, but since the first partial rise, *i.e.* since September 4, they have been so abundant as seriously to interfere with one's comfort. When a person can keep in motion, they do not settle much on him, but when I am obliged to be still, as in writing or working with the microscope, their torment is scarcely bearable.

Though I wear socks and tie my trousers round my ankles, and often put on gloves, they still find out vulnerable spots, and are especially persevering in biting my neck, breast, and forehead. In my visit to Solano (Oct. 2 and 3) I was surprised to not meet a single mosquito, and people who have come from the centre of the Casiquiari tell us that mosquitoes are much fewer there than here. It is said that many years ago San Carlos was as much plagued as any part of the Casiquiari, but for a good while mosquitoes have been scarce here. This year it promises to revert to its original state. Similar alternations of healthiness and disease seem not infrequent in the tropics.

The mosquitoes here are chiefly two sorts, one of which, the píum of Brazil, is a small fly of a darkish colour, and bites throughout the day, rarely beginning earlier than 7 in the morning and leaving off shortly after sunset. It leaves a small pustule filled with blood, and to persons unaccustomed to it considerable inflammation is caused by its puncture. The Indians are accustomed to squeeze out the blood (extravasated), and towards evening it is common to see women passing in review each other's backs and squeezing the blood from each puncture with a pointed stick. By this means it is said ulceration is prevented. I have myself rarely seen ulceration supervene, and this only where the patient had scratched the punctures, especially if his flesh was in an unsound state from some venereal taint. The irritation excited is greatest about the wrists, ankles, and feet, but the thinnest stockings are a preservative against mosquitoes, while the zancudo pierces through the thickest woollen garments, and English sailors in Pará are accustomed to say that it penetrates even jack-boots! Mosquitoes when numerous get into the eyes, nose, and mouth, and thus interfere more with any sedentary work than by the pain they inflict.

Along with this insect is another of about the same size and with much light green about it, not unlike some of our biting forest flies. And there may be several other species when they are accurately examined.

There is one very distinct and larger mosquito, remarkable for its large red head, hence its name in Spanish, Mosquito colorado, and in Lingoa Geral, Píum-piraga. There were a good many of this kind at Saõ Jeronymo and Saõ Gabriel, but they are scarce at San Carlos. It sucks an enormous quantity of blood, hanging on till its abdomen is extended to twice or thrice its original size and then falling helpless to the ground; but its puncture causes less irritation than that of any of the others.

All the flies bite from morning till night, relaxing only a little when the sun is very hot, and always congregating most in shady spots, as under trees and within houses. In complete darkness they cease their attacks, hence a respite can always be obtained by closing the doors and windows and stopping any large chinks. On the Casiquiari the people are in the habit of putting a sort of mat (called parí in Brazil and cacuri in Venezuela) in the doorways during the day. It consists merely of strips of the Gravatána palm tied together with sipó, but with interspaces wide enough for a wasp to pass through, and yet it is said to be quite sufficient to prevent the entrance of mosquitoes and it allows a little light to enter.

About sunset on some days, but not every day, we are visited by a sand-fly, the maruím of Brazil, ehén of Venezuela, which, though but a mere speck to the eye, and when flying scarcely distinguishable from a grain of dust, inflicts a more painful wound than any of the others. With me it always causes more or less inflammation of the part.

In travelling along the rivers either of white or black water, the greatest plague by day seems to be the mutúca (in Venezuela called tabano). The species most frequent on the Amazon is not larger than the common house-fly; it is a deep, almost black-green with a few white dots, and its proboscis is short and broad so that it cannot penetrate through clothing; but it bites fiercely on exposed parts of the body, and from its abundance is a very great pest. On the Rio Negro two or three species are frequent, all closely resembling the horse-fly (indeed, I believe this is the name it bears in Demerara), and possessing long needle-like proboscides which are more penetrating than even those of the carapaná. Their wound is attended with swelling and causes great irritation, especially when on the sides and soles of the feet. Generally they are infrequent on land, but in Saõ Jeronymo they were abundant, and about nightfall used to bite most savagely, allowing themselves to be killed rather than relax their hold ere they had sucked their fill. There is a species with a reddish body, the mutúca-piranga, rather frequent in caatingas, and its bite is very severe. Generally speaking, the forests of the Rio Negro are not much infested by stinging flies. By the rivulets there are occasionally a few mosquitoes and long-legged zancudos, and in sandy, open, low forests there are at certain times a good store of the smaller flies and the above-mentioned mutúca-piranga.

On the Amazon as far as the Barra, the only plague by day is the mutúca. At a day or two below the Barra, a stray píum now and then visited us; but on the Solimoẽs the píum and mutúca by day and the carapaná by night leave the poor voyager scarce a moment's respite, and the farther one goes up the river the worse one finds it.

[The following letter, written a few days before leaving San Carlos for the Orinoco, well illustrates the difficulties and delays a traveller in these countries is exposed to, and also gives a good description of the boat which Spruce had built for the purpose of the voyage :—]

To Mr. John Teasdale

SAN CARLOS, *Nov.* 20, 1853.

. . . It cannot be less than three months since I wrote to Sir William Hooker to say that I was just on the point of starting! I knew, in fact, of nothing to detain me: my boat was completed and in two days might have been caulked and launched, and my goods were all packed up for embarking.

But there is no calculating here as to when a person will start on a voyage, or when he will arrive at his destination. Just at that time we were in a sort of interregnum here. The Comisario of San Carlos was displaced and another person who lives some way up the Guainia nominated to succeed him. The latter declined the honour, and many weeks were taken up in correspondence with the Comisario General, who lives at San Fernando. The Indians, finding there was no one to control them, betook themselves to their cunúcos,[1] and for at least three weeks the pueblo was quite deserted. During this time I might easily have died of hunger, for the forests near San Carlos are quite exhausted of game, but I had fortunately received a short time before the salted flesh of an ox which I sent for to the cataracts of Maypures, and I had brought up with me from Brazil a considerable quantity of rice and mandiocca. You know, I daresay, how flesh is cured in the tropics—it is cut up into thin strips and dried in the sun with very little salt on it, and when cured has much the appearance of leathern thongs—whether the latter would be so tough when boiled I cannot say, as I never tried them—however, such as it was it came very opportunely, and sufficed to keep the life in me. When I saw the Indians were packing off I applied to the caulkers, of whom there are five or six in the place, to caulk my piragoa, but though I offered them twice the pay they are accustomed to get, they refused to work, and I had no means of

[1] Cunúco (= sitio in Brazil) is the name given to the plantations of mandiocca, which are generally on the banks of some stream deep in the forest.

forcing them. My boat had not been built under cover, and after completion, lay some six weeks baking in the sun, which opened all the seams and split some of the timbers. As she rested on the ground (for we use no "stocks" here, except for the amusement of the Indians) the termite ants had found their way into her and began to eat up some ribs that were of softer wood than they should be. I did not find them out until after the boat was launched, and it has cost me infinite trouble to kill them, with boiling water, for they were in thousands on thousands.

At length, Don Diego Pina—an old gentleman who resides at Solano on the Casiquiari, and is perhaps the only racional in the Canton del Rio Negro who remembers seeing Humboldt—was appointed Comisario, but after removing to San Carlos, it took him two or three weeks to get the Indians back to the pueblo. When they came every one brought his stock of bureche (the cachaça of Brazil), for not an Indian of them but has his still and his cane-patch. Two caulkers were set to work on my boat, but as they were constantly intoxicated and only worked half the day, they spent a whole week over it, and their work was so ill done that the night after the boat was launched it went to the bottom. It had cost me three gallons and a half of rum to put it into the water, and I had to come forward with another half-gallon to get it out again. I then set the caulkers at work to stop the holes where water entered; but as they and all their assistants were intoxicated, the work was done very imperfectly, and at this very time my boat makes much more water than it ought to do; but I am in hopes the muddy waters of the Casiquiari will stop the seams completely.

This may give you a faint idea of what boat-building is here, and you may well suppose I am disgusted with it. The worst of it is that one cannot calculate on a boat's lasting more than a couple of years, for the timber made use of is almost entirely of the inundated banks of the rivers, cut when the latter are high so as to fall into the water and be floated away in rafts, and it speedily perishes. There are many excellent timbers of the terra firme, but they have no means here of getting the trunks to the water-side.

The name piragoa is given to vessels built on a curiará (the name given here to boats made out of a single trunk) as a foundation. Vessels of a larger description are built of boards from the very keel and are called lanchas. My piragoa is 11 varas (each 2 feet $9\frac{1}{3}$ inches English) in length, a little less than 3 varas in breadth where widest, and not quite a vara in depth. In the afterpart the carroza (cabin) occupies a length of 5 varas; it is entirely of boards and not thatched with palm-leaf as is most customary here. The flooring is about 6 inches below the edge of the boat, and the roof, which is nearly flat—very slightly convex—is so high that I can sit very comfortably within the carroza on a little Indian stool of about 6 inches high. There is a small square window on each side and one in the stern of the carroza which can be opened to admit the air when necessary, and it is entered by folding doors which can be secured by a padlock. The roof is very difficult to make watertight—I had it twice caulked and still the rain penetrated it—I then had several strips of strong cotton cloth sewed together to the size of the roof and anointed with the milk of a tree called Pondári, so as to form a sort of cerecloth; this I nailed on to the roof, and it seems to do its office effectually. I have further a large and nearly waterproof mat on the roof, which serves to temper the heat of the sun. In the fore-part of the piragoa are the benches of the rowers, and I propose depositing in the same place our provisions, such as sundry mapires (baskets) of mandiocca, and any other things which the Indians are not likely to steal; the whole will be covered by two mats. In the very prow is a large coil of cable, essential for dragging the piragoa up the raudales (cataracts or rapids), of which there are several smaller ones in the Casiquiari. The oars to be used are, as you may suppose, paddles, which are of various shapes, some having an oval blade and some quite round. My crew is to consist of seven men and a little boy. I think I have before told you that no work can be done in this country without paying for it beforehand. Thus most of these men have already received pay for the voyage (calculated at three months). The Indian carpenters are all in debt to some racional or other, and if a person needs one for the slightest job he must first pay the debt of some carpenter, and then the latter will not put hand to work without a further advance of goods. Thus I, for instance, had a couple of carpenters to "buy," and after they had finished my piragoa and made me some boxes, one of them still owed me forty dollars. If I have no more work for him when I return to San Carlos, then I must try to "sell him," which is quite another thing, for no one here has any money; and if I receive piassaba and boards (which is all they have to offer), then I must build a boat to carry them down to the Barra and sell them, which will perhaps be a worse speculation than losing the money. . . . The first notice we had at San Carlos of who was the new president of Venezuela came in an English *Times* which reached me by way of the Amazon.

[The following is the last entry in the Journal before leaving San Carlos for the Casiquiari:—]

Nov. 4, 1853.—This was the feast of San Carlo Borromeo, the patron saint of the church and village.

For many previous nights, and throughout the feast day and night, most of the Indians passed the time in dancing and drinking, and when morning broke on the 5th, not a few were quite helpless. At about 8 A.M. I was called on to visit an Indian called Maestro Conde who was said to be dying. He was the best carpenter in the place, and I had had him engaged for two months in making the carroza or cabin of my canoe, a task he had only just finished when the festivities began.

I found him in his hammock—senseless and speechless—his eyes and mouth firmly closed, breathing stertorously and with scarcely any pulse at the wrists; his face bloated. In this state he had been all night. I had him moved to near the door for more air, and with the help of two men raised him up, and with much difficulty forced his mouth open, and gave him hartshorn and water in small spoonfuls. Following this I gave sweet oil and warm water, but it was very difficult to get it swallowed and to avoid suffocation. We also tickled the throat with a feather to induce vomiting, but he seemed to have no strength to throw anything off his stomach. Cold wet cloths were applied to his head and warm ones to his body as well as hot stones to his feet; and with the help of a Portuguese I managed to cup him behind the shoulder, and after several attempts drew a good deal of blood. Next a fowl was killed and broth made, and given him at intervals, and having spent several hours in this way, without being able to induce vomiting or restore consciousness, I was obliged to retire, but directed the fomentations to be kept up. About 4 P.M. they came to tell me that, after a violent spasm and vomiting a clot of blood, he immediately expired.

Deaths from drink are very frequent at San Carlos, and a short time ago two young men died from this cause. Conde left two sons, fine stout lads of from sixteen to eighteen. Soon after his death I engaged the elder of them one day to cut me some firewood. When I asked him what he would be paid in, he said at once trago; that is, liquor. I asked him if he had so soon forgotten his father's death and the cause of it. "Oh!" said he, laughing, "trago never killed anybody; my father was embrugadi (bewitched)." When I returned from the Casiquiari at the end of February, I learnt that young Conde had fallen a victim to trago—died almost exactly the same way as his father!

[Shortly before leaving San Carlos for the

Orinoco, Spruce wrote two botanical letters—to Mr. Bentham and Sir William Hooker. The former is mainly devoted to a general account of the vegetation of the Upper Rio Negro, interspersed with remarks on the amount and character of his collections, and on various matters bearing upon his past and future explorations, so that I think it will be interesting not only to botanists but to all lovers of natural history in its broader aspects. The other letter is chiefly occupied with an account of a fairly rich locality for Mosses and Hepaticæ, groups which were especially interesting to his correspondent, and which he himself made the chief study of his life. This will, I think, interest all who have any knowledge of these exquisitely beautiful little plants.]

To Mr. George Bentham

SAN CARLOS, *Nov.* 23, 1853.

My collections are very poor, and I am leaving a single case to be dispatched to the Barra by the first opportunity. Even had not my time been so much taken up by hunting up lazy and drunken Indians to their work and seeing they kept at it, I could not have done much in the wet season, when scarcely any trees flower, and there are not ferns here as at Saõ Gabriel, to keep me in work. Besides, the river-side vegetation has hardly any plants not already gathered, either on the Rio Negro or the Lower Uaupés. But among the few plants I have gathered, there are several interesting for their anomalous structure. The other day on the Casiquiari I gathered a tree, allied perhaps to Ochthocosmus (Ternstromiaceæ), but approaching also Humiriaceae, Olacaceæ, and Ebenaceae. I have some others which have something in common with the three orders just named, but do not very clearly belong to any one of them: a new genus of Rhizoboleæ allied to Anthodiscus, but scarcely combinable with it; a fine series of Dimorphandras apparently all undescribed; more nutmegs and Commianthi; and several other things which I doubt not will interest you, if they only reach your hands in safety.

. . . .

I did not look into the flower of *Caraipa paniculata* (murá-piranga), but set it down as a Myrtacea from the habit. On the Alto Rio Negro and Uaupés there are other murá-pirangas, apparently all Rubiaceæ, remarkable for the wood, and especially the bark, turning red when cut. I have by me just now some sticks which I found the other day in the house of an Indian; when the grey cuticle is scraped off these, the inner bark is exposed of the finest crimson. From this bark a brilliant red dye is prepared, far superior to that of the anatto and carajurú. I should like it to be tried in England, though nowadays chemistry has quite revolutionised the art of dyeing. I gathered two species on the Uaupés in fruit—I shall be glad if you find them to belong to Sprucea, though they are perhaps only Amaiouas (Cinchonaceæ).

. . . .

In *Hooker's Journal*, January 1853, there is a letter from D. C. Bolle, in which, speaking of the rainy months in the Cape Verd Isles, he says: "Even within (doors) how could plants be dried where clothes, shoes, furniture, everything is covered with its appropriate mucor?" Well, this and worse may be said of the Rio Negro all the year round, and yet plants can be dried. Were I a fixture here and could build a house such as experience has taught me to be requisite for keeping things dry and sound, I have no doubt I could dry plants here as well as they have been dried in any part of the world—I do not say with the same ease, for the manual labour under any arrangement would be great. Since I left Pará I have not inhabited a house through the roof of which heavy rains did not find their way. At the Barra I was much annoyed by a small, red, virulent-stinging ant which got into my boxes and made its nests among my clothes and dried plants. On passing in review a parcel of the latter I have sometimes found several thicknesses of paper soaked through with formic acid, and some of the plants in such a state that I was obliged to throw them out.

[In a letter written from Tarapoto, three years later, Spruce again refers to this subject as follows:—]

At San Carlos the dampness exceeded what I had experienced at Saõ Gabriel and on the Uaupés. If I were writing and chanced to drop a piece of paper on the ground, if I did not take it up for five minutes it was so moistened as not to bear writing on. Specimens well dried and put away in a box would be covered with mould in a month's time; but if left on the table, a single night sufficed to mould them. Any article of metal or ivory left all night on the table would be wet in the morning.

To Sir William Hooker

SAN CARLOS, *Sept.* 17, 1853.

In the angle between the Rio Negro and Casiquiari I have got some Mosses and Hepaticæ that have interested me much. As my predilection for these tribes is known to many, you may perhaps have been asked whether I was doing anything in them, and if I intended to distribute the species. I have hitherto avoided alluding to Mosses in my communications to you because the number was so few that I had no idea of their ever summing up to a quantity worth the trouble of distribution.

On the Alto Rio Negro I have been more successful, and I now think that some day or other I may make up sets of those Mosses and Hepaticæ which I have gathered in sufficient quantity. Of Mosses the number of species is still small, considering the space of ground passed over, and how sharply I have looked for them during four years of travel. I suppose that in all this time I have not gathered more Mosses than I could have gathered in a month in the space of fifty miles diameter in any part of Europe. Yet all are interesting and a good many will be new. The general character of the cryptogamic vegetation on the Amazon and Rio Negro seems to be quite that of Demerara and Surinam, and to bear little resemblance to that of the rest of Brazil. The Mosses are mostly pleurocarpous, and comprise a great number of minute Hypnums and a good many Hookerias. A pretty species of the latter genus, frequent on logs in the moist forest near San Carlos, seems to be the *Hookeria pallescens* which you described in *Musei Exotici* from specimens gathered by Humboldt at Esmeralda. I shall endeavour to look up all Humboldt's species from this region. Among acrocarpous Mosses the commonest and perhaps the most beautiful is *Octoblepharum albidum*, which grows everywhere on trees, both in wet and dry situations. *O. cylindricum* is much less frequent, and I have mostly seen it on palm trunks. I expect I have one or two new species of this genus! There are a good many minute Fissidens whose habitat is chiefly on termites' nests on the ground or in trees. The genera Macromitrium, Syrrhopodon, and Calymperes have all representatives, but they are far from being so abundant as I expected to find them. On the other hand, I have met with species of some genera considered peculiar to cooler climates, as, for instance, an Anacalypta at Santarem and a Phascum at Saõ Gabriel. On the Rio Negro a very common and a very handsome moss is *Leucobryum* (*Dicranum*) *Martianum*; it grows on wet logs, and has the additional merit of fruiting copiously. I have been somewhat disappointed that since I set foot in South America, now more than four years ago, I have not once seen *Funaria hygrometrica*—the moss which, as some one has said, more poetically than truly, "springs up wherever the wild Indian has lighted his fire." I have seen hundreds of places in Amazonian forests where Indians, wild and tame, have lighted fires, and the plants which spring up in such places are not mosses. I shall some day be able to tell you what they mostly are. There is a moss which seems partial to charred trunks; it resembles *Hypnum tamariscinum* in miniature, and I take it to be *H. involvens.* *Ceratodon purpureus* is an almost constant companion of *Funaria hygrometrica* in Europe, and has, like it, the reputation of being cosmopolite, but I have never seen it here.

The Hepaticæ have been everywhere much more numerous than the Mosses, and will, I hope, comprise much that is new. The great mass belong to the genus Lejeunia, but there are several species of Omphalanthus, Phragmicoma, Mastigobryum, Plagiochila, Aneura, etc. One of the commonest Hepaticæ on the Rio Negro is a Sphagnoecetis, quite like our *Jungermannia Sphagni* in aspect, but smaller, and fruiting abundantly towards the end of the rainy season. I have a good many new species allied to common European forms, as, for instance, to *J. bicuspidata* and *trichophylla*; and a series of several species, apparently all undescribed, intermediate between foliose and frondose Hepaticæ.

Very few Mosses grow on the inundated margins of the large rivers, and they are species that recur everywhere. It is necessary to plunge into the heart of the forest and to seek out rocky rivulets and the trunks of fallen trees which lie in or near them. Hence when I ascended the Rio Negro in November 1851, when the river was low, although there were abundance of trees in flower, the Mosses on the banks were so much dried up as to appear almost non-existent. The contrary was the case when I came from the Rio Uaupés to San Carlos in March last, when the rivers were rising and the rains were frequent and violent. The trunks of the inundated trees were in many cases clad with a green coating of Mosses and Hepaticæ, but the trees themselves were almost without exception destitute of flowers.

I shall do my best to explore the mountains at the back of Esmeralda, but I do not expect much from them. The great peculiarity of the mountains I have hitherto visited is that they are hills without valleys—lumps of granite sticking up out of the plain. They seem all destitute of water, and this is probably the reason why they are quite uninhabited, there not being, so far as I can learn, so much as an Indian's hut on all the mountains of the Rio Negro and Alto Orinoco.

I am glad to find that my specimens, both for the herbarium

and museum, have given you satisfaction. It is the certainty that my friends in Europe will appreciate my labours that enables me to bear up under the hardships of travel in this region. I have no doubt that a stronger man than I might do more, but even the strongest must be content to lose a great deal of time among a people so lethargic as this, as Mr. Wallace can better inform you. As to my health, about which you so kindly inquire, it is much what it was in England, easily disordered, but (with care) rarely seriously affected. I suppose I am so thoroughly acclimated to the tropics that I shall take ill to a cold climate again.

CHAPTER XII

IN HUMBOLDT'S COUNTRY: VOYAGE UP THE CASIQUIARI TO ESMERALDA ON THE ORINOCO, AND UP THE RIVERS CUNUCUNUMA AND PACIMONI

(*Nov.* 27, 1853, *to Feb.* 28, 1854)

INTRODUCTORY NOTE BY THE EDITOR

[THE Journal of this expedition is unusually full, and Spruce himself had always looked forward to it as one of the most interesting portions of his travels. In the first place, it traversed a large extent of ground visited by the early botanical travellers, Humboldt and Bonpland, and partially by Schomburgk; and in the second, because Spruce ascended, as far as conditions permitted, two rivers never before explored by a European traveller, and became acquainted with some little known and interesting tribes of Indians. I have therefore felt bound to present this Journal to the public almost in its entirety, only omitting such ordinary details of travel as are of no special significance, while I retain all that may serve to illustrate the difficulties and dangers of travel in this little known region, which is, I believe, to this day in almost exactly the same condition as that in which he found it. I have inserted in its proper place a vivid description of Esmeralda (given by Spruce in a letter to his friend, Mr. John Teasdale) which supplements the more technical description in the Journal, and I have distinguished such portions of the latter as are almost wholly botanical by printing them in smaller type. At the conclusion of the Journal itself I give a rather lengthy letter to Sir W. Hooker containing a connected and very readable account of the whole of this expedition, which is useful as explaining why the traveller was unable to carry out his whole programme. The specially botanical portion of this letter is also printed in smaller type. With the exception of these botanical passages, the whole of this chapter will, I think, be found generally interesting.]

Journal

On Nov. 27, 1853 (Sunday), I embarked for the Casiquiari. After infinite delays from drunken, unwilling Indians and other hindrances, I saw myself fairly under weigh about 10 A.M. At 4 P.M. I reached the raudal (rapid) at the mouth of the Guainia. The fury of this was much abated since the high-water of the river, yet it is still difficult to surmount for anything larger than a curiara. We crossed to the west bank, my pilot being of the opinion that it was more easily passed on that side. After two hours' toil my cable, which was a new one of piassaba, four inches diameter, broke just at the time when the boat was in the middle of the fall. She whirled round three or four times, and barely escaped being dashed in pieces against a projecting point of rock. As it was, a hole was opened in the keel through which we could distinctly hear the water hissing, though there was not light enough left to find out where it was. We made fast at the river-side, and the men were kept all night baling out water. I did not venture to sleep a moment, and roused the men in turn to their necessary task. In the morning we found out the leak and stopped it with clay, and when we reached Solano, on the 29th, we caulked it roughly without pitch, and the mud suspended in the waters of the Casiquiari soon made it completely water-tight.

On examining the broken cable I found that it had been previously cut more than half through with some sharp instrument, otherwise a new cable of that size could not have been thus broken. It was not until after my return from this voyage that the Indians let out that this had been the work of my pilot Carlos, a merry, lazy scamp, who had calculated on nothing less than the destruction of my boat on the rocks, which would have saved him the toil of a voyage for which he had already received pay. He and his companions would have easily saved themselves by swimming when the boat foundered, for they think nothing at any time of plunging into furious rapids.[1]

I took with me on this voyage some large

[1] On my next voyage towards the cataracts of the Orinoco, Carlos deserted me, taking with him an easy, quiet lad named Antonio, who had long been my personal attendant both on land and water. This was the only instance of an Indian running away from me during the whole time of my stay in South America, and I could not be surprised at it, for the Upper Rio Negro—one of the hungriest regions of the world at the best of times—was then in a state approaching positive famine. We were waiting at the village of Tomo trying to get provisions to enable us to push on to the Atabapo and the Orinoco, but could with difficulty procure a daily subsistence. When I again reached San Carlos, Carlos and Antonio came to me very penitently, and each one accused the other of having induced him to desert me; but they both honestly repaid me the articles I had advanced them for the voyage, to the great astonishment of the white residents, who said such a thing was never known of an Indian.

Portuguese frascos (square bottles of dark thick glass) for preserving succulent fruits in spirits; and for the same purpose I had had made a large demijohn (containing about six gallons, equal to twelve frascos) of resecado, that is, double-distilled cane-spirit. I took, besides, two narrow-mouthed frascos of the same spirit for drinking. During our first sleepless night, when the leak kept us alert, I dispensed of this spirit liberally to the men. The pilot was especially thirsty, and got so large a share that he became tolerably "well drunk." On the two following evenings he was clamorous for more, and two or three glasses made him fractious and impertinent. I saw then that the possession of this liquor would probably be a daily source of disquiet to me, and that even if I put poisonous fruits into it, it might be impossible to prevent the Indians from drinking of it. So, on the fourth night, I arose at midnight, took out the demijohn, and as quietly as possible poured its contents into the river. The men were sleeping in the prow, but when they woke up in the morning I overheard one of them, as I lay in the cabin, say to his neighbour, "What could it be the patron was pouring out of the demijana in the night-time? Did you not hear it, pop, pop, pop, pop, po? Surely it was not bureche!" The other thought it must have been oil that had become rancid, for I had with me two small demijohns of turtle oil for frying fish, and for my lamp. That there might remain no doubt on the subject, when we halted for breakfast I placed the demijohn on the cabin roof to drip. One and another approached it stealthily and smelt at it, and I could gather from their whispers the horror they felt at my having wasted a substance so precious. It relieved me, however, from all further importunity for bureche.

I retained only one frasco of spirit, into which I put two ounces of powdered rhubarb, hoping thereby to render it distasteful, as well as to increase its medicinal virtues. When my hunter—my best man—was taken ill with chills and fever, I gave him a strong dose of it, and it set him to rights. "That medicine of yours, patron," said he, "is bitter, bitter, but it's very, very good." I feared he would want it all, but he was put off by my assurance that strong medicines could be taken only in small quantities, or they became sure poisons; and I, of course, repudiated the notion that there could be spirit in it.

On the broad bed of granite, of which a wide extent is now dry, I had an opportunity of witnessing the exits of swarms of bats from beneath certain large flat slabs which lay upon the rock. Just after sunset they issued out in a continuous stream, which lasted two or three minutes. From beneath a single large stone not less than two or three hundred must have issued out. But on the evening of the 30th I witnessed the same phenomenon on a much grander scale when anchored near the rock of Guanári. I had just turned out of my cabin, after eating my evening meal, when my ears were saluted by a deep roaring sound in the forest in the direction of the rock (which, though not more than two hundred paces off, was hidden from view by intervening trees), quite like that of a coming thunderstorm. I ordered the Indians to gather up some linen which was drying on the top of the cabin, but they laughed and told me it was not rain but bats that were coming, and pointed out a long streak of black cloud extending across the river and far over the forest on the opposite side. I could not at first persuade myself it was a living mass until, by looking attentively, I distinctly perceived the movements of the little animals who thus sallied forth in an army to chase the nocturnal insects constituting their food. I am almost afraid to estimate their numbers, possibly they were not under a million!

When I have occasionally in the daytime sat down by one of these flat, incumbent rocks, my senses have been saluted by a warm and by no means odoriferous blast from beneath it, and if the ear be applied to the edge of the rock an unceasing whispering and fluttering noise is heard. I have seen children poke bats out from beneath these stones with long sticks.

My crew consisted of nine men: the pilot, seven oarsmen, and a little boy.

On the 29th, at 8 A.M., we reached Solano, a rather smaller pueblo than San Carlos, and the only ancient settlement on the Casiquiari. Here is an old man named Silvestre Caya Meno who recollects the Jesuits, and must have seen Humboldt. He is quite deaf, but his wife, who is of about the same age, has perfect use of her faculties. Both speak Spanish much better than any Indians of the present generation. Don Diego Pina, who resides at Solano as governor, supposes them to be not less than a hundred years old.

In the afternoon of the 30th we reached the rock of Guanári, which is a Cocuí on a smaller scale. I suppose it is less than 300 feet above the river. It consists of one large abrupt mass and three or four small broken ones to the right, of which two erect ones side by side, and each broken across above the middle, are called "varong hembra." There are also, as at Saõ Gabriel, many large blocks strewn about the base, under which nestle the hordes of bats. Amongst these rocks climb Arums in such quantities that it is scarcely possible to thread

FIG. 33.—ROCA DE GUANÁRI, CASIQUIARI, AS IT APPEARS FROM SOME MILES LOWER DOWN. (R. S.)

through their pendulous roots. There are a great many Paxiúba palms, but I found nothing new.

I stayed here till noon of December 1 to make a stage (trocha) for the rowers. At 5 P.M. we reached Buena Vista, a small place of some six houses.

On December 2 we reached Santa Cruz about sunset. This pueblo is nearly as large as Solano. There are a good many people in it, but neither fish nor fowl to be bought.

On December 3 we reached Quirabuena after sunset. There are about eight houses and a church. The soil is a red loam as at Marabitanas, the port of Tomo, and partly at Saõ Gabriel. There is much lofty forest quite near, and I saw several Seringa trees. There were great numbers of Piassaba palms, with dead panicles looking like those of Jará. There were also beds of a pretty Lepidocaryum (a palm) with vermilion flowers.

We stayed here all Sunday to kill and salt a pig. My men caught a large fish, a tambaki (murucúto, Venez.), in the port. It is the first I have seen since leaving Barra, and seems to be quite unknown in black waters—even at San Carlos it is never taken. In the Orinoco, as in the Amazon, it is abundant.

Dec. 5.—On this day we passed the mouth of the Siapa a little past noon. . . . There is a raudal on the opposite side, and a little way farther up is another raudal at an angle where the river is much contracted; this extends across the river, and was passed with some risk and trouble.

. . . . . .

The Siapa enters by a single narrow mouth (perhaps not 150 yards wide), yet it is a much larger river than the Pacimoni. The water is whitish, and the water of the Casiquiari was whiter towards its mouth than below Quirabuena.

Dec. 6.—Passed another raudal this day, and also two points where there are raudals in the height of the dry season. Above this there is a marked change in the margin of the river, which is, besides, considerably narrower. The land is low and inundated, often with beds of Jará-assu palm, and with small lakes opening out of it in places. . . . I saw within the gapó a Sassafras tree, about 4 feet in diameter, and certainly over 100 feet high. It had been tapped last year by cutting out a wedge with an axe reaching to the very heart, where there was a hollow as large as an arm. A small quantity of gum was coagulated within the wound.

At nightfall there were a good many birds crying

in the forest, especially socos (herons) and curucurús. . . .

Dec. 7.—About 4 P.M. we came up to a place where several blocks stood out of the river, some with trees on; and on the left bank, a short distance within the forest, rose a black rock to a little above the tree tops; it is called Cerro de Canumata. There are high banks, and terra firme again.

Dec. 8.—This morning before sunrise the whole air between the forest on each side of the river was filled, as with snow-flakes, by a white-winged insect allied to the mayfly. As the sun rose the mass gradually descended till it reached to within three or four yards of the river, numbers of insects falling exhausted into the water until by 9 o'clock not one was to be seen. We rounded an inundated point this morning which recalled some parts of the Amazon. The surface was clad with a low Inga, over which trailed Convolvulæ and other twiners so as to form an impervious mass, and out of it stood several slender Cecropias, 15 to 30 feet high, with smallish, not deeply-lobed leaves. For the last two days my men have taken two lablab (fish) each day—one being eaten by us fresh, and the other salted.

Dec. 9.—Humboldt says the Casiquiari, as far as to the mouth of the Vasiva, is from 250 to 280 toises wide (*i.e.* 530 to 600 yards), and therefore as wide as the Rio Negro at San Carlos. . . . We reached the entrance of the Lago de Vasiva at 2 P.M. The mouth is perhaps 150 yards wide, in direction continuous with Casiquiari. Some way in it narrows to 50, or even 30, yards. The forest is low, quite like that of the Guainia. Two hours of slowish movement brought us to the lake, of which I took rough bearings. There were great beds of Balsa-wood on the sandy beaches left dry in the summer. The forest is low and contains much novelty, but hardly anything in flower, for it was still *winter*. I found several new Melastomaceæ, two small-leaved Swartzias, etc. The water is black, its junction with the yellow water of the Casiquiari is very conspicuous, but the current is barely perceptible. On the farther side of the lake the river is continued in a broad shallow channel. I am told it runs a long way up, and that its course is nearly parallel to that of the Casiquiari. From near its head-waters Duida is very distinctly visible. Much turtle and cabezon are taken in this lake when the water is low.

Dec. 11.—At 3 P.M. we reached the Pueblo de Ponciáno on the left bank. Its founder Ponciáno was brought up at Solano by Padre Juan, and left it some thirty or forty years ago to establish himself in this place, where he died about two years since. On the same spot there had previously been an Indian settlement, and it is called Yamádu-báni, *i.e.* the land of the Yamádu, a fabulous animal resembling a man in size and aspect, but with long skinny legs and arms, which now and then shows itself in the forest to the terror of women and children. I have found a belief in the existence of this sort of wolf-man current among the Indians throughout the Amazon and Rio Negro. . . .

Ponciáno led with him several of his countrymen (Pacimonares), and they seem to have multiplied more than customary among Indians. The six or eight houses appeared each to have several families in it, and a great many children were running about. His widow is living, and she also had been brought up by the Padre Juan, in proof of which she still knows all the prayers of the Church, and speaks a pure Castilian which contrasts strongly with the imperfect speech of the modern Indians. She recollects two travellers coming down the Casiquiari when she was a very little girl—the one a German and the other a Frenchman (Humboldt and Bonpland)—who occupied themselves with gathering flowers by day and gazing at the stars by night. She did not herself see them, being absent in the cunúco, but Ponciáno did, and used frequently to speak of them.

There is much Piassaba at the back of the pueblo, but the trade of the inhabitants is chiefly in timber and turtles. Passed the Caño Itiniuini, along which there is a passage from Guainia to Casiquiari at high-water.

Dec. 13.—Early this morning we reached the site of the deserted Pueblo de Capibara, where a bed of grass sloped down into the water. About half a mile back in the forest are large, flat, naked beds of granite interspersed with low caatinga. There is much picture-writing here, of which I copied the principal figures; they are usually very perfect, but in some places are obliterated by the shaling of the rock.

Just above Capibara stand two rocks out of the river at a distance of 3 or 4 feet, which have obviously been riven apart. At this time they stood some 15 feet out of water. (See next page.)

Dec. 15.— . . . This afternoon we reached an angle whence we got an actual, though dim, view of the Cerro Duida. . . . The same day was rendered memorable by having just before found two new nutmegs.

Dec. 16.—We had a dim view of Duida this morning. Mosquitoes were terrible to-day, especially at 4.30 P.M., when we stopped to cook supper. . . . Looking into the cabin afterwards, it was like a beehive.

FIG. 34.—A RIVEN ROCK IN THE BED OF THE CASIQUIARI. (R. S.)

Dec. 17.—Early this morning we reached the pueblo of Monagas, called Camaciáno, from a raudal just above. I here met a Guaharibo, caught by Monagas about thirty years ago, and as at that time he was a young man apparently of twenty, he must now be near fifty. He speaks scarcely any Castilian, but through Monagas as interpreter I was able to converse with him. His name in his own land was Kudé-Kubúi, but he has been baptized José Miguel. In personal appearance he is low of stature (five feet), pot-bellied and knock-kneed (peculiarities of the vegetarian Macús), fair-skinned, and with light hazel eyes. His hair was black with a very slight tendency to curl over the forehead, where it had been left longer than on the rest of the head, in conformity to the custom of the

Rio Negro. He seemed very good-natured, but much less intelligent than the Barrés, etc., and when those around me laughed at the words of his language (of which I wrote down as many as I could get from him), he laughed more heartily than any.

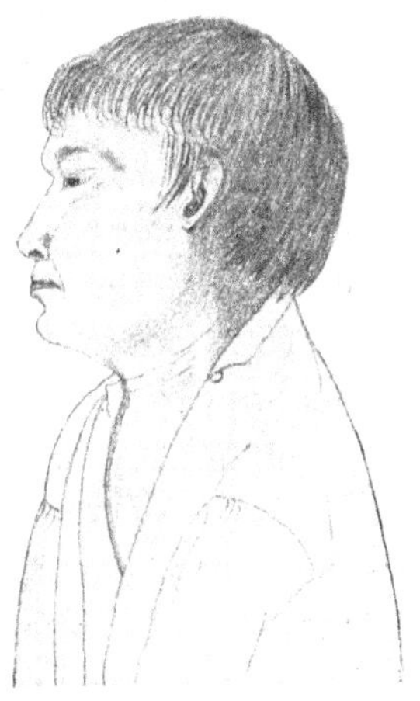

FIG. 35.—KUDÉ-KUBÚI, a Guaharibo Indian (50 years old).

Monagas, with six others, were gathering nuts of Juvia on a river which seems to be the Manaviche, and had gone very far up when they came upon a cleared space in the forest which constituted a pueblo of the Guaharibos. The houses were annular, the low roof sloping slightly outwards and being only two or three varas in width, while the whole of the centre was open to the sky. The roof and outer wall were made of the long, broad, simple leaf of a palm, apparently like the Bussú of Pará. Under the roof were slung the hammocks of several families. Several broad, clean paths led from the houses into the forest. In one house were two young men with three young women. One of the men fled, but Monagas and his companions captured the rest. After binding the captives, they were attacked by a party of returning Guaharibos, but escaped in the dense forest after killing one of them, and got safely back to their boats. . . . All three women died a few years afterwards of scarlatina.

According to Kudé-Kubúi, there are several

pueblos of his nation all the way up the Orinoco to its head-waters. He has never been to the latter, but knows that they lie on one side of a serrania, and that when one crosses these the Rio Branco is reached in about a day. Above the Raudal de Guaharibos there are mountains higher than Duida. All the way up there are many mosquitoes and zancudos. Little could I learn of their customs. Only one wife was allowed to each man. They burn the bodies of their dead, collect the calcined bones, and pound them in a mortar, and keep them in their houses in globular baskets of closely-woven mamuri. When they move their residence or travel, they carry with them the bones of their ancestors. Monagas found several of these mapires (baskets) in the house he entered.

When Monagas revisited the same place two or three years afterwards with several companions, hoping to catch more Guaharibos, the pueblo had disappeared and the roads were grown up.

When Schomburgk descended the Casiquiari, Monagas was residing in Quirabuena and had the Guaharibo with him, but he says the traveller did not land there.

Dec. 18.—We left Monagas a little before noon. In about an hour and a half we passed the mouth of the Caño de Dorotomuni. The Indians assured me there is no lake in this caño.

At the Caño de Dorotomuni nearly all the Rio Negro plants have disappeared. *Campsiandra laurifolia* and *Outea acaciæfolia* hold their way throughout and appear also on the shores of the Orinoco. There are also two or three twining Phaseoleæ which I cannot distinguish from the species gathered on the Rio Negro, and which occur here and throughout the Casiquiari. *Swartzia argentea* is as frequent as on the Rio Negro until somewhere about the mouth of the Vasiva, where I finally lost sight of it. . . .

Japura—*Erisma japura*, Spruce (Vochyaceæ)—scarcely occurs in the forests of the Casiquiari, and Uacu seems to disappear a little way up, but Cumirí is found throughout in terra firme. Chiquichiqui (the Piassaba of Brazil) is exceedingly frequent throughout. At the back of Ponciáno and Monagas are noble groves of it mixed with arbusculæ and shrubs, but with scarcely any lofty trees, and the effect is exceedingly novel and striking.

At one day's journey above Dorotomuni the shores put on quite an Amazonian look, being in some places sloping, sandy, and clad with tall rank grasses (chiefly a Panicum with the habit of *Paspalum pyramidale*) mixed with *Mimosa asperata* and a couple of weedy Ipomœas. Amidst this mass rise the slender soft stems of a Polygonea to the height of 30 feet, and on the water's edge I saw a few plants of a real Polygonum resembling one from the Solimoẽs. An Inga with broadly-winged petioles occurs in long continuous beds, and the species, so frequent lower down, almost disappears (but reappears on the Orinoco). There are great quantities of a bushy, narrow-leaved laurel, apparently one of the white-flowered species, such as are frequent on low shores of both black and white waters. A narrow-leaved, cedar-like Xylopia (Anonaceæ), very frequent on the Upper Uaupés, Rio Negro, and on the Uaupés, and from its singular habit very conspicuous and ornamental, scarcely passes the mouth of the Pacimoni, and the same may be said of *Heterostemon mimos*. On the Upper Casiquiari and Orinoco another Xylopia with similar habit is frequent, but it has fewer, smaller, and less rigid leaves, and the tree is generally loftier. If there is no Heterostemon on the banks, another and more remarkable species (*H. simplicifolia*, Mgf.), gathered also at Saõ Gabriel, is very frequent in the forest all the way up. Nutmegs are tolerably frequent on the banks. The commonest species had just gone out of flower, and was laden with a profusion of rudimentary fruit. My glass showed it to have leaves rounded at the apex as in a species gathered on the Uaupés. A very remarkable species with leaves sometimes nearly 2 feet long had so much the habit and form of leaf of *Vismia macrophylla* that until I came near enough to see whether the leaves were alternate or opposite I could not distinguish them, for the Vismia was also very frequent. I gathered four species which seemed new, and I avoided everything which looked suspiciously like *Myristica sebifera*. I was curious enough to notice, wherever I entered the forest on terra firme (as I did nearly every time we cooked our meals), the families which constituted the vegetation, when I could certainly ascertain them from the leaves and mode of growth, and in every place I saw at least one species of nutmeg, and there were some three or four species unknown to me. Of the four gathered only one was of the terra firme, the rest were of the gapó.

Below Monagas are a good many trees of a stout but low Anacardium, which seems the same in its leaves as *A. giganteum.*

The most curious feature of the Casiquiari is the occurrence throughout its course, though sparingly, of a Crescentia (calabash-tree) in the gapó; the first I have seen wild, but there was no flower or fruit.

This morning, December 18, I came on a small patch of Pontederia with inflated petioles, caught in the immersed branches at the margin of the river. It had evidently come from the Orinoco and was the first of the tribe I had seen since leaving the mouth of the Rio Negro.

Dec. 21.—A little after noon we reached the Caño de Calipo; at a point just above but on the opposite side of the river (the right) are beds of rock with numerous deeply-graved figures, but most of them under water.[1] Between three and four o'clock we entered the Orinoco. The Casiquiari for the last two days had had chiefly steep banks of clay and sand, and the Orinoco has the same. In both are here and there rocky points and sometimes exposed sandbanks. The Casiquiari upwards is much narrowed, but about its mouth it is a little wider. It seems to leave the Orinoco nearly at right angles. There are two Jagua palms in the entrance on the right bank. The Orinoco is about equal in width to the Rio Negro at San Carlos, but above it spreads out to a great width with beaches emerging in various parts, and we had some difficulty in finding a passage for the piragoa. . . .

We supped at a rocky point on the left bank where numerous bamboos and other marks indicated

[1] [These were dry on his return, and some were copied. See end of Chapter.—ED.]

an ancient Indian settlement. (It is now known as the Pueblo viejo, and was formerly inhabited.)

Dec. 23.—To-night we are anchored on a playa (beach) in sight of Esmeralda, but as we could not have reached it by daylight, and there is no travelling here by night, we preferred leaving the short remnant of the voyage for moonlight in the early morning. Duida looks down on us from the left and has seemed close by since entering the Orinoco; nor has our change of position much changed its aspect till late this afternoon, when rounding a point the southern end came in view deeply cloven into four abrupt ridges. At sunset the mountain was very grand, the ridges assuming a purple hue, while the interstices were veiled in impenetrable gloom, and a stratum of white fleecy cloud was floating below the summit. The conformation was much like that of the Serra de Curicuriarí, but less picturesque. My telescope shows that, except in a few places where the rock is very steep (whitish, sometimes streaked with brown), the mountain is forest-clad to its very summit. Yet so clear does it stand out to view, and so much nearer does it seem than it is in reality, that one would affirm its sides to be clothed with fern. Two flat summits to the north of the middle of the mountain seem to be the highest points, judging from the height of the clouds floating over them. The space below these is singularly hollowed out, and is said to be occupied by a laguna. The north extremity is a subconical peak.

Last night we supped on a large playa and to-day have come on several more. Sometimes so large a space is left dry that there is only a narrow channel left on each side of the river, and the water must be still more contracted when the river is at its lowest. . . . The vegetation has quite the same aspect as that of the Solimoẽs, though possibly all the species are different. There is much steep bank with no gapó. Where the shore is sloping and inundated, as on some islands (of which there are several), there are the same two Ingas as on the Casiquiari, but palms are scarcely so numerous.

We reached Esmeralda about 10 A.M. on the 24th.

The village consists of six houses scattered round a square plaza. One is the casa real (guest-house). In the centre is a cross, and there is a taller cross to the northward on the shoulder of the Cerro de Zamurro. (This cross was erected a few years ago to ward off thunderbolts, which have several times done much damage to Esmeralda.) This cerro is a ridge of fantastically piled granite blocks forming a cirque at the back of the village; it extends from S.E. 10 S. to N.W. 5 W., as seen from the cross in the centre; and nearly reaching the river on each side. Its highest point is three or four hundred feet above the pueblo. . . .

The inhabitants of Esmeralda assure me that nearly every summer fire is seen to issue from the summit of Duida, illuminating all the heavens above and emitting a considerable degree of smoke but nothing more. It is not the forest that is burning, for that only occurs on the sides.

In winter large pieces of rock are detached by torrents which are seen foaming down furrows in white lines. They are sometimes accompanied by a thundering noise which alarms the sleepers in their hammocks.

[I here insert Spruce's very vivid description of Esmeralda, both from a picturesque and residential point of view, as given in a letter to his friend Mr. John Teasdale.]

To Mr. John Teasdale

SAN CARLOS, *May* 22, 1854.

On the Orinoco I visited Esmeralda at the foot of the lofty mountain Duida—about 8000 feet high—you will find mention of it in Humboldt's *Narrative* and *Aspects of Nature.* This village, reduced now to six miserable huts, stands on the most magnificent site I have seen in South America. Between the Cerro Duida on the west and the mountains of the Guapo and Padamo on the east extend wide grassy savannas in which almost the only trees are scattered, fan-palms (Moriches). On the side next the Orinoco a semicircular ridge of fantastically-piled granite blocks, in whose crevices grow a few scattered shrubs, cut off a small savanna on which stands Esmeralda. All up and down the Orinoco, and on the margins of the savanna, rise hills of granite and schist, some nearly naked, others forest-clad, and at the back (to the N.W.) rises the abrupt and frowning mass of Duida. If you can fancy all this seen by a setting sun—the deep ravines that furrow Duida on the east buried in nocturnal gloom, while the salient edges glitter like silver (the rock is chiefly micaceous schist)—you will realise in some degree a scene which has few equals. Looking up the savannas to the northward, from the top of the before-mentioned granite ridge, reminded me of a view in the drive from Killarney to Kenmare, where on reaching the summit of the Pass of Cairn-a-Dhur one looks down on the valley where are 30,000 acres of as fine bog as any in Ireland. But Duida is 8000 feet above the sea, while Macgillicuddy's Reeks are only some 3000.

You will credit me when I say that to the sight Esmeralda is a Paradise—in reality it is an Inferno, scarcely habitable by man. When I stood in the middle of the small square, round which are built the houses at Esmeralda—the straw doors all carefully closed and looking as if nothing human ever came forth from them—the warm east wind fanning my face and raising the sand in the plaza, but bringing no sound of life on its wings—no bird or even a butterfly to be seen—amid the luxuriance of vegetable life, animal life almost extinct—I thought the scene inexpressibly mournful. But the utter absence of living things was only apparent, not real. If I passed my hand across my face I brought it away covered with blood and with the crushed bodies of gorged mosquitoes. In this you have a key to explain the unearthly silence. The apparently tenantless houses had all inhabitants in them who, bat-like, drowse away the day, and only steal forth in the grey of morn and evening to seek a scanty subsistence. Throughout the day the very air may be said to be alive with mosquitoes, from which even with closed doors one can only imperfectly escape. I constantly returned from my walks with my hands, feet, neck, and face covered with blood, and I found I could nowhere escape

FIG. 36.—CERRO DUIDA (8000 feet), FROM THE CROSS NEAR THE VILLAGE OF ESMERALDA. Looking north. (R. Spruce, Dec. 1853.)

these pests. If I climbed the cerros, or buried myself in the forest, or sought the centre of the savannas, it was the same, but it was worst of all on the river. At San Carlos they were often bad enough; in ascending the Casiquiari every day brings an increase of mosquitoes, until, towards the upper mouth and out on the Orinoco, they are an indescribable annoyance. Many times there is no sitting down to eat a meal, but one must walk about, platter in hand, and be content to eat one's food well peppered with mosquitoes. I found working at my plants very difficult, although I put on gloves and tied down my trousers over my ankles. The face and neck were necessarily exposed, and my gloves and sleeves were constantly streaked with blood from brushing away the little insects. Most of these minute flies leave a small clot of blood in the place where they have been sucking, and with me the wounds often bled considerably.

[The Journal now continues :—]

Duida as seen from Esmeralda seems a cubical mass, one face parallel to the Orinoco and another to the Guapo. . . . Throughout it is forest-clad (in the ravines to the very summit) save where the rock is nearly perpendicular. The south-east angle seems to be micaceous schist and glitters like silver when the sun shines on it. Most of the rock about Esmeralda is schistose, and where the stones are placed with the lamina perpendicular they are worn by the action of rains and the atmosphere into close-set sharp edges which treat the naked feet very cruelly.

The inhabitants of Esmeralda are a quite different race from those met by Humboldt. When it had become depopulated, a few Indians from the caño above San Miguel (Uarikena) came and settled in it. . . . When by death or migration the population had again become reduced to an old woman with her daughters, grand-daughters, and her nephew, several Manaca Indians came and married the women. There seem to be now eight or ten families of mixed Manacas and Uariquenas. The old woman speaks excellent Castilian. The men all speak Castilian imperfectly, but nearly all know something of Lingoa Geral; this is accounted for by the captain of the Manacas being a Brazilian (an escaped murderer from Barra), and also by the Manaca Indians trading with Brazilian merchants in salsa, passing from their river (the Manaca) by the Castaño and Marárí to the Padauirí.

. . . . . .

[Leaving Esmeralda on Dec. 28, Spruce descended the Orinoco to the mouth of the Cunucunúma river, which enters the former from the north about as far below the mouth of the Casiquiari as Esmeralda is above it. It is a rather shallow black-water river, somewhat smaller than the Casiquiari, but full of small rapids, several of which can be ascended when the river is full; while the river has its source among the lofty Marayuaca mountains at the back of Duida. On the Ist of January 1854 he passed the first fall, a ledge of rock extending quite across the river, and on the second had an uninterrupted passage till the evening. The Journal tells the rest of the story why he was unable to prosecute the ascent of this unknown and very promising river as he had intended.]

Jan. 2. . . . In the evening when cooking our supper there came down to us a curiara with seven Maquiritares whom their chief (Ramon Tussarí) had sent to assist in passing the second fall (Uarinăma). I had met him at Esmeralda, coming down with turtle from the Guapo, and he had promised to send me assistance by the Sunday, when we calculated we might reach the falls. Some of the men were tall; all remarkably fair (light red-brown) and long-nosed, but not so good-looking as the Uaupés. Only one, a brother-in-law of Tussarí, wore shirt and trousers.[1]

The rest had a large tanga (apron) of a rectangular piece of cotton-cloth with tassels at the corners, tucked in under a string encircling the loins, and at the back passing up over one shoulder or allowed to hang down. They buy this of the Piaroas, by whom it is manufactured. They have garters of many convolutions of their own twisted hair below the knee. The arm is much compressed below the shoulder by a ligature like the garter of the Uaupés. They have a thick mass of beads (mostly blue) round the neck, and waist-bands of white beads.

They were very noisy, and very curious in examining everything about the piragoa.

This morning at 8 we reached the second fall —a long rapid where the river spreads out wide and shallow and runs over a bed of rounded pebbles rarely larger than one's head. We struggled for two hours to find a passage, but the piragoa

[1] Only this one spoke a little Spanish; he was a tall, well-made man named Miguel. He was at Saõ Joaquim on the Rio Branco when Schomburgk left on his expedition to Esmeralda, and was engaged by him as a guide. He continued with Schomburgk for three months.

grounded and several times ran great risk of being swamped, nor did all our force suffice to drag it above half-way up the rapid. The river had dried much since we had entered it, and indeed since leaving the Pueblo de Monagas the drying of the rivers may be said to have gone on very rapidly. With sorrowful heart I gave the word to return, and we again took up a position at the base of the fall where a small caño enters on the left. Hastily gathering together a few trifles for the Maquiritares, I embarked in my curiara (small canoe) with five men and set off up the river to visit Tussarí. It was past 10 A.M. when we started, and it was near 5 P.M. when we reached the pueblo at the base of the third fall (Tauarupána). This fall is very difficult to pass, as the river is full of rocks among which the water tumbles about.[1]

The pueblo was established only two years ago; previously its site was much higher up the river, and Tussarí moved down on account of the dangerous navigation.[2]

In going from the second to the third fall we

[1] The raudales of Rio Cunucunúma, in ascending, are: (1) Casurúbi; (2) Uarináma; (3) Tauarupána; (4) Curiripána; (5) Urukarutfóri (the Raudal de Puerco); (6) Mapáku; (7) Matfipiríma; (8) Paikitú-púpe (Cabiza de Peceri), San Francisco; (9) Mauari-púpe (Cabiza de Culebra), San José; (10) Amékui; (11) Uamupatari, in front of Mount Marayuaca. Of all these raudales the eighth is the highest. The sources of the Cunucunúma are at the foot of the Cerro de Kuinéna. The sources of the Guapo are at the foot of the Cerro de Marauáka. The sources of the Padamo are at the foot of the Cerro de Arapami. According to Tussarí these mountains are nearly equal in height.

[2] The next pueblo above Sta. Ramona is San Francisco, which contains four houses in the style of the whites and one round house. It stands in the middle of a small savanna equal in size to that in which Esmeralda is situated. Directly to the north of the pueblo and apparently very near, but at more than half a day's journey, stretches a lofty wall of rock over which there are four or five waterfalls in winter, and two all the year round. A portion of the cerro above the wall seems to be woody, but little of it is visible from San Francisco. The wall is nearly bare of forest, there being only a little here and

passed in one place a lane which some squall had opened through the trees, extending westwards to the limits of vision; its breadth was perhaps 40 yards. The trees were rarely uprooted, but were broken off at about 15 feet from the ground as if some giant hand had passed over them.

. . . . . .

There were but two houses completed in the usual style of the Rio Negro and Orinoco; one is Tussarí's, the other the casa real. They are very neat—whitened outside and inside and painted in original devices by Tussarí's own hand—the colours being red and black. Inside I noticed some figures of men wearing coats, and some on horseback. I was more interested in the other houses (two or three) in the ancient style of the Maquiritares—from a circular base they are sub-hemispherical and tapering to the apex almost in the form of a Turkish minaret. They all consist of the broad fronds of Bussú palm fastened on

there on protuberances. Like other mountains, this shows many white patches (mica).

There is another pueblo — San José, above San Francisco, and there is only one raudal between them. There is also a passage by land from one to the other.

From San Francisco to the Ventuarí by land takes four days.

This information about the Upper Cunucunúma I derived from a Portuguese, Senhor José do Eirado, who went there in June 1854. During his stay with Tussarí there came a number of Maquiritares from Padamo, their captain and six others by water and fourteen by land. Their object was to make a Dabocurí for Tussarí, and they brought presents, ollas, guapas, and roasted frogs and grubs.

The articles which the Maquiritares trade with the whites are curiaras (cascos, of which they furnish the largest and best that appear on the Rio Negro), guapa, curari, and gravatánas, manioc, Aceite de Cuparba, Seringa (only lately begun), caraná, and taçamahaca, the two last only when particularly commissioned beforehand.

Senhor Eirado found the water of the Cunucunúma with a decidedly whitish tinge in the winter. Mosquitoes were few, as also on the Orinoco and at Esmeralda.

rafters meeting in the apex and fastened on a central pillar. One of these was 24 feet in diameter by 15 feet high.

Tussarí's house consisted of two large rooms and two smaller. I slung my hammock in one of the former. The utensils were similar to those of other Indian houses, with the addition of low stools cut out of a single piece of wood, rudely imitating an armadillo, but much clumsier and heavier than the stools of the Uaupés.

Fig. 37.—Ramon Tussarí, Chief of the Maquiritari Indians on the River Cunucunúma (Orinoco) (about 50 years old).

On the large trochas (stages or shelves) there were also evidences of the industry of those Indians in several mapires of mandiocca, masses of circular shallow baskets, and a sort of reticule much used in this region for carrying tinder-box, tobacco, and other indispensables. Suspended from the roof were quantities of camazas and tapáros; also a few gravatánas (blowing-tubes), paxiúba outside, bamboo inside, the latter brought from the head of the Guapo about the base of Marayuaca.

Tussarí is a remarkable man, and his wife is, for an Indian, a still more remarkable woman. She and her daughters manufacture mandiocca, guapos, etc., and she understands the selling of them quite as well as Tussarí, who makes no bargain without consulting her, and takes her with him to San Fernando and elsewhere when he goes trading.

The only branch of industry in which Tussarí employs himself is in the hollowing of cascos (canoes); those made by him have a great reputation. The wood is a heavy laurel, probably Paraturí.

He has travelled about much. Many years ago he, with his family, his two sisters and their husbands and families, went as far as Fortaliza do Saõ Joaquim on the Rio Branco to trade, crossing from the headwaters of the Cunucunúma to Padamo and thence to Parime. From the Cunucunúma to Padamo takes five days, and the path is rugged, passing over much high ground. From Padamo to Parime took three days. Instead of returning they settled down there, cleared a cunúco, built a house, and even became possessed of a few cattle. Here they traded much with the Macusis for articles the latter had bought of the English at Demerara. A few years passed over and the chief of the Maquiritares died in Cunucunúma. The Comisario of San Fernando sent for Tussarí to take his place, and the latter returned to the land of his fathers. A little daughter born shortly after his return seems to be some six years old. He was at Saõ Joaquim when Schomburgk passed that way.

FIG. 38.—MAQUIRITARI GIRL (14 years old).

The only garment worn by the women is a guayuco (small apron) of beads woven into a tasteful pattern on cotton threads. The beads are mostly red and white or black and white, and the minutest are preferred. From the quantity of these used it must be to them a costly article and weeks must be spent over its manufacture. (I saw one in process of formation, the frame on which it was woven being merely a small stick bent into a bow by a string attached to its ends.) I doubt, therefore, that the guayuco is not worn *en famille*, especially as the first woman we met was quite naked, and I presume it to be only put on at some feast or when some stranger visits them.

[The accompanying engraving from a French explorer shows a group of Maquiritari Indians from the Orinoco, above Esmeralda; and it will be seen that they agree very closely in costume and ornaments with those here described by Spruce. This traveller nearly reached the sources of the Orinoco, as noted at the end of this chapter.—ED.]

I stayed two nights with Tussarí, and bought of him a large quantity of mandiocca, guapos, etc. On the second night he invited all his people to drink jaraki and exhibit the native dances to the white man. Men came with bodies smeared all over with anatto.[1] Necklaces of beads, others of tiger's teeth, or peccary's or monkey's teeth. Pieces of arrow-reed a foot long were stuck through the lower part of their ears, and projecting in front of the face, looked like a pair of tusks. At their backs were hung skins of birds (such as macaws and toucans) and monkeys' tails, and he who was rich enough to possess a knife carried it either

[1] The illustration on p. 419 shows this beautiful shrub, cultivated all over the Amazon valley for red colouring matter in the arillus of the seed, called Anatto, or Urucú in the Lingoa Geral. Its native country is not accurately known, but is believed to be near the base of the Andes. The plant photographed (at Pará) was only three years old.

FIG. 39.—MAQUIRITARI INDIANS. (From *L'Orénoque et la Caura*, by J. Chaffanjon.)

slung at his back or in his hand. One had a small instrument of a conical shape in his hand, made of some heavy wood (apparently murá-piranga); he told me it was formerly used in war in close encounters, and he who wielded it sought to smite his antagonist below and behind the ear. The dance was unfortunately interrupted in the commencement by a serious quarrel, arising from a young woman, recently married, refusing to remain any longer with her husband, the brother of Tussarí. The young woman's part was taken by a stout fellow called Aranáu, brother-in-law to Tussarí, not, however, because he wanted the woman for himself (being already married to Tussarí's sister), but because, as I understood from the women, he was foremost in every quarrel. The young woman clung to her own father's arm and, though tearful, seemed resolute. The brother seemed to wish that she should follow the bent of her inclination. Tussarí tried to soothe all parties, and to induce the woman to return to her husband, but the quarrel grew more fierce, and suddenly Aranáu knocked the flambeau out of the hand of Tussarí's wife, knocked down Tussarí himself, and threw himself on the husband. The men shouted, the women screamed, we were in total darkness in a room not over 14 feet square and the combatants had long knives. At one step I could have laid my hand on my gun, which had both barrels loaded, but I thought to myself, if I seem to notice their quarrels it may serve as a pretext for turning their rage on me, so I walked quietly out of an opposite door, and when I got outside was quickly joined by my men, who were also afraid of being implicated in the quarrel. In a short time Aranáu was led out by his brother (the aforementioned Miguel), who had grasped him in his arms ere he could reach the forsaken husband. The storm was now over, but dancing was at an end. Drinking of jaraki (caxirí of Brazil) went on as before. Most of this is made of Yuca, but some is made from yams. It is prepared in large ollas (pots), into which calabashes were dipped, and, all slimy outside with the beverage, dispensed at once to the company. They drank enormously; at first some of them drank two or three full calabashes, one after another. At any time when their stomachs were inconveniently full they seemed to have the faculty of vomiting forth its contents, only to make room for its immediate repletion with jaraki. The floor was soon in a disgusting state.

The water of the Cunucunúma is black and clear, like that of the Guainia. The bottom is sandy with rocks sometimes standing out, but from the first to the second fall the bed is mostly rock. Above this the river is again tranquil and its bed sandy, till the third very rocky raudal, from which upwards the river would seem to run chiefly over rock.

There is mostly very little gapó, but in a few places where the shore is gently sloping and sandy I noticed the same Ingas as on the Orinoco and Casiquiari. . . .

The stones under water in the second fall are covered with a green leafy mass of vegetation, which, when it emerges by the drying up of the river, raises itself erect and bursts into flower. It is composed of two species, one a Hygrophila (Acanthaceæ), and the other a curious Eriocaulaceæ (Papalanthus). There is also a small quantity of Podostemon here and there, but at the third fall the rocks were covered with the same species, only just beginning to be exposed.

Game is as frequent as on the Orinoco, and fish nearly as much so. There were no turtle.

FIG. 40.—URUCÚ, ANATTO (*Bixa orellana*).

The current is rarely strong, and we had only once to use ropes (save on raudales) a little way within the mouth. The bed was mostly so shallow that we could get along with poles.

Jan. 4, 1854.—This morning early we left Tussarí's pueblo. He accompanied me to the piragoa, where I paid him for his goods. About noon I started on our downward voyage and proceeded safely till the first raudal. The curiara was sent ahead and the men reported the waters to be much fallen, and nowhere depth of water for the piragoa to float; still, it was thought that by keeping a firm hand on the helm, she might shoot the fall in safety though she scraped the rocks. We ventured, and reached the edge of the fall, where the impetuous current bore us irresistibly along. A scrape and a bump and we were down the fall, but unfortunately we leaped off one rock only to light on another. The vessel swung round and fell over first to one side and then the other amongst the roaring breakers which prevented us hearing one another's voices. We thought she would inevitably be swamped, but at length she righted with her prow to the falls, and there stuck. I took the helm and the men all leaped into the water and applied their shoulders to the prow, but could not push her off the rock—a smallish round-backed one which had caught her amidships while the prow and stern swung free. We had then to disembark the cargo by little and little in the curiara, and convey it with great risk to a flat rock on the right margin below the fall. After two hours of labour we succeeded in getting the piragoa off the rock, and fortunately her bottom had received no damage. By the time we got all embarked night was approaching, and we desisted from our voyage till the following day.

On Jan. 6 at 8 A.M. we got into the Orinoco, and about noon on the 7th reached the mouth of the Casiquiari. . . .

Jan. 10.—We reached the Pueblo de Monagas yesterday before noon, and as the people were all absent in their fields we awaited their return, as I wished to purchase some pigs which the people here are noted for rearing. Meantime I strolled into the forest. Chiquichiqui (Piassaba of Brazil) was exceedingly abundant, in some places quite gregarious, and here and there a magnificent tableau. When the trees grow high and the beard is not cut off, its own weight brings it down, but it still remains as a sheath to the lower part of the stem, and as the new beard is forming at the apex the stem has a very singular aspect. It was in young fruit, and from the ramification of the panicle I have no hesitation in referring it to Leopoldina. Along with it were an Aldina (Leguminosæ) and a Rhizobolea (gen. nov.) in flower, but the trees were so thick and lofty that not one could be climbed. To-day on the voyage down I gathered a small-leaved Connarus, which was everywhere in flower. A Bignonia with large yellow flowers is also abundant on these two days. Another bignoniad (*Arrabidæa inæqualis*) with smallish rose flowers was completely crowning a lofty tree (100 feet high), so that it appeared to belong to the latter, until an Indian with great peril ascended and brought down specimens.

When we re-entered the Casiquiari it had fallen about 2 feet. The same night we had heavy continued rain which did not pass off till 10 of the following morning. At daybreak the water was rising, and so continued. On the 12th towards evening we reached the Pueblo de Ponciáno, and found the Casiquiari higher there than it was when we ascended. We did not leave again till the morning of the 22nd, as I wanted to dry and pack away my plants before I ascended the Pacimoni. During this time it was very rainy, and except on one day (the 19th) the sun was scarcely ever seen clearly. The river rose steadily till the 19th, when it had risen above 4 feet (since the 7th). On the 19th and 20th it receded a few inches, but yesterday morning (21st) when we left it was again rising.

. . . . . .

We entered the Vasiva towards night of January 21, and left it on the afternoon of the 25th. The first three days and nights were dreadfully rainy, and as the waters continued rising I saw it was hopeless to wait in expectation of the sandbanks becoming exposed. Our position was gloomy and lonely in the extreme. A singular circumstance occurred here. Every day towards evening, say from 4 to 5 o'clock, we were startled by hearing the report of a musket in the forest on the opposite side of the river, which was here not more than eighty yards wide. It is scarcely possible to conceive the strangeness of such a sound, in so desolate a place, in forests which we knew scarcely any human being could penetrate, and especially one accustomed to use firearms. . . .

My sailors, not being able to explain it in any other way, concluded it to be the Yamádu, *in propria persona*, who was hunting near us, and predicted that he would send us a terrible rain, or some other calamity. In reality, on the first two days, we had rain from 4 P.M. to midnight, and on the two following days from 7 or 8 P.M. throughout the night.[1] . . .

Ascent of the Pacimoni River

On January 27, a little after noon, we entered the mouth of the Pacimoni. The river was wide, black, and still, and so continues for a long way up.

[1] This remarkable sound is explained later on in Chapter XXV.

At the very mouth, and especially on the opposite shore of the Casiquiari, a long range of lofty mountains is visible (Aracamuni). It was not until the 2nd of February that we again caught sight of it. Towards evening of the fifth day (January 31) we reached the lower or principal mouth of the Bariá, a large caño coming in from the south, from which the Cauaborís may be reached by a short portage. . . . As far as the mouth of the Bariá, and for a day's journey above it, the forest is all low (30 to 50 feet), and generally inundated for a great breadth, so that it is difficult to find a piece of dry ground whereon to cook; and we went one day till after midday ere we could prepare our breakfast, having started considerably before daybreak. Higher up there is land not inundated, and higher forest, but still caatinga predominates.

At the time of my ascent nearly everything was out of flower. . . .

The vegetation was very similar to that of the Guainia, and almost identical with that of the Vasiva, as nearly all the plants of the Vasiva were repeated on the Pacimoni. Perhaps nothing was more abundant on all three than a large-leaved Terminalia, not yet seen in good flower or fruit. *Parkia americana* (Mimoseæ) was exceedingly frequent, always hanging over the water's edge, and very ornamental from its large pendulous crimson tassels. I only saw one palm to above the mouth of the Bariá, viz. Jará, apparently the species common in the islands of the Rio Negro. . . .

At the wide mouths of the lagoons were beds of Palo de Balsa and clustered Jará. No doubt these are left dry when there is a regular summer.

. . . .

On February 2 I met the first nutmeg, with slender acuminate leaves. Palms appeared simultaneously with nutmegs—Bacába, Inajá, Assaí. In the caatinga I saw only Jará. Both palms and nutmegs are signals of better soil, the forest is loftier, and there is little gapó. Up to this place there was no soil suitable for Yuca. On the 3rd and 4th we passed three cunúcos, the owners of which reside in Santa Cruz.

. . . .

About 4 P.M. on February 4 we reached a new pueblo established a year ago by a mulatto named Custodio who, many years ago, escaped from slavery in Brazil. He counts some sixty souls (of whom a very large proportion are children) in his township, the other families, besides his own, being relatives of his wife (a Yabahana Indian from the source of the Marania), and one Baria Indian. The ground is high—perhaps rising to 150 feet from the river—and the soil good, but cold winds sweep over it from the cerros, especially by night, and squalls come with such force as to threaten to overthrow the houses.[1] To this point my piragoa ascended without difficulty, but a little higher up the stream narrows considerably, and many caños and lagoons open into it. At a short day's journey above, and at the foot of an abrupt conical cerro (Araucana), is a small pueblito where Custodio first established himself. . . .

I stayed a day with Custodio, and then leaving the piragoa, proceeded in my curiara to visit the pueblo of Sta. Isabel.

The Pacimoni above Custodio narrows considerably and winds more. Several caños and lakes communicate with it There are small islands here and there. Clustered Mauritia is frequent, and on the second day the stream winds as if it would never find its way out of a Morichal (*M. vinifera*). . . .

Midway up we encountered a Posoqueria (Cinchonaceæ), 18 to 25 feet high, bending over the water, and clad with a profusion of white odoriferous flowers. At the bottom of a long tube was a quantity of honey which my Indians sucked with great relish.

. . . .

It took us two days to reach the caño of Sta. Isabel (Uaranaka), which branches off to the left

[1] These are real hurricanes like those of the West Indies, but of brief duration, and apparently not spiral.

(as one ascends). It is white water, while the Pacimoni continues to be black. The latter is slightly larger, but both are insignificant streams, swelling with every rain, in many places not wide enough for a curiara to turn, and in the dry season so shallow in parts that the smallish canoes have to be dragged over sand. At all times of the year it is necessary to be furnished with an axe and cutlass to clear away the trees which are constantly falling into it. Hardly a day passes without a strong squall from the cerros, which never fails to overthrow such decayed and insecurely rooted trees as lie in its course, and during my stay in the Pacimoni I heard frequently the crash of their fall. I was furnished with a cutlass, but, unfortunately, not with an axe, as I knew not previously that the latter was necessary, and we were put to serious straits in consequence, for we encountered two fallen trunks stretching across the caño and standing out of it 1 to 3 feet, far too stout to be severed by the cutlass. With much difficulty we dragged the curiara over them, and with great risk of precipitating the cargo into the river, for the dense brush allowed nothing to be landed. . . . It is only when the sun is nearly vertical that it penetrates the overhanging trees and climbers. Logs and branches of trees were hanging into the water, and sometimes stones covered with large Hypnum, having quite the habit of *H. riparium*, but more closely allied to the common Rio Negro species. . . .

Starting with the earliest dawn, it was midday when we reached the port of Sta. Isabel, and we had then a portage of at least two miles through the forest to the pueblo. The track was easily found, and logs had been laid over caños and hollows filled with water in the winter, but as all the ways along which heavy goods can be carried are by water a pueblo is ill-situated when remote from easily navigable waters.

Sta. Isabel is inhabited principally by Cunipusana Indians, of whom there are still a good many from

FIG. 41.—PUEBLO DE STA. ISABEL, WITH CERRO TIBIALI, RIO PACIMONI, NOT FAR FROM THE SOURCES OF THE ORINOCO. (R. S.)

the head-waters of the Pacimoni towards the Siapa, and on the Caño Castaño. There are also a good many Mandauaca Indians, who seem to have been the original inhabitants of the Upper Pacimoni, a few Manaca Indians, and a family of Yabahánas brought by Custodio from Maraniá.

There are fourteen houses (of which one is the casa real), and every house contains at least two families. They are built principally round a plaza slightly sloping to the cañito, which runs to the Uaranaka. The ground is sandy, rising at the back (to the N.W.) to a low hill. To the N.E., and apparently quite close, though really so distant that the forest can scarcely be distinguished on it, rises an abrupt cerro (Tibiali, the name given by the Cunipusanas to a little bird of bright blue), falling almost perpendicularly at its right (E.N.E. as seen from Sta. Isabel). Bearing S.E. by S., rises into a lofty cone the northern extremity of a long range of high mountains called Imei (the wasp). The Venezuelans limit the name Cerro de Abispa to this cone; another peak about midway is called Cerro de Danta, and the southern extremity (very distant, and only visible by ascending the hill at the back of the pueblito) the Cerro de Mono.

We lost some time at Sta. Isabel through my own fault in forgetting to take with me my shot-bag. My men had also left their fishing-lines, but indeed the fish we saw were scarcely larger than minnows. With a basket of farinha and my gun I am generally independent in the matter of provisions, especially when (as on this occasion) I have with me an Indian who is a good shot. Fortunately my gun had both barrels loaded, and in the evening of my first day my hunter shot with it two cojubims (Penelope sp.), on one of which we supped and had the other for breakfast on the following morning. After this I ate no more till near five in the afternoon of the third day. At Sta. Isabel we found only two families of women with two youths. One of the women owned a fowl, which I was glad to purchase, for I was near famishing. Several other fowls were running about, but their owners were away in the cunúcos. As these fowls were the only thing I could eat, I had no alternative but to send on the following morning to call the capitan, who owned most of them. The capitan's cunúco was a long way off, and after waiting through that day and the next, till evening, he made his appearance, and I purchased a few fowls of him. When I have nothing to eat I find it impossible to work, and, besides, I had been able to take in the curiara only two small bundles of drying paper which I wished to reserve for the plants of humble growth I hoped to meet in the cerros. For this reason I had left several interesting plants in the caño, the specimens of which would have been so bulky as speedily to fill the papers. The weather proved showery, but in these two days I gathered a few plants near the pueblo as interesting perhaps as any got elsewhere, and took a sketch from the casa real looking towards Tibiali.[1]

On the morning of February 11, having caused a fowl to be roasted to eat on the way, I started for the Cerro Imei (Cerro de Abispa), accompanied by a young man, as guide, and by two of my Indians.

[1] Most of the other inhabitants came to the pueblo along with the capitan when they heard a white man had arrived there. As it was a fine dry moonlight evening, I got them out into the plaza and set them a-dancing. In a place so remote from civilisation, and where the people, since they were gathered together into something like a Christian pueblo, had not been visited by any missionary to baptize them, I expected to see and hear something quite new to me in their dances and songs; what then was my astonishment when, to the sound of a kind of guitar made from an internode of bamboo, the dancers began to caper wildly about and to throw their legs high into the air in a way quite foreign to the grave and stolid Indian, and to sing in good Portuguese, "Vamos à ver, vamos à ver, vamos à ver a Mai de Deos!"—precisely the song and dance of the negroes at the Barra de Rio Negro when, during the festival of Christmas, they go about visiting the altars on which is exposed a figure of the Virgin and Child set up usually at the corners of the streets. When I asked them to change the "figure" it was still a nigger dance and song. "Oh," said I to myself, "my friend Custodio has been here," and I afterwards ascertained that the Indians had really derived their novel accomplishments from the Brazilian slave. However, I was highly amused, and praised their performance as it deserved.

[This is probably the steep conical mountain shown to the right of the tall tree in the drawing.—Ed.] I had been deceived by false information of the distance, and calculated on returning to Sta. Isabel before sundown. Instead of this, though we started just after sunrise, it took us till after midday to reach a cunúco at the base of the cerro. In this space we crossed streamlets forty-three times, without including the pools of standing water in which we sometimes walked a quarter of an hour together, for the forest where the ground was lowest was almost turned into a lake. Poles had been laid across most of the caños, but some were rotted away and nearly all were covered with water, so that it was critical work traversing them. We crossed the caño Uaranaka three or four times, once with water up to the waist. It was the only considerable stream of water we encountered.

After reposing for a while, we started for the cerro, but without any hope of reaching a height where good plants might be expected. We crossed a low hill and descended a steep valley, and then commenced ascending the slope of the mountain, which seemed to continue uninterrupted and clad with lowish forest till about midway, where (as could be seen from below) there began to appear abrupt exposed rock. We continued along more than an hour, but there was nothing in flower, and I saw scarcely any trees which I did not already know. The soil was dry, yet a good many ferns began to appear, consisting solely of two tall species, one or both of which I had previously gathered. It came on to rain, and a thunderstorm was brewing up to the northward of the cerro, so that I judged it best to return, and we again reached the cunúco, having reaped nothing but a wetting.

. . . . . .

This mountain could only be climbed to the summit (if that be practicable), or even to any considerable height, by sleeping two nights at the cunúco and devoting the whole of the intervening day to the ascent. But we had no provisions and there was nothing to eat in the cunúco but cassave, so that I passed a miserable night, for I had no supper and I tried in vain to sleep in a tiny hammock of very open texture, shivering with cold and tormented by zancudos, which are said to be abundant all along the base of Imei.

We started to return next morning without breakfast save a little cassave soaked in tucupí. I had torn my naked feet on the previous day and I contrived shortly after starting to deepen one wound by treading on a sharp stump, so that, what with bleeding feet and an empty stomach, I found the journey sufficiently toilsome. But this did not prevent me gathering such plants in flower as I had noted on the previous day.

There was one large Tiliaceous tree of which we could find no individual whose branches were accessible, though the yellow flowers and discoid prickly fruits everywhere strewed the ground. It is of the genus Luhea, and from Pará upwards I have found these fruits strewed in forests, but never in a good state, and never accompanied by flowers.

We got back to San Custodio on the 14th

February, and the next morning, in company with Custodio, visited the low cerro Tarurumari, which rises a little north of the village. At the base it slopes gradually and is well wooded; higher up it is steep and the furrowed rock is bare save strips of peculiar dwarf vegetation in the hollows. It is only about 500 feet high, but a most extensive view is obtained, including the whole of the mountain ranges called Aracamuni, Tibiali, and Imei. The first of these runs up between the Siapa and Pacimoni, inclining more to the Siapa at its western end, and to the Pacimoni at its eastern. It is of nearly equal height throughout, and may rise about 4000 feet above the plain. Over the west shoulder of Aracamuni we could barely distinguish in the distance the cerros which rise from the very banks of the Siapa. Tibiali is situated at the back of Sta. Isabel, and still farther east a gradual rise conducts to the fine cone of Abispa, which forms the northern termination of the long serrated ridge of Imei. Abispa bore S.E. ½ S. and must be about 6000 feet high, and there are other peaks closely approaching it in elevation.

In front of these noble mountains stretched the forest plain, like an immense heath, its surface unbroken save by a slight winding depression nearly at our feet marking part of the course of the Pacimoni. I had hoped to make more extended observations, but when we reached Tarurumari a heavy shower was passing over the cerros, and until it should clear away I occupied myself in culling the interesting plants which grew around. Unfortunately, the shower took our mountain in its course and gave us a thorough wetting. Other showers followed the first, and when these were over we had only just time to get home before night.

[At the end of this chapter I give the story of how Custodio, a mulatto slave from the lower Rio Negro, became a powerful Indian chief and valued official of the Venezuelan Government.—ED.]

The vegetation of these lower cerros, which alone I was able to reach, have quite the same character as that of the Cocuí mountain near Marabitanas. It is supported on declivities by a margin of Bromeliaceæ—perhaps of the same species—and there are the same two Orchises. Mixed with these was a Pandanacea (or perhaps two species) with fronds like those of a young Assaí palm. The most curious plant—which occurred in considerable quantity —was a shrub about 5 feet high, with fleshy shoots and leaves, and a few tubular scarlet flowers. It is a Rutacea allied to Galipea. A much-branched Sipanea was 8 feet high, though still herbaceous like others of the genus. A Holly was frequent; and I was surprised to meet again the same Remigia (Rubiaceæ) as I had gathered at Esmeralda.[1]

In descending the Pacimoni from San Custodio in the afternoon we reached a gently sloping granite rock at the base of a low bare cerro, rising not more

[1] [I will here give a list of the few plants which Spruce gathered on these mountains near the head of the Pacimoni, as the locality has probably not yet been visited by any other botanist. For reasons already stated he collected nothing but what was new to him.

On Mount Imei (at foot of the rocks).

Cephaelis sp. (Rubiaceæ); Miconia sp. (Melastomaceæ); Badula sp. (Myrsineæ); Davya sp. (Melast.); *Swartzia grandiflora*, Boug. (Cæsalpineæ); Faramea, n.s. (Rubiaceæ).

On Mount Tarurumari.

Sipanea rupicola, Spruce (Rubiaceæ); Aspidosperma sp. (Apocyneæ); *Galipea oppositifolia*, n.s. (Rutacea); *Echites anceps*, Spruce (Apocyneæ); Myrcia sp. (Myrtaceæ); *Liriosma micrantha*, n.s. (Olacineæ); Ruyschia sp. (Maregraaviaceæ); Cupania sp. (Sapindaceæ).

It must be remembered that only a few hours were devoted to either of these mountains, and that he had already spent two or three weeks in an examination of the forest plain which surrounded them.—ED.]

than 200 feet. As we decided to pass the night here, I climbed the rock, when I was astonished at the magnificent scene that burst on me, exceeding that from Tarurumari both in extent and distinctness. A rain-cloud streaked with lightning was passing between Tibiali and Imei, which added to the picturesque effect. The whole horizon was visible except from W. towards N.W., which was shut out by trees on the top of the cerro. As this was nearly in the direction of the Casiquiari, I do not suppose there were any hills to be seen even had the forest been cleared away. Not only were all the mountains seen which had been visible from Tarurumari—especially Imei in its entire length—but by moving one's position a little a distinct view was obtained of Cocuí (S.W. $\frac{1}{2}$ W.) and the cerros below San Carlos, besides a set of low hills extending between S.W. and S., and in the extreme distance at the back of these we could dimly distinguish Pirá-pukú.

[The Journal of the return journey ends here; but the short record of botanical excursions shows that Spruce reached the mouth of the Pacimoni on February 24, 1854, remained a day collecting at the junction of the Pacimoni and Casiquiari, and arrived at San Carlos on the last day of the month. Here the whole of March was occupied in sorting and packing his collections and dispatching them to England, while April and half of May were spent in further botanical excursions around San Carlos, till he started on his journey to the cataracts of the Orinoco by way of the Guainia and Javita. The following letter to Sir William Hooker gives a connected sketch of the interesting voyage just concluded, and explains some matters not touched upon in the Journal. It is also a good example of Spruce's style of writing and serves to illustrate his general interest in scientific inquiry; and though it may contain some repetition of facts in the Journal, I do not think any readers of his travels will consider it out of place.]

To Sir William Hooker

SAN CARLOS DEL RIO NEGRO, VENEZUELA,
March 19, 1854.

. . . I calculated on spending a month in the voyage up the Casiquiari, but after passing the mouth of Lake Vasiva mosquitoes began to be so abundant that my Indians became very impatient of stoppages. So long as we continued in motion, comparatively few mosquitoes congregated in the piragoa, but when we stopped to cook or gather flowers they were almost insupportable, and the cabin especially became like a beehive. You will easily understand that, however much my enthusiasm as a naturalist might conduce to render me insensible to suffering and annoyance, I could not help occasionally participating in the feelings of my sailors, and was not sorry to get along as quickly as possible. The weather was unusually fine and dry for this region; hence the abundance of mosquitoes. The same circumstance was favourable for preserving specimens, but the trees of the riverside had mostly shed their flowers and had fruit too young to be worth gathering. Still I found enough to keep me occupied. In the afternoon of December 21, 1853, we got out of the upper mouth of the Casiquiari. I could not look for the first time on the Orinoco without emotion, and I thought of the illustrious voyagers who more than fifty years previously had explored its course, and the vegetable products of its shores; not without hope of being able to collect again some of the latter in the places where they were first discovered. My original intention (as you already know) was to explore the river Cunucunúma, which flows along the western side of the mountains Marayuaca and Duida, and enters the Orinoco a little below the mouth of the Casiquiari; but first I had resolved to have a peep at Esmeralda. We started, therefore, up the Orinoco, and in the morning of the 24th reached Esmeralda, having experienced no small difficulty in finding a way for the piragoa, for the Orinoco was falling fast, and in certain places where it spreads out to a great width we could hardly anywhere find 3 feet of water, all that was necessary to float my little vessel. As my provisions were falling short, I had to devote some time to hunting up the Indians of Esmeralda, and setting them to work to bake cassave. With this exception every moment of daylight during my short stay was given to collecting the plants of the surrounding cerros and savannahs. I suppose I mentioned to you that the Comisario General of the Canton del Rio Negro (residing in San Fernando de Atabapo) had invited me to accompany him on an exploratory expedition towards the sources of the Orinoco, and appointed to meet me for that purpose in Esmeralda on Christmas Day. As above stated, I arrived at the rendezvous a day earlier than agreed on, but

I already knew that everything was in confusion at head-quarters in Venezuela, and that it was probable nearly all the officials would be changed throughout the country; though I found that orders had been given by the Comisario to prepare a quantity of mandiocca in Esmeralda, Cunucunúma, and in other places higher up the Orinoco—proof that he was sincere in his proposal. Some time afterwards, when I was on the Pacimoni, I received a letter from him, informing me that he was no longer Comisario, and that he could not leave his post until the arrival of his successor, which, in fact, has not taken place until within the present month (March). I would willingly have waited some time in Esmeralda, but the Orinoco continued to fall rapidly and I began to fear I should not be able to enter the Cunucunúma; so after a stay of four days I bade adieu to Esmeralda and its mosquitoes. It occupied us through the 28th and till noon of the 29th to descend the Orinoco as far as the mouth of the Cunucunúma. We entered the latter, which may be compared to the upper half of the Casiquiari for breadth and depth; but the water is black, not white, and yet notwithstanding this, mosquitoes are quite as plentiful as on the Orinoco. The Indians inhabiting the river Cunucunúma are Maquiritares, and I hoped to be able to conduct my piragoa as far as their first pueblo, which is at the foot of the third raudal (rapid).

We reached the first raudal on the first day of the present year (1854). There was just water enough for my piragoa, which we dragged up with some difficulty. At 8 on the following morning we reached the base of the second raudal, a long rapid where the river spreads out wide and runs over a shallow bed of rounded pebbles, of all sizes up to that of a man's head. For two hours we struggled to drag the piragoa up this rapid, but found it useless to attempt to go farther, and with a sorrowful heart I gave the word to return. I had calculated on spending at least a month among the Maquiritares and exploring their river by means of small boats up to its sources, which are on high land towards the head-waters of the Ventuarí and Caura; but this was impracticable unless I could get my stock of paper and goods to some station which I could make my head-quarters, for the lower part of the Cunucunúma is embosomed in forest so dense that we had difficulty in finding a spot of ground whereon to cook our victuals. . . .

After visiting the pueblo in a small canoe and staying a day there, I returned to the piragoa, where I found the river had sensibly fallen, and it was evident there was no time to be lost, for the first raudal, passed with difficulty on the ascent, might now be impassable.

. . . . . .

On January the 6th we emerged from the Cunucunúma, and I had now to decide whither I should next bend my course. There was little chance of getting much farther up the Orinoco, from the small depth of water. In my way up the Casiquiari I had entered Lake Vasiva, and though it had dried so little that we could nowhere find a spot of land whereon to light a fire, the adjacent forests seemed to contain a peculiar vegetation. There were large playas covered with Palo de

Balsa,[1] now several feet under water but left bare in the dry season, and my pilot, who had spent a summer in Vasiva catching turtle, told me that at that time the sand was covered by thousands of little annual plants. I determined, therefore, to explore Vasiva thoroughly, and I pictured to myself the numbers of new Burmannias, Utricularias, Ptychomeriæ, etc., I should gather on its shores. It was necessary to use all expedition, for when the Casiquiari is at its lowest, only small boats can navigate the upper part. We re-entered it about noon on the 7th and commenced our downward course. It rained every day, and, instead of falling as we expected, the water rose again. On the 12th we reached an Indian settlement a little above the mouth of the Vasiva. . . . Hoping the rise of the waters might be only temporary, I waited in Yamádu-báni until they should again go down. On the Casiquiari and Alto Orinoco the driest months of the year are considered to be January, February, and March, and in the last-named month the rivers are expected to be lowest. This year, however, the turning-point was on the 8th January, and the swelling of the streams has gone on continuously, with the exception of a very slight subsidence in the middle of February, until the present time, when they are as full as usually at the end of June. Hence every one says there has been no vasante this year, and the consequences are disastrous. No turtle oil could be collected on the

[1] Spruce says that this is a species of Amyridaceæ, an order of highly resinous trees and shrubs, some of which produce the myrrh and frankincense of the East, while many South American species produce gums, oils, and balsams, especially those of the genera Icica and Hedwigia. Spruce was not able to obtain flowers of the species here referred to.—ED.

Alto Orinoco and Casiquiari—no turtles caught and no fish salted. . . . I left for Vasiva on the 22nd, and in the evening of the same day took up a position within the outlet of the lake, on the only piece of land that was not inundated. During the four following days, which were dreadfully gloomy and rainy, I explored the lake in my curiara, and then, seeing that I could do no more there, again continued down the Casiquiari. I was not content to return to San Carlos without adding considerably to my stock of dried plants, and my best plan now seemed to be to explore the Pacimoni. This I was enabled to execute partially. I entered the Pacimoni on January 27, and in the space of a month explored it to nearly its head-waters, which are in the midst of magnificent mountains, the latter uninhabited and all but inaccessible, and scarcely known to geographers even by name.

I have not time to write in detail of the plants collected. Those from the Pacimoni include the most novelty, but perhaps the small collection made at Esmeralda will be looked on with more interest by Mr. Bentham and yourself, although I suppose all the species have been gathered previously either by Humboldt or Schomburgk. The low cerros near Esmeralda—the debris of Duida—have a scanty scattered fruticose vegetation, among which one of the most prominent plants is a Commianthus, apparently *C. Schomburgkii*, Benth., though a smaller form than I gathered nearly two years previously on a small sandy campo near the Barra. It is so abundant within a quarter of an hour's walk from Esmeralda that I can scarcely credit its not being among Humboldt's plants. Another shrub or small tree growing along with it in great quantity is a stunted form of *Humirium floribundum*; the same widely-distributed species accompanies the Commianthus near the Barra. Equally frequent was a Remijia with densely pilose capsules, shorter than usual in the genus; I was surprised to meet afterwards the same species on a small granitic mountain by the Pacimoni, especially as none of the plants accompanying it in the latter locality were identical with those of Esmeralda. Other shrubs were a Byrsonima, apparently a form of *B. spicata*, a Guatteria, a Pagamea, etc. Under large stones grew the most delicate little fern I have ever gathered, looking at first glance like a miniature *Allosorus crispus*, but in reality more allied to Schizæa, and along with it a small grass with broad truncate-cuneate leaves, which I had gathered abundantly in similar situations by the cataracts of the Rio Uaupés. Rooting into clefts of the rocks and twining on adjacent shrubs or over the rocks themselves, grew an Asclepiadea, with narrow leaves and minute white flowers, looking not unlike *Galium saxatile*. In moist, rocky places I found a shrub of about 4 feet high, with long pinnate branches, minute rigid leaves ending in an arista, and solitary axillary fruits the size and colour of haws. It is quite new to me and seems to me to be a capsular Myrtacea, but I have not examined it closely. There were also a few Melastomaceæ and other things.

The savannas near the pueblo (Esmeralda) were mostly dried up by the heat. The grasses showed only withered culms, but I recognised among them several species of Paspalum, Setaria, Andropogon, Trichopogon, etc. I crossed the two first savannas in the direction of Duida, but found scarcely anything in flower. It is curious that on the second of these the only tree, besides the Moriche palm, is a Qualea, which seems to me identical with one gathered on a low campo of quite similar character opposite the Barra, and which Mr. Bentham has called *Q. retusa*. The tree at Esmeralda had neither flower nor fruit, and if it was in the same state at the period of Humboldt's visit, most probably he did not gather specimens.

On a savanna which extends towards the Guapo there were still some moist places left, and in them I gathered several interesting little plants. They include two Burmanniaceæ (perhaps the true Burmannias), one of them with a violet flower far larger than I have seen in any other species of the tribe; four Gentianeæ, of which two are Lysianthi, the one a small species with a bright blue flower, exactly resembling *Campanula rotundifolia*, the other a tall plant with green flowers; the other two species are minute things allied to Schübleria, three or four Xyrideæ, two Asclepiadeæ, two minute Rubiaceæ with yellow flowers, species of Perama, one of them *P. hirsuta* (gathered also at Santarem), three Polygalæ, in one of which I recognise *P. subtilis*, H. B. B., and several others.

I gathered also all I could on the banks of the Orinoco, including the *Palma Jagua*, whose beauties are so highly and so justly eulogised by Humboldt in his *Aspects of Nature*. It is an undescribed Maximiliana, and I brought away with me specimens and notes on the living plant which will enable me to describe it. There were two splendid trees of it in the mouth of the Casiquiari. I had one of them cut down and a frond and a spadix embarked in the piragoa, where I could examine them at my ease and also continue my voyage. The frond measured 34 feet long and was composed of 426 pinnæ. The spadix bore about a thousand fruits and was a load for two men. Several spadices are matured simultaneously. These statistics will alone suffice to give you an idea of the magnificent aspect of the Palma Jagua, which is one of the chief ornaments of the Upper Casiquiari and Orinoco.

About half-way up the Casiquiari, where the water begins to be unmistakably white, the rocks by the river-side and the overhanging inundated branches of trees begin to be clad with a moss having exactly the aspect of *Cinclidotus fontinaloides*. It is so abundant on the Upper Casiquiari and Orinoco, that I think I could in an hour have laden a small boat with it. This moss you were the first to describe, under the name of *Grimmia fontinaloides*, from Humboldt's specimens gathered on the Alto Orinoco. If it be pleasant to discover an undescribed species, the pleasure is at least equal (and it is free from any selfish admixture) when, after the long lapse of years, one gathers again a plant in the spot where it was originally discovered by another. I can fancy Dr. Hooker's gratification at gathering again the mosses discovered by Menzies in New Zealand.

One of the most notable things in the Pacimoni was a tree which was conspicuous from afar by certain white cones thickly scattered among the deep green foliage. These cones my telescope revealed to be fruits, but my Indians insisted they were wasps' nests, and even when we came directly under the tree, which was not more than 40 feet high, not one of them would venture to climb it until they had first poked one of the cones with a long stick. Nor did their caution appear to me ridiculous, for on the Casiquiari we had had feeling proof that wasps' nests occur of all shapes and sizes. I expect this tree will constitute a new genus of Clusiaceæ, allied to Platonia.

In returning from one of my long expeditions, I always feel a sense of humiliation at the little I have been able to effect for other sciences besides botany, and especially when the country traversed is perhaps more interesting to the geographer than to the botanist; nor does it console me to reflect that one person cannot do everything, that the preserving of plants in this climate involves great mechanical labour, and that the daily cares and *contretemps* of a voyage where one's only workmen are Indians, and where food must be sought from day to day in the rivers and forests, consume no little time. In my late voyage, in addition to my botanical collections, I brought away with me rough maps of the rivers Pacimoni and Cunucunúma, with materials for constructing them more accurately at a future day; a few sketches, including a good deal of picture-writing; and vocabularies, more or less complete, of six different languages, including that of the Guaharibo Indians. But there are persons who would have done much more, and some one will come after me, possessing more health and strength, aided by industrious hands, and with resources of every kind at his disposal, who will complete whatever I have left imperfect.

[The following is the story of Custodio, the Comisario of the Pacimoni river on the Casiquiari, as told to Spruce by himself.—ED.]

Custodio is a dark mulatto, nearly black, apparently from forty-five to fifty years old, tall, stout, and good-looking. He was born a slave in the village of Barraroá[1] on the Rio Negro. His master treated him well—even as though he had been his son—he had no son of his own. When Custodio grew up he accompanied his master in his expeditions on the Uaupés, Maraniá, etc., in quest of salsa and other products of the country, and was often entrusted to trade alone with a quantity of goods. He thus visited the Maraniá in eight successive years and became well acquainted with Yabahána Indians, who inhabit the sources of that river. His

[1] [Sometimes called San Thomas, situated about midway between the Barra and Marabitanas.—ED.]

master had a caseira (or housekeeper)—a woman by no means young—whom Custodio declares he looked on with as much respect as if she had been his mother; but being on one occasion left at the sitio whilst his master was absent a few weeks, evil tongues put it into the head of the latter that the slave and the caseira had played him false. He entered the house and laid his loaded gun by him, and when Custodio shortly after entered to bid his master welcome, he without saying a word presented the gun at Custodio and pulled the trigger. Fortunately it merely flashed in the pan, but Custodio, though totally unable to account for such a reception, saw, not only from the act but from the expression of his master's countenance, that the latter was bent on killing him, and needed no second warning to flee for his life. He was soon deep in the forest, and at night came down to the river-side, seized a montaria, and set off up the river. When he reached Sta. Isabel he ventured to go ashore, and entered the house of some Indians whom he knew, where he sat down to eat and recounted to them his story. But he did not remark that among those who listened to it were two half-whites who resolved to gain a reward by placing him again in slavery. They accordingly waited for him outside the house with loaded guns; but Custodio was made aware of their intention. His only weapon was a long knife fastened to the end of a stick; he grasped this and, waving it right and left, leaped out of the doorway. His assailants gave back, and, ere either of them could present his musket, he had rounded the corner of the house and plunged among the Coffee trees and other brush, of which there is mostly no lack near an Indian village. Thence to the forest was only a few steps, and he was soon safe from his pursuers, for the nonce. He resolved to venture no more on inhabited places, and painfully made his way through the forest till he reached the mouth of the Maraniá, swimming across the mouths of the caños that lay in his path. Here, with only his knife, he stripped the bark off a tree-trunk in a piece and made of it a canoe—an art he had learnt from his friends the Yabahánas, whom it was now his object to reach. With his knife he made also a paddle, and thus equipped set off up the Maraniá, subsisting solely on wild fruits and procuring fire when he needed it by rubbing together fragments of Cocurito. In this way he succeeded in reaching the land of the Yabahánas. Here he remained in safety two or three years, and took to himself a wife; but wishing to inhabit some place where he could turn to account his skill as a blacksmith, crossed over to the Pacimoni, and down this river and the Casiquiari to San Carlos. By the constitution of Venezuela, slaves from Brazil who cross the frontier are free, but by some treason a Portuguese, who was going to the Barra, seized Custodio by night and carried him away bound all the way down the Rio Negro. This was some time in 1836, towards the end of the cabana revolt, and there was in Barra a sloop of war on board which were placed a number of cabanos (prisoners). Custodio's old master was dead. but he was claimed as property by his executors, and placed on the sloop in irons along with the rebels, until the country should be quiet again. The sloop shortly afterwards sailed down the Amazon, but had not been gone many days when an express was sent to recall it to assist in repressing a new outbreak of malcontents. Arrived again in the Barra, its services were not needed. It was Christmas time, and sailing down was deferred till the end of the festival. By an extra act of grace the captain of the sloop allowed the prisoners to leave the sloop every evening when there was dancing, on condition of their returning to sleep on board by 10 o'clock. On the two first nights that Custodio profited by this licence he returned punctually to his prison, but on the third night when he left the gay throng to betake himself to his miserable lodging in the hold of the sloop, he by chance found himself alone in the street. It was a bright starlight night and one cannot wonder that he should be seized by an irresistible longing for liberty, or that thoughts should rush into his mind of his Indian friends' home, of his wife and his two little ones. More than a thousand miles of forests and rivers separated him from them, and he had no friend to aid him, but he was familiar with every part of the way, and he was accustomed to live in the forest. In an instant his resolution was taken. He ran down to the port. There was not a single montaria to be seen save one laden with water-melons, etc., which an old woman had that instant brought to land. "Boas noites, senhora," said he, "you come heavily laden; allow me to help you to land your cargo." "De bon ventade" (willingly), "my son," said she; "your aid will be most acceptable." Her house was close by and Custodio had soon stowed in it the contents of the montaria. The old woman was pleased and gave him a pataca for his trouble. This was, however, not what he wanted, and he had now to forge an "historia" in order to attain his object. "Mother," said he, "will you not lend me your montaria an instant in order to visit my friends in the sitio across the mouth of the igarapé" (the Igarapé da Cachoeira). "Take the montaria," said she, "but fasten it up again securely in the same place when you return." This was readily promised; Custodio leaped into the montaria, and with a hasty "ate logo" (good-bye), shoved off, nor has the poor woman from that day to this ever set sight on either one or the other.

At the northern extremity of the Barra is a small peninsula bounded by steep cliffs of earth, called San Vicenti, where there is a sort of fort. Custodio crept along the base of the cliff as silently as possible, but did not escape being challenged by the sentinel, who, however, allowed him to pass on the same plea which had imposed on the old woman. It wanted but a few minutes of ten, his absence would soon be remarked in the sloop, and he plied his oar with all his force. Approaching the mouth of the igarapé, he heard within it splashing of oars and numerous voices laughing and talking. These sounds proceeded from several montarias which were on their way to the town. It would not do to be seen by the occupants of these, so he stood out wide and did not again approach land until he could not distinguish either human sight or sound. Warned by former experience, he determined to trust himself rather to tigers than to men, and to avoid every appearance of a habitation. For greater security he kept always on the left bank of the river, along which vessels scarcely ever pass either ascending or descending, nor is there any village on this side until Sta. Isabel. He rowed night and day, and allowed himself very little sleep, and this chiefly in the middle of the day, when he pushed his canoe deep into the gapó, and closed his eyes in security from the assaults of his fellow-men, but not without risk of being strangled by some water-snake. His food all the way up the Rio Negro was the fruit of a curious twiner (Gnetum) with jointed branches, the joints tumid and bearing a pair of leathery leaves: it is called Ituán in Brazil, and the species which furnished Custodio's food is common all the way up the Rio Negro, where I have gathered it in flower and fruit. I have met with it also on the Casiquiari and Pacimoni. He gathered these wherever they appeared in quantity, and at any convenient place lighted a fire and roasted them, afterwards piling the roasted fruits in the prow of the canoe, and when they were eaten up he roasted others. Thus he went on until he reached the mouth of the Maraniá, up which he had decided to ascend. A far shorter and easier way would have been to continue right up the Rio Negro to San Carlos, where was his family, but he could not hope to pass the garrisons of Saõ Gabriel and Marabitanas, especially as they were known to be on the look-out for fugitive escaped slaves and cabanos. Up the Maraniá he therefore proceeded, and here was no more Ituán, but its place was imperfectly supplied by eating the thin pulp of the Mirití palm, which is barely sufficient to sustain life, though it is very insipid. At the end of thirty-five days, counting from his leaving the Barra, he safely reached the friendly Yabahánas. During all this time he had never spoken to a human being, nor had he once tasted cassave or farinha, or indeed any sort of food which is produced by human industry. After a short repose among the Yabahánas he once more started for San Carlos, by the same route as before, and had the happiness to rejoin his family just as he had left them.

For several years after this he remained in San Carlos working at his trade; until the Comisario General of the Canton, who knew that he was well acquainted with the Indians of the Maraniá, Pacimoni, and Siapa, and had great influence over them, offered to him to undertake the management of the river Pacimoni. He closed with the offer and immediately removed his family to the Pacimoni. There was already a pueblo near the source of the Pacimoni, called Sta. Isabel, which had been founded a few years before with a few families of Mandauaca and Cunúqurana Indians. This pueblo Custodio augmented, and founded another pueblo some distance lower down the river, at a point which can be reached by boats of considerable size, and therefore better situated for commerce than Sta. Isabel, which can only be approached within two miles by boats of the smallest size. Not content with this, and being joined by several families of his wife's relatives, he moved a day lower down the river, some two years ago, and commenced another pueblo which has apparently taken firm root and has been named San Custodio. So that Custodio—the mulatto—the slave—the captive—now figures as "Comisario del Rio Pacimoni y fundador de los pueblos de Santa Maria y San Custodio" (Chief of the river Pacimoni and founder of the villages of Santa Maria and San Custodio !).

Note on the Sources of the Orinoco

[In a volume on *L'Orénoque et la Caura*, by J. Chaffanjon (Paris, 1889), the author describes his voyage up these rivers reaching to the sources of the Orinoco. Unfortunately, he appears to have had no means of fixing any positions, and his small, very sketchy map is evidently quite untrustworthy. This is shown by its making the distance from Esmeralda to the highest point reached, at a mountain which he names after Ferdinand de Lesseps, rather more than that from Esmeralda to San Fernando de Atabapo at the mouth of the Guaviare, which would bring him to a point far beyond the Sierra Parima as shown on all the maps.

Leaving Esmeralda on December 2, 1886, in a large canoe with eight rowers, and ascending numerous rapids, he reached the Raudal des Français on the 13th. Thence, in 11½ days, in a small curiara with two men, he reached a point where the river became impassable on account of rocky obstructions. A walk of two hours brought them to where it became one of several small rivulets with inclined rocky beds on a mountain slope, which may fairly be considered as the sources of this great river. It still remains, however, for some competent observer to fix the position of the point reached.—Ed.]

THE RIO PACIMONI

From the point where the Cano Uaiauaka begins to be navigable down to its confluence with the Casiquiari

Febr. 1854

Reduced from Spruce's Original Sketches

From a comparison of time occupied as compared with that on the Casiquiari, whose length is approximately known, the scale of this map appears to be about 10 miles to an inch, and the total distance following the windings of the river, about 200 miles (A.R.W.).

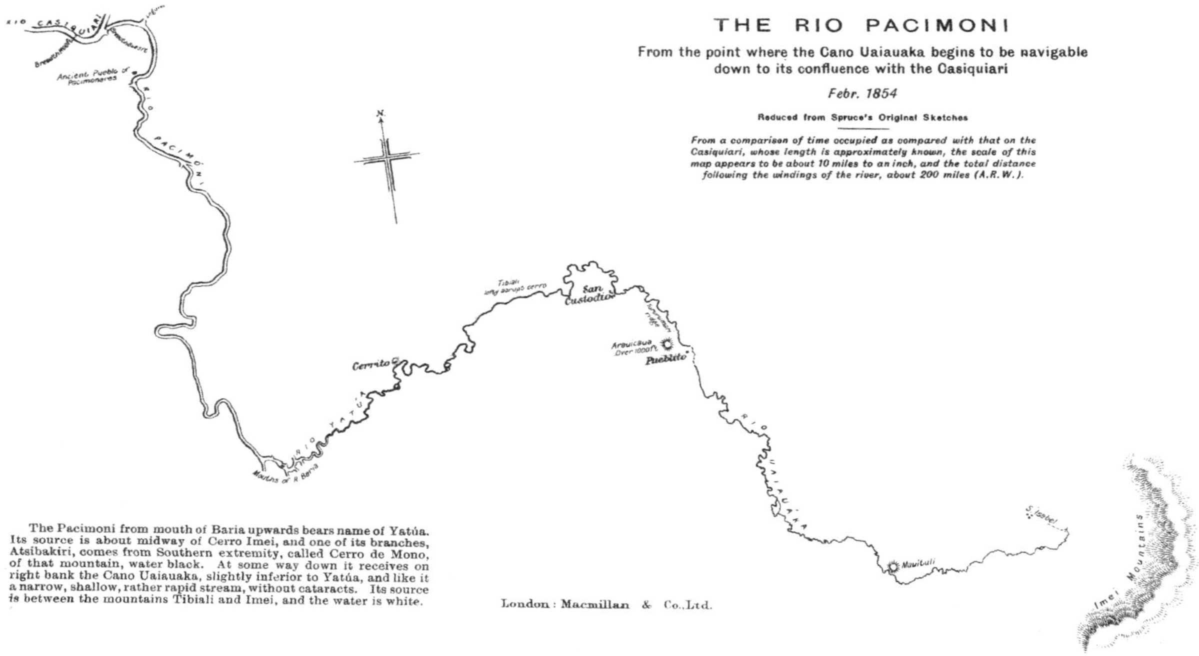

Rio Casiquiari
Ancient Pueblo of Pacimonares
Rio Pacimoni
N.
Mouths of R Baria
Rio Yatúa
Cerrito
Tibiali
San Custodio
Pueblito
Rio Uaiauaka
Mauituli
S. Isabel
Imei Mountains

The Pacimoni from mouth of Baria upwards bears name of Yatúa. Its source is about midway of Cerro Imei, and one of its branches, Atsibakiri, comes from Southern extremity, called Cerro de Mono, of that mountain, water black. At some way down it receives on right bank the Cano Uaiauaka, slightly inferior to Yatúa, and like it a narrow, shallow, rather rapid stream, without cataracts. Its source is between the mountains Tibiali and Imei, and the water is white.

London: Macmillan & Co., Ltd.

CHAPTER XIII

TO THE CATARACTS OF MAYPURES BY WAY OF JAVITA, AND RETURN TO SAN CARLOS

[As this particular route has been more often described by other travellers, I have thought it advisable to give only an abstract of the greater part of the Journal, while printing the account of Maypures and the cataracts, and the notes on the vegetation of the river-banks, in full. San Carlos was Spruce's head-quarters during a year and eight months, and he actually resided there on three separate occasions for periods of five and a half and three months (twice), making altogether only a fortnight less than a complete year. He there made himself familiar with the Spanish language, as well as with the most common Indian dialect, the Baria; and through intercourse with the Venezuelan officials, as well as with many traders and Indians, he obtained an extensive knowledge of the country and its productions, as well as of the people, their government, and their past history. His Journal and some of his Letters contain many short notes and essays which he no doubt intended to elaborate into a systematic account of this interesting and still little known region. All I am able to do here is to give a few of the more generally interesting of these letters and notes, to form the latter portion of this chapter.]

VOYAGE TO MAYPURES

(*Abstract of Journal by Editor*)

May 26, 1854, Spruce left San Carlos in his large canoe and travelled slowly up the Guainia, collecting plants on the way. On Sunday, June 4, he reached Tomo, and as the weather was very rainy he stayed there four days to dry and pack the

FIG. 42.—TOMO, ON THE RIVER GUAINIA OR UPPER RIO NEGRO. The palm is the *Guilielma speciosa*. (R. S.)

plants he had gathered, having to leave his boat here till his return from the Orinoco. This, he says, was a dreadfully hungry place. There was no fish to be had, and a couple of toucans formed his only fresh food while he stayed there. On the 9th he left in a much smaller boat for Maroa and Pimichin, reaching the latter place on the afternoon of the next day.

. . . . . .

Spruce observes that the road from Pimichin to Javita is kept clear and in good order, being about 12 feet wide; but the bridges of trunks of trees across the numerous streams are often in bad condition and dangerous to cross, care not being taken to make them of good and durable timber. With the exception of a short breadth of caatinga at Pimichin, and another about three-fourths of the way to Javita, the forest is all lofty. Jebaríe is very abundant, and there are some very fine specimens. There are also fine rubber trees (*Siphonia lutea*) from which the people had lately begun to collect the gum. In some parts the road was covered with large patches of *Leucobryum Martianum*, and at one place were several tufts of a white species looking like *L. glaucum*, but with more elongated points to the leaves. About midway a cross is erected. As he only found two young men to help his own Indians, they were all rather heavily laden, and it took them the greater part of the day to reach Javita.

Spruce remarks that Javita and Balthazar (the first village on the Atabapo river) are the neatest in the whole district, and their inhabitants the least demoralised, due chiefly to the teaching of an old man, a Zambo, resident in Balthazar, whose talent for singing masses and litanies and strict attention to religious observances have given him great influence, and gained for him the name of Padre Arnaoud.

Having obtained the use of a boat belonging to a trader who had come up from the Orinoco, Spruce was able to go on at once from Javita, and in three days reached San Fernando de Atabapo, situated at the junction of that river with the Guaviare, one of the largest western tributaries of the Orinoco and only a few miles from its confluence with the great river. It is surrounded by very low lands, flooded in winter, when it is inaccessible in any direction except by water. It is very unhealthy, and June, July, and August are the worst months.

The village much resembles Maroa on the Guainia, but is larger and less neat. It has an ancient church and convent, and a few good houses. The inhabitants seem to be the scum of Venezuela —few whites, mostly half-Indians and Zambos. Many are fugitives from distant provinces. While staying here for two days Spruce examined the registers in the convent where travellers enter their names, hoping to find some record of Humboldt and Bonpland, but all entries before 1842 have disappeared through neglect, and much of what exists is ruined by damp and insects.

On the Orinoco, a little above the mouth of the Guaviare, is a national hacienda, at a place called Menicia, where a small quantity of coffee and sugar-cane are cultivated, the latter for the fabrication of a coarse rum, the former barely sufficient for the consumption of the village.

Spruce's Notes on the Vegetation of the Rivers Temi and Atabapo

Most of the plants are identical with those of the Pimichin and Guainia. The palms noticed are the tufted Mauritia (so abundant that even the Indians remarked it), Caraná, a Bactris with bunches of scarlet fruits, a pretty Desmoncus, and a Jará with solitary stems. . . . A very frequent tree on the Atabapo is a Henriquezia, rarely exceeding 15 feet high. It was in fruit as I went down (June), and as I came up (August) it was beginning to flower, but I was too weak to gather and preserve it. The corolla was purple, and quite as in the Rio Negro species. When the river is full the only land accessible on the margin is certain rocks which lie at a considerable distance apart, and which travellers exert themselves to reach for cooking and sleeping thereon. They are black, irregular masses with a little earth only in hollows, which, being humid, produces a crop of small annuals in the wet season.

Among these are some Xirids, Utriculariæ, Polygaleæ, the same large blue-flowered Burmannia as at Maypures and Esmeralda (*B. bicolor*, Mart.), etc. Of shrubs, very frequent is a Melastomad with copious roseate flowers, and a pretty slender Cassia growing 2 to 4 feet high.

The Atabapo is a counterpart of the Pacimoni, a little broader in its lower part, but shallower. In the summer there are a few small rapids, and sometimes it dries so much that all but the smallest canoes have to be dragged over sandbanks. As in the Pimichin, it has a broad caatinga gapó, bounded by lofty forest.

The Voyage down the Orinoco to Maypures

(*Abstract*)

[On June 18, about noon, Spruce left San Fernando in company with Senhor Lauriano, a trader, to whom he lent two of his men so that they might travel together. The water of the Guaviare is whiter than that of the Upper Orinoco, and for a considerable distance below the junction the Orinoco is dark on the right and white on the left side. The general aspect of the river is like that of the Solimoẽs (Upper Amazon), but that there are no willows on its margins. There are only two small Indian villages between San Fernando and Maypures, at the second of which, Marana, they stayed for the night. There was here a small cane-field and rum distillery, and the refuse cane had attracted millions of biting ants which swarmed everywhere in the village and down to the port, so that it was impossible to walk anywhere without being overrun and bitten. No kind of food can be saved from them. The travellers slung their hammocks in the open shed where the cane was crushed, and were glad to embark early the next morning.

At 9 A.M. they stayed to cook a fowl at the

Cerro de Mono, on the left bank, where a rock slopes down into the water. Where the rock is hollowed so as to accumulate soil, there is a dense shrubbery whose upper edge is protected by Bromeliaceæ tenaciously adhering to the rock, and whose lower margin is formed by a dense mass of a shining Selaginella bearing some resemblance to a closely cut box-edging.

Owing to the trader crossing the river to sell some goods, it was dark when they approached Maypures, and only Lauriano and one Indian had been there before. The creek leading to the village was not easy to find even in the daytime, missing which the boats would be carried down the falls. Some torches, specially made to resist rain, were lighted and carried in the foremost boat, but a violent wind, with heavy rain, often extinguished them, and only after much trouble and anxiety the entrance was found and the port of Maypures reached. Thence in total darkness they had to walk over a partly-flooded savanna for about 300 yards, which, in the absence of any track, seemed to Spruce to be a mile, and the exposure to wet in the canoes and afterwards probably led to the serious illness which soon attacked him.

Maypures only contains half a dozen families of permanent inhabitants, all of mixed blood—white, Indian, and negro. But there are other occasional residents or visitors, mainly Indians of two tribes, the Piaroas and the Guahibos; a number of the latter at this time occupying some open sheds on the ancient site of the village farther away from the river. Spruce here made one of his characteristic drawings of a very old woman of this tribe, of whom he remarks—" Her only article of clothing was the strings of red beads I threw over her neck to induce her to keep still while I took her portrait."

The pilot of the falls, named Macapo, was a Piaroa Indian, and it was in his house that Spruce lodged during his ten days' stay at Maypures. This man had in his possession an old oil-painting of San José, the patron saint of Maypures, which was formerly in the church, but was removed for safety to the pilot's house when the former building fell into decay. It represented a sitting figure of life size, very much battered, and looking more like that of a woman than of a man. Of this Spruce says:

FIG. 43.—A VERY OLD GUAHIBO WOMAN SEEN AT THE CATARACTS OF MAYPURES.

"Talking one day to Macapo of the very great number of times he must have passed the falls, he said to me: 'If I have conducted so many vessels over these cataracts without an accident ever happening either to them or to myself, it is not because of my skill or dexterity, but because, before leaving my house on such occasions, I have never failed to devoutly beseech the aid and protection of San José'—pointing to the tattered picture, and laying his hand on his heart with an expression of the most profound gratitude. To myself the picture was an object of deep interest, not so much on account of the veneration with which Macapo regarded it, as because it was almost the only relic I had seen of those devoted missionaries who sowed the germs of civilisation on the wilds of the Upper

Orinoco; and I thought how, fifty years previously, Humboldt had probably looked upon the same picture, which at that time revolutionary troubles and sacrilegious iconoclasts had not so defaced as to render its identification impossible."

I will now give Spruce's description of Maypures and the falls.]

Extract from the Journal

No part of the river is visible from the village, a narrow fringe of forest concealing it. To the westward extend wide savannas with interspersed clumps and patches of forest, while on every side, and sometimes rising in the midst of the savanna, are bold black cerros only partially covered with vegetation. I climbed the cerro bounding the savanna of Maypures on the west; it rises out of the plain to a height of about 1000 feet, its sides abruptly swelling and naked save in hollows, its summit crowned with low dense forest, among which the foliage of Corozito is very conspicuous. From this cerro a magnificent view is obtained. At its foot, to the north, is a broad black river (the Tuparo) which enters the falls, and is seen winding round a broad sheet of granite at the base of the mountains; while its upward course can be traced, first, among broken hills soon subsiding into a level plain, and then across the latter through an alternation of savanna and forest to the uttermost limits of vision. Looking directly west, not a single elevation appears upon the horizon. To the eastward of the Orinoco appears the whole range of the mountains of Sipapo, of which a remarkable triune peak, called the Troncon, forms the southern termination. These mountains scarcely yield in elevation to those of Esmeralda, and are equally picturesque. Looking north, the course of the Orinoco can be traced nearly to Atures.

Both in ascending and descending the rapids, boats are unladen and the cargo carried overland. From the upper part to the lower the distance may be about four miles. Approaching the river near the lower end of the rapids we pass over a moist turfy plain quite resembling a peat moor in England; it is traversed by small streams, on the banks of which are a few marsh plants, the most frequent being an Aracea with erect, long, lanceolate leaves. Instead of our heaths, but not near so pretty, we have tufts of a procumbent purple-flowered Cuphea. Between this plain and the river the track passes over a low bald cerro on which the scattered vegetation is very interesting, and it is from near the summit of this that (as Humboldt mentions) a view is obtained of nearly the whole course of the rapids. One of the most interesting plants on the cerro is the slender bamboo of which the Indians make their carizos or pipes on which they play to their dancers. Here too, and especially on the higher cerros near Maypures, there are considerable quantities of a Barbacenia with dichotomous stems 3 to 6 feet high, long pungent leaves at the apex of the branches, and solitary white, tubular, very odoriferous flowers 4 to 5 inches long. It is the first I have seen of the tribe in a wild state.[1]

[1] [It is allied to the Vellozias (Hæmodoraceæ), curious arboreal endogens allied to Bromeliads, and especially common in the highlands of Brazil.—Ed.]

A little above the mouth of the Tuparo is, perhaps, the highest fall of the rapids between the mainland and a small island, but it is impassable for canoes of any kind; another channel along which they are conducted lies on the other side of the island. By the right (eastern) margin a number of large blocks are rudely piled up jutting into the fall on which one may creep out so as to have a splendid view of the cataract, the spray from which dashes in one's face, and whose roar drowns one's voice. They are clad with vegetation both arborescent and herbaceous, the latter principally Aroideæ and Orchideæ, among which is an orchid exceedingly like *Peristeria Humboldtii.* Trunks of trees and moist overhanging rocks are clad with mosses, among which is the same Hypnum with compressed stems as is frequent on the Upper Rio Negro, and another species which I have not seen elsewhere. I looked in vain for *Grimmia fontinaloides*, Hook., both at the falls and all the way up to San Fernando. In some places circular holes are worn into the rock as on the cataracts of the Rio Negro. The remarkable ones visible near the summit of the island above the present flood surface of the river are called by the inhabitants ollas de xamuro (turkey buzzard's pans).

I arrived at Maypures on the night of the 19th, and hoped at once to have proceeded with the killing and salting of an ox, but the Contratista was away at Atures and did not return till the 24th, which, being the feast of San Juan, was a day devoted to feasting and pleasure, and no one was to be found to search after cattle. Early on the following morning two or three men were

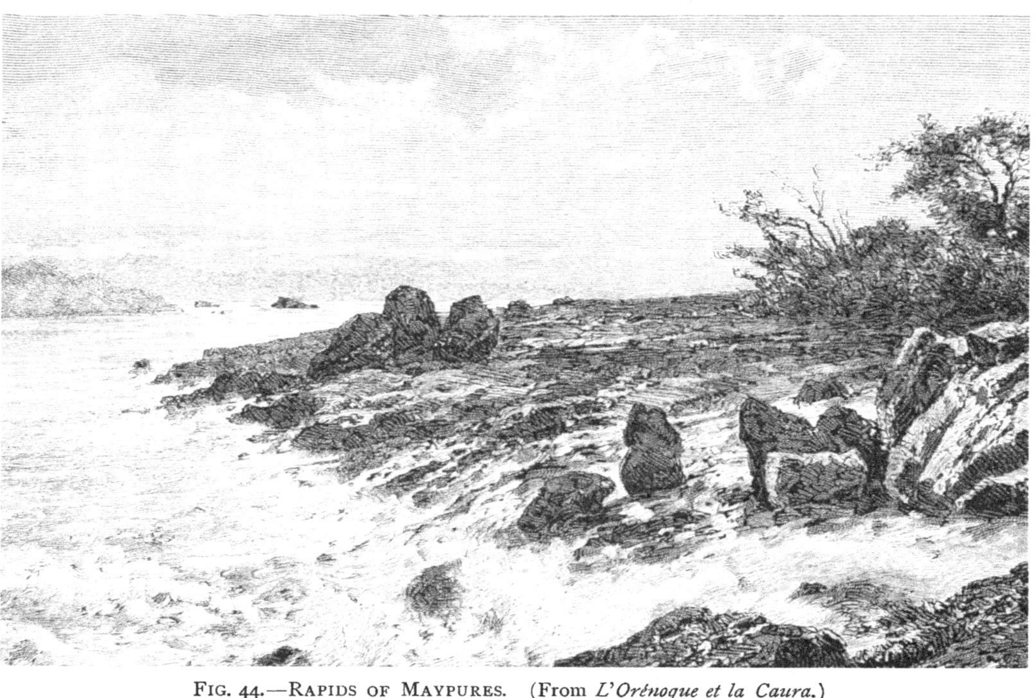

FIG. 44.—RAPIDS OF MAYPURES. (From *L'Orénoque et la Caura.*)

dispatched on horseback, and towards nightfall brought up a drove of cattle to the corral. I had an order from the Comisario to pick the best steer I could find at Maypures, but as among perhaps 100 cattle there were only one or two steers, the choice was soon made. During the period of waiting I had made good use of my time, and had collected a few specimens of everything (save the commonest weeds or widely-distributed species) which I could find in flower or fruit; but my Indians were becoming very impatient of the zancudos of Maypures, and after the first night in the village (as they declared without being able to sleep) they moved out into the river every evening at sunset, fastening the canoe to a small island, sleeping huddled up under the tolda—exposed to the rain beating in, but unannoyed by the bloodthirsty zancudo.

FIG. 45.—A LLANERO AT MAYPURES. (R. Spruce.)

[During Spruce's stay at Maypures he engaged one of the Llaneros to be his attendant, and took a sketch of him, here reproduced. The features show his Spanish descent unmistakably, when compared with any of the Indian tribes.]

When the ox was killed and salted, I had scarcely a moment's respite through the day; the frequent showers and the thousands of maggots bred in the flesh as it hung to dry demanding my constant vigilance. I could place no dependence whatever on my Indians for turning the pieces of flesh as they needed or for seeking out the maggots, as they never half did the task. The smell of fresh beef was to them very disagreeable, although they had all learnt to eat it years ago. The rainy season is always bad for drying beef, but as it is the time of greatest hunger in this district, more cattle are killed at Maypures than in the summer.

Exposure to the hot sun for several hours each day in the occupation of turning the flesh and clearing it of maggots did not produce any beneficial effects on me, and my previous rambles over the heated black rocks on the cerros and by the falls, together with the wetting I sustained on my arrival at Maypures, had probably already sown the seeds of fever in me. In ascending to San Fernando, which took me four days and five nights, symptoms declared themselves unmistakably. The canoe I had borrowed at San Fernando was small, as the current is so strong, especially in the wet season, that larger vessels often spend two weeks on the voyage. My stock of dried plants and of beef, together with the few necessaries of a voyage, occupied so much space in the tolda that I was compelled to half-sit, half-lie in a very uneasy posture at the entrance, where it was impossible to protect myself completely from sun and rain, although at night I fortified myself with blankets as well as I could. Every day was rainy, and the nights were worse than the days. When I reached San Fernando I had been for two days nearly helpless with continued fever, and to have proceeded farther as I then was would have been almost certain death. But even in San Fernando I very narrowly escaped this result, and I was unable to resume my voyage until thirty-eight days had passed.

Indians are sorry nurses, and are ever more ready to flee from the sight of a sick man than to help him. When they desert even their own sick relations, it can hardly be expected of them to abide by a stranger in that state. My Indians did not leave me, but I might as well have been alone. I had violent attacks of fever by night, with short respites in the middle of the day, and on the second night, on stepping out of my hammock, I was seized with vomiting, which symptom being desirous to encourage, I called to my men to heat water for me to drink. They were all so completely stupefied with rum that not one of them was able to help me. Although I had given them a bottle of rum to keep them in good-humour, I found they had sold some of my beef to obtain more. I passed a dreadful night, and in the morning I resolved to seek better aid. A friend wanted men to go to San Miguel on the Guainia, so I lent him my Indians on condition that he would find a woman who would undertake to nurse me. In the afternoon he brought with him an elderly woman who agreed to act as my nurse, but on condition of my moving to her house, where she had a family which she could not leave; and I had no choice but to agree.

This woman—Carmen Reja by name—I shall not easily forget. She was a Zamba—that race by which nine-tenths of the most heinous crimes are said to be committed in Venezuela—and when young she had not been ill-looking, but when out of temper (which for the most part occurred without any reason that I could possibly assign), her face put on a scowl which was almost demoniacal. I was already very ill and almost helpless, and nearly all I could do was to ask for what I wanted, yet my every slightest word or action was interpreted as a complaint or an accusation against her. When I needed to send to the shops (of which there were two or three) for anything, her little grand-daughter was the only messenger I could procure, and as the child was unable to ask for more than the simplest thing, I used to give her a slip of paper with any desiderata written on it. These billets the old woman (who could neither read nor write, and had a mortal hatred of these acquirements) was sure could contain only complaints against herself. She did not say so to me, but she would converse for hours together with her daughters (two grown-up young women) on the subject in the next room, and never fail to work herself up to a high pitch of indignation, and to mutter not a few curses against the *foreigner*. She was exceedingly fond of rum, and when, having ascertained this, I took care constantly to have a bottle on the table, her temper was a little mollified but still only very partially.

[Here follows a very detailed account of the various symptoms of his long and dangerous illness, which I will briefly summarise. He had no medicine with him but quinine, and some ipecacuanha which he took to produce vomiting but without effect. The Comisario and his nurse advised certain local pills always taken for fever with good results. One of these was a violent purgative, and he was persuaded to take it repeatedly, but it produced no good effects. The fever increased in intensity and duration, he got absolutely no sleep for days and nights together, he was unable to take any food but a spoonful or two of arrowroot water-gruel daily, and he was reduced to the extremity of weakness and exhaustion. He had an unquenchable thirst, great difficulty of breathing, with occasional violent sweats, and for some days he himself and those around him were nightly and almost hourly expecting his death. He had given the Comisario instructions as to the disposal of his plants and few other belongings, and then waited the end in a state of almost complete apathy.

During this period his nurse would often leave the house empty for six hours at a time, evidently expecting and hoping to find him dead on her return. In the evening, after lighting his lamp and leaving a supply of water on a chair by his bed, she would often fill the house with her friends and spend the time in discussing or abusing him; calling him all the vile names in which the Spanish language is so rich. Among other things she would call out: "Die, you English dog, that we may have a merry watch-night with your dollars!" One night when the symptoms were very bad she shut up the house and did not return till long after midnight. On another evening she invited her son-in-law and other friends to spend the night with her, in the expectation (as Spruce heard her whisper to them) that the Englishman could not last out the night. Another night, when a similar termination was expected, she scolded him because he was going to leave her responsible for the safety of his goods, and one of the men whispered to her that he thought it would be necessary to give the white man some poison. At length, on the nineteenth day of the fever (July 23), a change for the better occurred, partly, he thinks, owing to his leaving off the purging pills which he had taken too frequently. He now slept better, was able to eat a little, and obtained some good red wine which he took daily.

On August 13, though still excessively weak, a Portuguese trader from Tomo, Senhor Antonio Diaz (the chief manufacturer of the celebrated feather hammocks), being on his return to the Rio Negro, Spruce determined to travel with him. He had to be carried in a hammock from Javita to Pimichin, and reached Tomo (where he had left his large canoe) on the 20th. Here he remained to recover his strength somewhat till the 26th, when he descended to San Carlos on the 28th.

As Maypures is a very interesting locality, being one of Humboldt's collecting stations, I give here a list of the plants collected by Spruce during his four days of leisure, so as to enable botanical readers to form some idea of the characteristic features of the vegetation. The numbers are those of Spruce's botanical register, and I have added the natural orders for the benefit of those who may not be familiar with the generic names.]

LIST OF PLANTS COLLECTED AT MAYPURES

3568.	Perama hirsuta, Aubl.	On the Campos	Cinchonaceæ
3569.	Desmodium adscendens, DC. . .	On rocks at falls	Fabaceæ
3570.	Polygala gracilis, H. B. K. . .	On damp rocks	Polygalaceæ
3571.	Cuphea Melvillæ, Lind. . . .	Banks of Orinoco (flooded)	Lythraceæ
3572.	Arrabidæa carichænensis . . .	Shores of Orinoco	Bignoniaceæ
3573.	Bignonia . . .	„ „	„
3574.	Turnera . . .	On Campos	Turneraceæ
3575.	Phlebodiûm . .	On rocks at falls	Polypodiaceæ
3576.	Ixora capitellata, Benth. . . .	„ „	Rubiaceæ
3577.		„ „	Melastomaceæ
3578.	Aægiphila (20 ft.) .	„ „	Verbenaceæ
3579.	Herpestes Salzmanni, Bth. . . .	On moist rocks	Scrophulariæ
3580.	Dioscorea . . .	„ „	Dioscoreaceæ
3581.	(twiner)	„ „	Asclepiadeæ
3582.	Tocoyena velutina, n.s.	On rocks and mts.	Rubiaceæ
3583.	Faramea odoratissima, DC. . .	Banks of Orinoco	Rubiaceæ
3584.	Evolvulus linifolium, Linn.	„ „	Convolvulaceæ
3585.	Xylopia salicifolia, Dun.	„ „	Anonaceæ
3586.	Echites	„ „	Apocyneæ
3587.	Declieuxia herbacea .	„ „	Rubiaceæ
3588.	Jussieua acuminata, Pl. Am. . .	On moist rocks	Onagraceæ
3589.	Cassia prostrata, H. B. K. . .	On Campos	Cæsalpiniæ
3590.	Dichumena pubera, Vahl. v. . .	Rocks of falls	Cyperaceæ
3591.	Polygala variabilis, H. B. K. . .	On Campos	Polygaleæ
3592.	(tree, 20 ft.)	„	Samydeæ (?)
3599.	Echites . . .	Banks of Orinoco	Apocyneæ
3600.	Plumiera . . .	Cerros	„
3601.	Schieckia orinocensis, Meisn. . . .	Campos	Liliaceæ
3602.	Smilax . . .	Rocks	Smilaceæ

3603.	Cleistes rosea . .	Moist rocks	Orchideæ
3604.	Manihot . . .	Granite rocks	Euphorbiaceæ
3605.	Ipomœa sericea, n.s. .	Granite mountains	Convolvulaceæ
3606.	Barbacenia . .	Granite rocks	Hæmodoraceæ
3607.	Paullinia capitata, Bth. (shrub) .	Campos	Sapindaceæ
3608.	Cipura paludosa, Aubl. . . .	„	Irideæ
3609.	Phaseolus monophyllus, Bth. . .	„	Fabaceæ
3610.	Echites . . .	Rocks	Apocyneæ
3611.	Helicteres guazumæfolia, H. B. K. .	„	Sterculiads
3612.	Tussacia . . .	Rocks at falls	Gesnereæ
3613.	Ditassa glaucescens, Dene . . .	Campos	Asclepiadeæ
3614.	Rhynchospora . .	„	Cyperaceæ
3615.	Rudgea . . .	Mountains	Rubiaceæ
3616.	Tabernæmontana (tree)	„	Apocyneæ
3617.	Aspidosperma . .	„	„
3618.	Arrabidæa Chica, var. Thyrsoidea .	Orinoco (banks)	Bignoniaceæ
3618*.	Eriope nudiflora, H. B. K. . .	Campos	Labiatæ
3619.	Couma oblonga, n.s. (tree)	„	Apocyneæ
3620.	Sipanea radicans, Endl. . . .	At falls	Rubiaceæ
3621.	Selaginella . .	„	Lycopodiaceæ
3622.	„ . . .	„	„
3623.	Wulffia stenoglossa, DC. . . .	Orinoco (banks)	Compositæ
3624.		Moist campos	Orchideæ
3625.	Apeiba Tibombon, Aubl. . . .	Campos (stony)	Tiliaceæ
3626.	Mimosa microcephala, H. B. K. .	Mountains	Mimoseæ
3627.	Mimosa microcephala, H. B. K., var.	At falls	„
3628.	Randia . . .	Mountains	Rubiaceæ
3629.	Rhexia leptophylla, H. B. K. . .	Moist campos	Melastomaceæ
3630.	Allamanda . .	Mountains	Apocyneæ
3631.	Stachytarpheta mutabilis, Vahl. (sm. tree)	„	Verbenaceæ
3632.	Elaphrium („)	„	Amyridaceæ

3633.	Isertia parviflora, Vahl. (sm. tree)	Orinoco banks	Rubiaceæ
3634.	Amasona genipoides, Spr. (tree) . .	Campos	„
3635.	Swartzia microstyles, Bth. (tree) . .	Rocks	Fabaceæ
3636.	Panicum latifolium, L.	Mountains	Gramineæ
3637.		R. Guaviare	Amarantaceæ
3638.	Byrsonima nitidissoma, H. B. K. .	Rocks	Malpighiaceæ
3639.	Isolepis leucostachya .	Damp rocks	Cyperaceæ
3640.	Hyptis dilatata, Bth. .	Campos	Labiatæ
3641.	Neea (sm. tree) .	Woods	Eleagnaceæ
3642.	Cordia (sm. tree) .	Campos	Cordiaceæ
3643.	Cordia interrupta, DC. (sm. tree) .	„	„
3644.	Utricularia . .	Moist campos	Lentibulariæ
3645.		„ „	Turneraceæ
3646.	Cassia . . .	Campos	Cæsalpinea
3647.	Borreria tenella, Cham. et Schl. . .	„	Rubiaceæ
3648.		Streams	Halorageæ
3649.	Pleroma . . .	Moist campos	Melastomaceæ
3650.	Habenaria . . .	„ „	Orchideæ
3651.	Abolboda pulchella H. B. K. . .	„ „	Xyridaceæ (?)
3652.	Sipanea acinifolia, Pl. Amaz., 299 . .	Rocks, falls	Rubiaceæ
3653.	Vitex orinocensis H. B. K. (tree) .	River banks	Verbenaceæ
3654.	Mimosa, n.s. . .	Campos	Mimoseæ
3655.	Psychotria limbata, Pl. Amaz., 1723 .	Woods	Rubiaceæ
3656.	Lantana . . .	Rocky places	Verbenaceæ
3657.	Decheuxia chioccoides, H. B. K. .	Campos	Rubiaceæ
3659.	Icthyothera Cunabi, Mart. . . .	„	Compositæ
3661.	Buttneria pentagona, Spr. . . .	Falls	Byttneriaceæ
3662.	Hypoxis scorzoneræfolia, Linn. . .	Campos	Liliaceæ
3663.	Centrosema angustifolium, Bth. . .	„	Fabaceæ
3665.	Sipanea glomerata, H. B. K. . .	Rocks	Rubiaceæ
3666.	Isolepis . . .	„	Cyperaceæ

3667.	Hydranthelium callitrichoides, H. B. K.	Rocks (overflowed)	Scrophulariaceæ
3668.	„ „ var.	Deeper	„
3669.	Platycarpum orinocense, H. B. K. (tree)	Campos	Bignoniaceæ
3670.	Mayna laxiflora, Bth.	Forest	Flacourtiaceæ

The Decadence of the Canton del Rio Negro under the Republican Government of Venezuela

To Mr. John Teasdale

San Carlos del Rio Negro, Venezuela,
July 2, 1853.

In most parts of the world, and especially the New World, some progress has been made during the last fifty years, but could Humboldt revisit bodily (as I know he often does in imagination) these scenes of his early wanderings, he would find in every respect a lamentable falling off. The missions which existed in the time of the Spanish rule, and which effected so much in drawing the Indians out of the forest and keeping them together in pueblos, have all disappeared. There is still in San Carlos a building called the convento, but no padre has resided in it for the last twenty years, nor has there been for that space of time a minister of religion in the whole Canton del Rio Negro—an immense territory comprising the whole of Spanish Guayana above the cataracts of the Orinoco. There is an equal destitution of doctors, lawyers, police, and military; we are therefore (you may suppose) in a state so primitive that Jean Jacques would have delighted to form one of our community. How I wish he could have made trial of it for the space of a few months only!

The Canton del Rio Negro is governed by a special code, of which I have not yet seen a copy, and can only judge of its edicts by what I have seen of its working. The whole community is divided into two classes, the racionales, who are descended from the whites, and the peones, who are of Indian descent. All the manual labour is performed by the latter—the peones (pawns!); while the racionales (whom you may consider the players) have only to sit still and direct their moves on the chessboard. The prime mover in the game is represented by a Comisario General, who lives in San Fernando de Atabapo, and with whom resides the power of nominating a comisario particular for each pueblo. The comisarios of pueblos appoint among the peones one as captain and another as his lieutenant, whose office it is to seek up Indians (*i.e.* peones) for the service of any racional or of the community, when required; and also generally to execute the behests of the Comisario. In the case of any crime being committed, every adult male is a policeman for the nonce, and can be called on to assist in the capture of the criminal, and afterwards (if necessary) in administering the punishment awarded him. In the more weighty cases, such as murder or robbing with violence, the law requires that the accused be remitted to Angostura (the capital of Spanish Guayana, now Venezuela), there to take his trial before a regularly constituted court; but this is very rarely done. In the plaza of San Carlos—and when I speak of plaza, I must caution you against making any mental comparison with the plazas of old Spain, as for example the Plaza de Torós at Seville, and conjuring up the multitudes assembled to witness the combat, the shrieks of delight of gaily-dressed ladies when a wounded torero is carried out, and all that sort of thing—seeing that, in point of fact, the green in my native pueblo of Ganthorpe, whether considered in itself or with respect to the houses that environ it, far exceeds in magnificence the plaza of San Carlos. In the said plaza, then, and towards the southern side of it, standeth the casa real. . . . As you have now a distinct perception of the casa real, I may venture to admit you to a view of its interior, and you will find that it is divided into two compartments, both on the ground-floor—literally the *ground* floor, there being no other sort of floor in San Carlos—whereof one is appropriated to the administration of justice, and the other to the confinement of those who are accused and the punishment of those who have been convicted. On entering the latter, one is struck by the sight of a ponderous machine of sinister aspect extending the whole length of the wall on one side, and looking at first glance something like a recumbent guillotine, but when more accurately examined proving to be nothing more than a monster "stocks," and called by the people of the land a sépo. This is perforated with sundry round holes, serving for the confinement of the ankles, and if necessary the wrists, of those whose fingers are too light, and wits so heavy as after committing a theft to allow themselves to be found out and seized, in a country where concealment of crime and escape are so easy. There is also a larger hole in the centre of the sépo for confining the neck of any very refractory subject, who is laid on his back and has his wrists also confined in the two holes adjacent the central one. I am told that a quarter of an hour's confinement in this way is sufficient to reduce the most stubborn to gentleness and submission.

With the map before you, you will have some idea of the extent and markings of this enormous chessboard, and what I have said above will give you some idea of the players and the pieces, and especially of the mode in which the pawns are "taken." If I may be allowed to vary my metaphor a little, I would say that while on your chessboard the pawns exactly equal the other pieces in number, and are therefore far inferior in collective strength, *here*, on my chessboard, the pawns are at least as twenty to one to all the knights, bishops, queens (queans?), etc., put together, even if we include in the latter category the few foreign pieces (English, French, and Portuguese) scattered over the chessboard; and you may naturally inquire if the game is carried on with tolerable quietness, and if the pawns allow themselves unresistingly to be pushed about by the superior pieces, seeing that were they to unite their forces, there can be no doubt of their being able *to floor*—*i.e.* to shove off board, table, and all, the kings, queens, and all their abettors. In some of the pueblos the system is said to work well, especially in those on the Atabapo. There is the prestige of the old Castilian prowess, and the fortifications erected by the Spaniards to overawe the Indians and defend their frontiers still exist, though decayed and untenanted, as for example one by the Rio Negro opposite San Carlos. There is also the natural docility (or if you will, apathy) of the Indian. He allows himself to be ill-treated and thinks not of revenge. He is overwhelmed with kindnesses and caresses and deserts his benefactor without scruple. Such is the character of all the Indians I have seen in South America. . . .

To the causes above cited for the submission of the Indians to their governors and oppressors may be added that the captains and lieutenants have a sort of pride in their office and in the maintenance of order. They owe their elevation to that rank to their good conduct and the influence they possess over their brethren. Very often they are descendants of ancient chiefs of tribes. Notwithstanding these reasons for the maintenance of order, the fabric of society stands here but on slippery ground, and is perhaps daily becoming less secure. Nowhere is this more apparent than at San Carlos. The memory of the rigours of Spanish sway is becoming indistinct. The so-called Indians have in many cases no small proportion of "white" blood in their veins, and with the mixture they have become proud and revengeful; and the captains are often appointed so arbitrarily and changed so frequently by the comisarios that the Indians hold them in little account. At San Carlos there is another cause for disquiet—nowhere else have I seen Indians so demoralised by the immoderate use of ardent spirits. Without rum no work of any kind can be done. A good many boats are built at San

Carlos, and the other pueblos above, sometimes of considerable size, and since I came to the Rio Negro, a schooner of 145 tons was built in Tomo and sent down to Pará. About an equal number are built every year for the Rio Negro and Orinoco, and it is only when these rivers are full that large vessels can pass down the cataracts; of course they never come up again. This branch of industry necessitates the sawing of a good many planks, and without rum it is impossible to land logs from a raft on the river (most of the timber being cut on the Casiquiari), or mount them over the saw-pit, or launch a vessel when it is completed; each of these operations having its stated price in gallons of rum. Sometimes the whole Indian population will be drunk together, and when this is the case they enter without ceremony the houses of the whites to ask for more rum, quite prepared to take it by force if refused. Such a circumstance occurred the very day I reached San Carlos, and made me almost repent of having come.

. . . . . .

On the Growth of San Carlos

(*Journal*)

San Carlos seems to have grown up into a Pueblo since 1830, in consequence of the boat-building that is carried on there. Anteriorly there were but two or three houses besides the fort, which is on the opposite side of the river and a little lower down.

. . . . . .

The villages on the Amazon and elsewhere have in the first instance been formed of the Indians in the immediate neighbourhood who were induced or compelled to gather themselves together, build houses, and plant mandiocca. When the population began to fall off, expeditions of soldiers were sent towards the head-waters of the tributary rivers to attack the Indian settlements by night, kill all who resisted and carry off the rest, especially the women and children. As the Portuguese taught the Lingoa Geral to all the Indians, this language soon absorbed all the rest; but in the Spanish territory, as no other medium of communication has been used between whites and Indians but Spanish, and the whites have not taken particular pains to teach this language to the Indians, the latter still constantly speak the native languages in conversing with one another; and the language of any pueblo is that of its first inhabitants.

In every pueblo where there is no commerce its duration is necessarily brief and uncertain; for when all the land suitable for cultivation is exhausted in the vicinity of any settlement, the inhabitants betake themselves to some other spot, either bodily or more frequently by one or two families at a time. Of the pueblos on the Rio Negro, on the Venezuelan side, only two, San Miguel and Maroa, date from the time of the Spaniards.

The agrarian Indians, who of their own accord, and before any whites visited them, had settled down to the cultivation of the soil (their settlements in Brazil are called mallocas), must also have frequently migrated from the same cause;

nor have they always kept together, as on the Uaupés we find sections of the same nation inhabiting distant spots, and often mixed up with offshoots of some other nation. Probably the individuals of any nation were more united before the proximity of the white man compelled them to abandon their intestine wars.

. . . . . .

The nomadic tribes seem to know no other limits to their movements than the meeting with white or red enemies. Such are the Macús, who roam over the forests between the Rio Negro and the Japurá, ascending to the Uaupés, and sometimes, I am assured, descending nearly to the Barra. Rarely they are seen on the opposite side of the river. The Guaharibos, who inhabit on the Alto Orinoco above the Raudal de los Guaharibos, rarely descending below it; and the Guahibos on the Meta and Calcanapára. All these tribes are ignorant of the construction of canoes, and, when they have to cross a stream which is not fordable, make use of rafts. Their food is chiefly fruits, eaten raw.

On the Patauá Palm called Uarúma by the Barré Indians, and by Spanish Settlers Séje, which is a general Name for all Palms whose Fruit is used for mixing with Jueuta

(*Extract from Journal*)

. . . .

There are two species at San Carlos. One, which is the same as the Barra Œnocarpus, whose beard is used for arrows of blowing-canes (called here Sarabatana), a tall, noble species with large oblong fruit; the other (which I have not seen) has much smaller subovate fruit (not globose as in Bacába), and the drink prepared from it has a distinct reddish tinge almost like that of the Bacába;

while in the larger Patauá it is nearly white—very slightly flesh-coloured.

That with the smaller fruit is probably the smaller species spoken of in another place, distinct from *Œ. Minor*, Mart., by the fastigiate pinnæ. Patauá-yukisé, L.G. (yukisé is the general name for extracts produced by cooking either vegetables or meats, and is equally applied to gravy of flesh or fish, and the juice extracted from fruits, roots, etc.), jukúta de seje (Venezuela), is one of the most wholesome and delicious drinks in nature. I still think the assaí of Pará the most delicious of the palm drinks, but to me it is not palatable without plenty of sugar, while patauá, now that I am accustomed to it, drinks better by itself. The taste is exceedingly rich, resembling more that of new milk than of anything else. It is prepared in the same way as assaí, either by scalding the ripe fruit, or still better, by slightly boiling it, then breaking it up by hand in water, when the thin light-coloured pulp mingles with the water and the brittle purple skins fall with the stones to the bottom. The liquid is either poured off or the whole is passed through a sieve which retains all the grosser parts. A small quantity of mandiocca is added, as in making xibé, and when it has softened the whole is ready for drinking. Sometimes, instead of mandiocca or cassave, mingau de farinha (mandiocca boiled in water to the consistency of thick oatmeal gruel) is mixed with the patauá, and the whole drunk warm; in this way it is very delicious, and is an excellent mess the first thing in the morning. Instead of mandiocca, boiled ripe plantains may be mashed up with the patauá, but the compound, though very sweet and pleasant to drink, is rather windy.

Patauá contains perhaps the same quantity of oil as Bacába. The oil is extracted occasionally near Pará, but on the Alto Rio Negro only that of Bacába is sometimes to be met with in the sitios. It is perhaps owing to the presence of this oil that patauá-yukisé is rather aperient. When I have abstained for some time from drinking it and return to its use, it always produces a slight looseness in the bowels, but this effect passes off in a day or two, and is rather beneficial than otherwise.

There was a small quantity of Patauá ripe when I left the Uaupés in March, and we have had it at San Carlos all through the months of April, June, July, August, and September. The trees are very abundant in dense forests on the west side of the river from the pueblito of San Felipe (by the fort of San Carlos) towards the Guasié.

The Indians in their sitios grow exceedingly fat during the season of Patauá, and there can be no doubt of its being very nourishing.

On Vegetable Oils

(*Extract from a Letter to Sir William Hooker*)

San Carlos del Rio Negro, Venezuela,
March 19, 1854.

. . . .

Vegetables yielding oil abound in this region, but with the present scanty population, and their listless lazy habits, it is exceedingly difficult to get together even a small quantity of the oils, resins, etc., which in Europe would be highly esteemed. Nearly all the palm fruits yield oil in greater or less quantity. You are aware that very pleasant drinks are prepared here by triturating the fruit of the Assaí and other palms in water, and adding a small quantity of sugar and farinha. The Portuguese give the name of vinho to these drinks, though totally different from the palm wine prepared in other parts of tropical America (and I believe also of Asia). . . . All the palm drinks are exceedingly nutritive, and several are slightly purgative, owing no doubt to the oil they contain. By allowing the liquid to stand a short time in a basin the oil rises to the top, and an idea is obtained of the quantity yielded by any particular palm fruit. Of all that I have seen, the Caiaré (*Elæis melanococca*, an actual congener of the African palm) yields oil in the greatest quantity and in appearance exactly like the oil of *E. guineensis*, but I have never heard of its being collected and put to any use. The Caiaré palm is abundant all about the mouths of the Rio Negro and Madeira, but I have not seen or heard of it anywhere up the Rio Negro. I sent you a spadix with fruit from the Barra do Rio Negro. Why it was called "melanococca" is hard to say, for the fruit is of a bright vermilion colour. Perhaps Gaertner had only the nut.

After the Caiaré, as to quantity of oil, come the various species of Œnocarpus (*Œ. Bacaba*, *pataua*, *disticha*, etc.). The oil of these is apparently of finer quality than that of Caiaré; it is colourless and sweet-tasted, and not only excellent for lamps but for cookery. The shopkeepers of Pará buy Patauá oil of the Indians, and mix it in equal proportions with olive oil, retailing the whole as "olive oil," from which indeed even the best judges can scarcely distinguish it. I can bear testimony that for frying fish, oil of Bacába is equal to either olive oil or butter. The various species of Œnocarpus abound on the Amazon and Orinoco, and on their tributaries. I have lately seen the Patauá in the greatest plenty throughout the Casiquiari, Alto Orinoco, and Cunucunúma. Near the Barra it is frequent, but less so than the Bacába. The forests opposite San Carlos, extending from the Rio Negro to the Xié, are literally sown with Patauá. The fruit is in season nearly all the year round. We are just now beginning to make use of it, and we shall have it (in unlimited quantity if there were always Indians to climb the trees) all along until November. I am passionately fond of patauá-yukisé, and it is the only thing I shall regret when I leave San Carlos. When I have passed a long time without drinking it and recommence, I always find it slightly aperient, but this effect passes off in two or three days.

. . . .

Among oil-yielding Dicotyledons of equatorial America, I suppose the Andiroba (*Carapa guianensis*) holds the first place. Andiroba oil has the great advantage (in a tropical climate) of being so bitter that neither ants nor any other insects will touch it. The tree is abundant near Pará, especially at the mouth of the Tocantins, and is met with all the way up the Amazon.

From the seeds of two trees, apparently undescribed, abundant on the Alto Rio Negro, Orinoco, Casiquiari, Pacimoni, etc., the Indians prepare a paste resembling cream cheese in appearance and taste. The seeds are first boiled and then steeped for some days under water, after which they are broken up by the hand. In the boiling a quantity of oil is said to be collected, but I have never been able to get a sight of it. These Indians are exceedingly shy in showing to a white man the edibles, etc., whose use is peculiar to themselves, thinking that his only object must be to ridicule them. I first saw one of these trees (the Cunuri, a Euphorbiacea allied to the India-rubber tree, but with simple leaves) near San Gabriel above two years ago, and though I have since that time continually come upon it, it is only very lately that I met with its flower and fruit on the Casiquiari, and still later that on the Upper Pacimoni I came upon some Indians eating Cunuri cheese (if I may so call it). From them I obtained a small quantity which I wish to send you, but have at present nothing to put it in. For Cunuri oil I must still wait with patience. It is said to be as bitter as Andiroba oil, but to afford an excellent light. The other tree, whose products are quite similar to those of the Cunuri, is called Uacú. It is a leguminous tree with pretty pink flowers of very curious structure, and I sent Mr. Bentham two species of it from the Rio Uaupés.

There are numerous other trees and palms of this region yielding oil, and I have only particularised a few of those which are so abundant that their oil might be procured in any quantity *were there only industrious hands to collect it.*

Of resins also there is no lack, but I doubt if any of them would come in for candle-making. The Venezuelans make a flambeau, which they call mechon, of the resin of various species of Icíca, poured when melted into the decayed stem of the blowing-cane palm from which the soft interior has fallen away, or into a bamboo. It emits rather too much smoke (as Mr. Wilson remarks of resins), but the odour is very agreeable.

. . . .

FIG. 46.—CUMARÚ (*Dipteryx odorata*)
The "Tonquin bean" of commerce.

[I give here a photographic print of the tree that produces the well-known Tonga or Tonquin bean, valued for the beautiful scent produced by its seeds. It is a lofty forest tree bearing red papilionaceous flowers, and pods with a single large seed. It is abundant at Santarem and almost equally so in the Upper Rio Negro, whence the seeds are exported in considerable quantity, though I do not think Spruce found it in a wild state or even mentions it in his Journals. The essential oil is used by perfumers, and especially for scenting snuff and tobacco.]

On Insects used for Food

(*Journal*)

Indians of the Rio Negro, Uaupés, Casiquiari, Orinoco (and perhaps of the Amazon) eat the large grubs bred on various growing palm stems, but especially in Pihiguas. They are said to be of the size of the forefinger, and the mode of eating them is this. By a sudden twist of the head, it is pulled away along with the intestinal canal, and the animal is then roasted on the budari or mandiocca oven. There is another grub or caterpillar found on Marima trees which they are very fond of. When this insect is in season, it constitutes a principal part of the food of the Maquiritari Indians, and Don Diego Pina related to me that, travelling once on Alto Orinoco with a crew of those Indians, he was near perishing of hunger, for they would neither fish nor seek after any sort of food but these caterpillars, and wherever they stopped by the way they climbed into the Marima trees in search of them.

I have many times seen Indians eat the saúba ant (called bacháco in Venezuela). The large kinds only are eaten, and at those times when the bachácos pour from their holes in great numbers (probably sending forth colonies after the manner of bees), if it be near any pueblo all the unoccupied Indians in the place turn out to collect them. The head and thorax is the part eaten, the abdomen being nipped off (at San Carlos I constantly see them eaten entire), and it is eaten uncooked. The taste to me is strong, fiery, and disagreeable, but those who have eaten the bacháco fried in turtle oil tell me it is quite palatable.

Certain frogs called Juí are also eaten wherever I have been. They seem most abundant in the wet season, and when they begin to croak much by night in the gapó it is certain that the waters have risen considerably and that winter has set in. The Indians fill a pan with them, all alive and entire, and set it on the fire to boil.

There are at least two species on the Uaupés; one very large, of which I made trial, taking care to have the intestines well cleared away and the residue roasted on a spit. No chicken could be more delicate.

Remarkable Thunderstorms at San Carlos

(*Journal*)

Sept. 27, 1853.—Last night after sundown much lightning was seen to the east, and a little after 7 P.M. the squall commenced at San Carlos with such force as to threaten to upset the houses (a house had been blown down a few days before). The thunder was exceedingly close and loud, keeping up a continuous roll, and the glare of the lightning almost ceaseless. Not much rain fell here and by 9 o'clock all had cleared away.

It is difficult to ascertain the frequency of the flashes when you are in the midst of the explosions, but in coming from Marabitanas last month, when we did not reach our sleeping-place until long after dark, a heavy thunder-shower passed near without giving us a drop, and I could see beautifully the flashes issuing from the cloud. I counted the pulsations of my wrist between consecutive flashes. They varied from two to eight, giving an average of five pulsations, and fifteen flashes to a minute.

In the middle of the rainy season, though the amount of rain that falls is greater, these very violent thunderstorms are of rare occurrence. . . .

When I was at the Jauarité caxoeira on the Uaupés in October 1852, the lightning struck a house at the mouth of the Paapurís and prostrated the inmates, but injured no one save a young man in a hammock, who was deprived of the use of a leg (whether permanently I did not learn). This was late in the afternoon, and on the following morning every person I met had his face and arms streaked with red carajurú, intended as a protection against the pajé whose incantations had brought down the thunderbolt on the wounded man affecting them in the same manner.

Near electrical discharges are always followed by augmented fall of rain. This is peculiarly evident where there is a lull or a slight abatement in the shower and a vivid flash of lightning comes, restoring the fall of rain in all its force. Generally several seconds elapse ere this augment occurs, and the report is mostly heard first. As the last few drops of a heavy thunder-shower are falling, there are often two or three very loud and close reports like parting salutes, which may sometimes be seen to bring the rain pouring down in the advanced position of the storm.

Sept. 30.—We had fine and tolerably dry weather for some days before the equinox, but when the sun had passed we had every day in the afternoon (say from 4 o'clock to half-past 4) violent squalls with heavy thunder and rain, which are said to have fallen heavier lower down the river at Marabitanas. It was thus up to the 28th, when the day opened with thick fog and was afterwards fair and hot throughout. On the 29th and 30th the afternoon showers returned, and this day the thunder has been particularly loud and frequent.

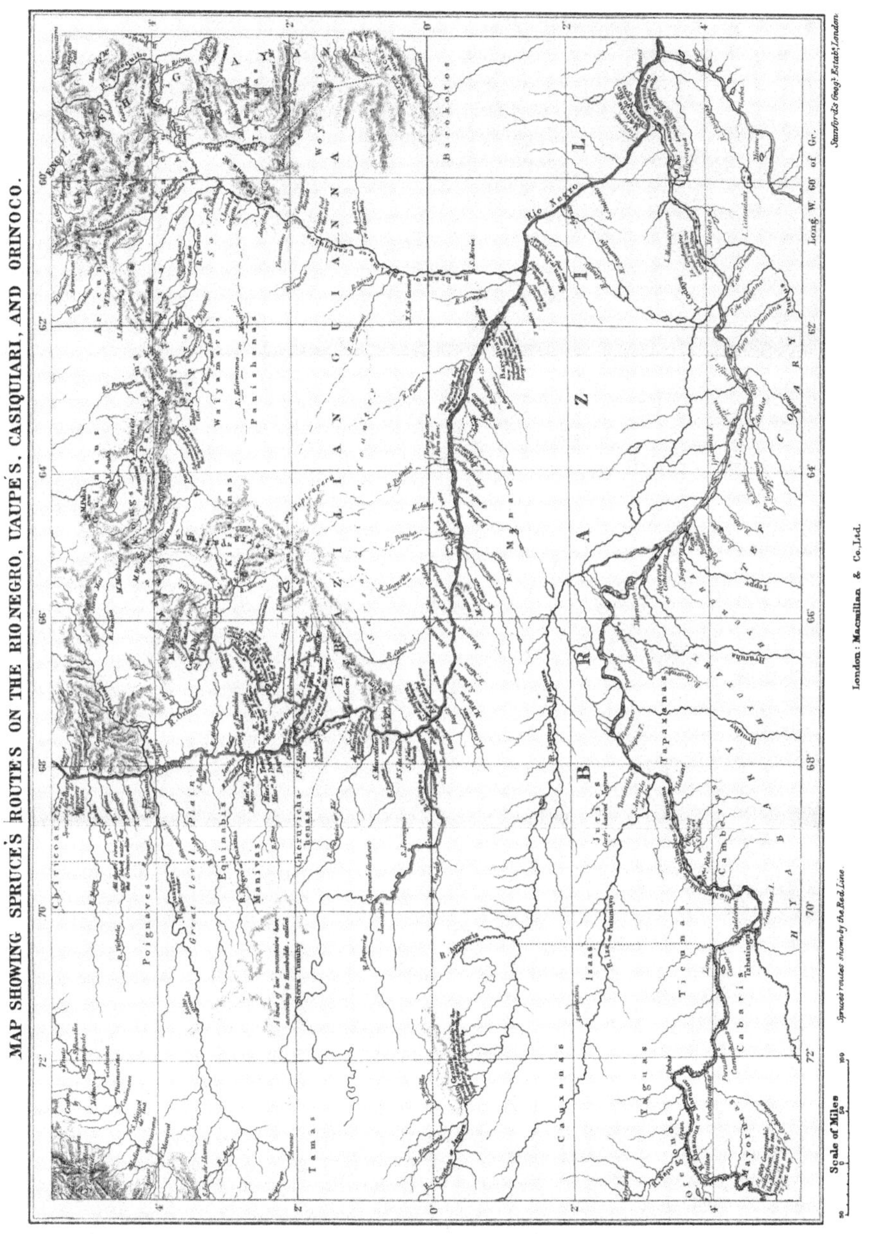

The material originally positioned here is too large for reproduction in this reissue. A PDF can be downloaded from the web address given on page iv of this book, by clicking on 'Resources Available'.

CHAPTER XIV

DOWN THE RIO NEGRO FROM SAN CARLOS TO MANÁOS

(*November* 23, 1854, *to March* 14, 1855)

[THIS chapter completes the record of Spruce's five years' exploration of the Rio Negro and Upper Orinoco, with several of their tributaries. It consists of a rather full Journal of his voyage down the river—at the very beginning of which he narrowly escaped assassination—an account of a botanising excursion from the Barra, together with three short articles on characteristics of the vegetation, and extracts from letters to Sir William Hooker and Mr. Bentham, which serve to bind together the rather fragmentary materials into a personal narrative. The "Charlie" mentioned at p. 496 appears to be the sailor again referred to in the following chapter, whose engagement turned out to be so disastrous.]

Nov. 23 (*Thursday*).—This day about noon I left San Carlos. My crew consisted of four Indians; two of them were sons of the pilot (Pedro Deno). On the same day at 4 P.M. we reached the pilot's cunúco, a little within a narrow caño on the left bank, and stayed the night. Here a plot was laid to kill me. There were several people at the cunúco, including the pilot's wife, other sons and daughters, a son-in-law, etc. They were engaged in distilling bureche, and my men on arrival began to test its quality, which, though not of the best, sufficed to turn their heads and set them vomiting, all except the son-in-law (Pedro Yurébe), who drank enough to make him noisy but not to render his movements unsteady.

The cunúco consisted of two sheds, open at the sides, in one of which the still was at work. The port where the canoe was anchored was perhaps some 80 yards distant, down a rather steep descent. I had my hammock taken up and fastened under one of the sheds, and when night fell, after eating a small quantity of the forequarter of an alligator which I bought of Yurébe, I turned in. The Indians were very noisy, but as nothing is more tiresome than the conversation of these people when intoxicated, I paid little attention to it, save that I noticed one of the pilot's sons was inviting his brother-in-law Yurébe to make the voyage with us to the Barra. After a while I heard them talk so much about "heinali"[1] that I could not help listening attentively to what they said, and it was well I did so.

Pedro Yurébe owed some forty-three pesos to the Comisario of San Carlos and others, but he had no scruple to leave this unpaid till his return from the Barra, and a brilliant idea had just struck him. "The man," he said, was going to his own country,

[1] "Heinali" means "the man." Spanish Indians in talking of their master call him el hombre, and when speaking their own language translate this by its corresponding term.

whence he would return no more. In the morning he (Yurébe) would offer his services for the voyage and get the pay beforehand (according to custom); they would then embark, and on reaching the mouth of the river Guasié, which they might do in three or four days, they would take the montaria while I was sleeping and make their way up that river, whence they could at any time return to their own territory, as it is but a short cut (a day overland) from the upper part of the Guasié to several tributaries of the Guainia. They would thus shirk the long, tedious voyage for which they had already received pay. This was largely discussed and approved by all. It then entered Yurébe's head to ask if "the man" had much merchandise with him. "Hulasikali! Wala!" ("He has plenty. He has everything") was the reply. But they deceived themselves, for most of my boxes were filled with paper and plants, and not with woven goods as they supposed. "Then," said he, "we must not leave him without carrying off as much as we can of his goods, and for this purpose it will be necessary to kill him." This also was approved of and the consequences discussed at length, it being considered that if they remained four months in the Guasié (where there were plenty of fugitives to bear them company) the affair would be quite forgotten. His genius seemed to expand as he talked the matter over and finally conducted him to what might be considered a climax. "Why should we not kill the man now?" said he; "we have him here sleeping in the midst of the forest, far removed from all observers. When he left San Carlos every one knew him for a sick man, and no one will be surprised to hear of his death." "Hena nu camisha" ("I have no shirt"), "and no sign will remain of violence having been used." (In fact, his only clothing was a strip of bark between his legs.) This he repeated a great many times, and all his companions applauded the idea. Three questions remained for them to discuss: the disposal of the body, of the goods, and of themselves. The first offered no difficulty, for on account of the climate the dead are mostly interred at the end of twenty-four hours, and they could say "the man" had died of his illness and they had buried him. As to the goods, they would leave a few in each box so as to give the latter the appearance of not having been disturbed. As to what they would do themselves there were various opinions; but at length they concluded that a better way than hiding themselves in the forest would be to present themselves boldly before the Comisario of San Carlos, tell their tale, and there would be an end of it, for "the man" was a foreigner a long way from his country and had no relations here to make any inquiry as to the mode of his death. It may be supposed that I listened to all this with breathless attention, and I could hardly believe that their acts would be conformable with their words, till I heard them begin to lash themselves into a fury by recapitulating all the injuries they had received from the white men, all of which they considered themselves justified in retaliating on my devoted head—though in my short intercourse with them I had shown them only kindness, and particularly to Pedro Yurébe, whose little daughter I had a short time before cured of a distressing colic, which for many consecutive days and nights had allowed her no rest. I had on me a slight attack of diarrhœa—this is mostly the case with me on the first day I embark, when the excessive heat causes me to drink a great deal of water—and I had been obliged to leave my hammock two or three times since nightfall. It was now past midnight, and just as I lay down the last time I heard them deciding that the best way would be to strangle me as soon as I should be asleep again, which Yurébe undertook to do, and one of the others undertook to ascertain when I had fallen asleep. The fires had gone out and only the dim light of the stars illuminated the interior of the cabins. Though reclining in my hammock, I kept my feet on the ground ready to spring up should I be attacked. The darkness prevented their noticing this, and as I kept perfectly still for some time the man who had placed himself to watch me reported I was sleeping. I heard them all whispering one to the other, "Iduali! Iduali!" ("Now it is good—now it is good"), and as Yurébe hesitated a moment, I got up and walked leisurely towards the forest as if my necessities had called me thither again; but instead I turned when I got a few paces and walked straight down to the canoe, unlocked the door of the cabin, which I entered, and having fortified the open doorway by putting a bundle of paper before it, I laid my double-barrelled loaded gun, along with a cutlass and knife, by my side, and thus awaited the attack which I still expected would be made. At intervals I could hear angry exclamations from the Indians, wondering that I did not return to my hammock;

and it may be imagined in what a state of mind I passed the rest of the night, never allowing my eye and ear to relax their watchfulness for a moment. However, they did not once stir to see what had become of me, and at length the break of day relieved me partly from my anxiety, but not entirely, for in that lonely place the dark deed contemplated might have been done almost as secretly by day as by night; and when shortly afterwards Pedro Yurébe came to offer to accompany me to the Barra, I took care while conversing with him never to move out of reach of my gun. Of course I declined his offer, excusing myself on the supposition that the Commandant of the Brazilian frontier would not allow him to pass on account of his name not being entered in the passport along with the others.

Though Pedro Yurébe was left behind, I took care throughout the rest of the voyage that the Indians should never approach me unarmed, and I never spent a gloomier time. On the very first night, at the mouth of the Guasié, after supper the Indians lay down on a slightly sloping rock which there formed the river's bank. The montaria was fastened to the poop of the piragua and to the rounded top of a rock which stood out of the water close by. A little past midnight I had occasion to turn out of the cabin; the moon was just setting, and I noticed that the Indians had left their first berth and were all sitting together on the top of the stone to which the montaria was fastened. I could have no doubt they had planned getting into the montaria and silently eloping up the Guasié as they had first proposed doing. I therefore took out my gun and laid it gently on the top of the tolda so as to point towards them. Then I went in again, satisfied that they would have noticed my movements, and would know that any attempt to unloose the montaria would lead to the death or serious wounding of some or all of them, as I could easily keep a watch on them from the cabin. At daybreak I found them all back on the rock where they had first lain down.

On the evening of November 30 we reached the mouth of the Uaupés, where I was fortunate in meeting two old acquaintances, the traders Amansio and Amandio, the former making rubber, the latter collecting salsa. They lent me four men, with whom the next morning I continued my voyage.

. . . . . .

[Reaching Saõ Gabriel on December 2, after an absence of more than two years, Spruce found the village somewhat improved in appearance. The church had been repaired and a school established under a "professor de primeras letras," who had twenty-eight pupils (Indians and half-breeds). But in other respects there was no change—no industry, no cultivation—and the people were as usual complaining of "passando muito fome"—being always in want of food. Here he was so fortunate as to find a negro mason who had been sent from Manáos to repair the church, and who begged a passage back, offering to take an oar when required. He was, Spruce says, a very decent, respectable man, and his company rendered the voyage a much less anxious one than it would otherwise have been, as the botanist could thereafter occasionally stray into the forest in search of plants without feeling uncertain whether his men would not have deserted him during his absence. The negro was a slave, and belonged to a widow lady at Barra, who, he said, treated him more like a son than a slave. He was, in fact, all the property she had, and the labour of his hands was all she could depend on for her maintenance. Spruce adds: "I found him a sensible, well-behaved fellow. He was tall, slender, and well-made, quite equal, in fact, to the run of journeyman masons of any colour or country. He might easily have freed himself by escaping across the Venezuelan frontier, but he had evidently a great contempt for the Spaniards and he loved his 'country,' as he called the Barra, where he was liked and respected, and where he had his little boy (for he was himself a widower)."

During the voyage to the Barra there were no further incidents beyond the usual sudden storms, more or less dangerous, and the ordinary inconveniences and incidents of a boat journey on these great rivers; but some interesting notes on the vegetation were made, showing that its novelties and curiosities were by no means exhausted.]

Peculiarities of Vegetation observed during this Voyage down the Rio Negro

(*November* 23 *to December* 22, 1854)

On the north bank above the Rio Branco a Terminalia (Combretaceæ) was frequent, which has a remarkable ob-conical mode of growth, sometimes nearly flat-topped, sometimes slightly convex; but the most curious feature is, that the short trunks and extruded roots are often nearly hidden by a quantity of black rootlets, the whole forming a mass the size and shape of a moderately large haycock.

On the south shore, where the land was somewhat elevated, the great Bertholletia (Brazil-nut tree) was frequently noticed in the lower half of the river, its stem and slightly convex crown rising a long way above the adjacent trees.

Diplotropis nitida (Fabaceæ) was one of the commonest trees of the gapó all the way down to Barra. When I started it was in full flower; it had flowered also in September. On nearing the Barra there was another burst of flower. At the first flowering many of the panicles are only in bud; these open later. *Dicorynea Spruceana* (near Cassia) was almost equally frequent as far down as the mouth of the Rio Branco.

Lecythis amara occurred also all the way down, but the great country of this group seems to be from the mouth of the Maraivia (above Sta. Isabel) to that of the Rio Negro, especially on the south bank and on the islands.

The fine new genus Henriquezia was seen all the way down where the soil was rich. It was in flower and very ornamental. Drepanocarpus (Fabaceæ) was also abundant all the way from the mouth of the Casiquiari to that of the Rio Branco and on to near the Barra. It was conspicuous from its whitish lunate pods hanging in bunches from the free portion of the stems that spring in graceful arches from the sides or summit of the forest wall. There are two species of the genus, differing in the number of the leaflets.

On a steeply sloping bank above Cabuqueno were several trees I did not know—some of them in flower and fruit. A little below Barcellos, and especially about Airaõ, we saw the Castanheiro (Brazil-nut tree) frequently, on ground that rises from the river and stretches away into a low hill. This tree is remarkable for its trunk rising naked above the surrounding forest, like the Samaúma, but it has a flatter-topped crown, easily distinguished from the hemispherical dome of the Silk-cotton trees.

[During his enforced stay in Barra waiting for the steamer to take him up to Peru, Spruce made a few botanical excursions, the most interesting being to a stream which enters the Rio Negro some fifteen miles above the city, and has on it what is said to be the highest waterfall in the whole district. His account of this visit, slightly condensed, is as follows:—]

Excursion from Barra, February 12, 1855, to the Rio Taruma

This small river enters the Rio Negro about five hours' rowing above the city, where the coast bends inwards forming an extensive bay into which the Taruma enters. It is fairly wide at first, but as it receives numerous small streams from either side it soon becomes narrower, yet its sources are said to be a long way off in the forest. At about an hour from its mouth a rather large igarapé enters on the east side and is celebrated for having the loftiest waterfall known on the Rio Negro. My object was to visit this; and I accordingly established myself at the only Indian sitio within this branch, tenanted by an old man named Nicolas (a Manáos Indian born at Barcellos), his wife, two sons—stout lads—two grown-up daughters, and a little boy, a grandson. Here I and my companion found a little room about eight feet square, whose walls were of woven Caruá leaves and roof of Bussú. Fortunately it contained a small table and a stage of Jará stems, both of which were very useful for depositing my boxes and plants.

The next morning, accompanied by Charlie and the old Nicolas, I started to visit the fall. We ascended the winding igarapé for nearly an hour. It was much obstructed by the gapó vegetation, and at last became so grown over that we had to leave our boat and make our way through the forest. A little more than an hour brought us to the fall, which we approached from above, but we scrambled down the rocks to the bottom, where we could obtain a perfect view of the fall. I have seen few finer things in South America, and it reminded me a little of the Irish "Turk cascade." This branch of the Taruma traverses a narrow valley, contracted to a ravine below the fall, which rushes over a concave cliff in an unbroken cascade of from 30 to 40 feet high. The upper stratum of the cliff is of hard whitish sandstone, and projects considerably beyond the lower, which are of softer stone with thin alternating layers of vermilion strong-smelling earth. It is thus easy to walk under the cataract without being wetted, though the rocks drip here and there and are everywhere thickly clad with ferns and Hepaticæ, but especially with Selaginellæ, of which I gathered four species not found in the adjacent forests. The water falls into a deep trough, from which spray dashes out and is borne downward by the violent wind caused by the rush of the cataract. The water winds away among mossy blocks and then is lost beneath them for a considerable distance. From among these blocks springs a tree to the height of some 100 feet, the spreading sapopemas (buttresses) at its base clad by *Micropterygium leiophyllum* and a Plagiochila (Hepatics), the trunk rough with termites' nests, on which Philodendrons (Araceæ) and a Carludovica (Pandaneæ) have established themselves. This tree bore numerous grey fruits the size of an orange, but I could not distinguish the form of the leaves, and my guide could not give me a name for the tree, as, he said, the fruit was not edible. It was probably a Caryocar (Rhizoboleæ). From top to bottom of the cataract hangs a thick rough rope of tangled black rootlets proceeding from a tree on its edge.

The whole aspect of this mossy cirque, with its broad riband of falling water, embosomed in dense luxuriant forest, in which was visible no palm, was something of an admixture of tropical scenery with that of temperate climes.

In the adjacent forest, however, there were several palms, including some smallish ones new to me, such as a small Bacába (Œnocarpus) of 15 feet with equidistant pinnæ. Above the cataract the forest became dwarfer—more of a caatinga in character—containing several Assaí-zinhas and Bussús; and at some distance up begins a caranásal where Indians are accustomed to cut fronds of *Mauritia Caraná* for thatch.

Several fruits strewed the ground, but these were difficult to dry at this rainy season, besides that the leaves of the trees from which they had fallen were inaccessible. One fruit the size and shape of a hen's egg, with thin greyish-green covering peeling off, and a thick woody endocarp with many radiating fibres, has a kernel tasting quite like that of Caryocars, yet the leaves were simple. The Indians call it Castanha-rana (wild chestnut).

The bed of the Taruma is so level that in the season of flood the water of the Rio Negro enters it and flows up to the base of the fall, but a day's heavy rain produces a downward current, while during dry intervals the water remains almost motionless.

To Sir William Hooker

BARRA DO RIO NEGRO, *June* 5, 1855.

. . . .

I reached San Carlos on the 28th of August. This was a good time for descending the Rio Negro, and I had an opportunity of making the voyage along with Senhor Antonio Diaz (the manufacturer of the feather hammocks), who shortly afterwards went down to the Barra with two large vessels;

but after having been nearly three years in the land of Piassaba, I did not like to leave it without seeing either flower or fruit of that remarkable palm, for which purpose I had previously made not a few unsuccessful journeys. I decided therefore to remain some time longer. The time of ripe fruit is about midsummer; this had already passed, and on returning to the Guainia I learnt that no fruit had been seen there, but that the trees on the Casiquiari had borne a little fruit. The year 1853 was a year of scarcity for fruit of the forest of all kinds. In 1852 the Patauá palm fruited so copiously that I drank the wine prepared from it nearly all the year round; while in 1853 I did not drink it once. In October 1854 I succeeded in getting flowers of the Piassaba at Solano on the Casiquiari. A few days after this I caught the virulent chilblains of this country by walking barefoot in the wet forest, and from this apparently simple cause I was confined to the house for five weeks, and great part of the time to my hammock. The skin of the sole of the right foot came off as completely as if a blistering-plaster had been applied to it, and this was followed by tumours which burst and sloughed.

[In the district above referred to—the angle between the Rio Negro and Casiquiari—which he fully explored during his long stay at San Carlos, Spruce found a new and elegant little palm of which he made a very accurate and beautiful drawing (here reproduced half size). He describes it as being about 18 feet high, with the stem a little over 3 inches diameter, and very closely ringed, the divisions of the leaves about 20 inches long and gracefully drooping. It was confined apparently to a limited tract of low forest, since in all his excursions in the surrounding country it was nowhere else met with.]

I left San Carlos for the Barra on the 23rd of November, with a crew of four Indians. As I was going down the stream I did not trouble myself to seek for more, and I thought myself fortunate in meeting with Indians who had previously made the same voyage, though three of them (an old man and his two sons) were quite new to me. The old man was my pilot, and at the first night from San Carlos we slept at his sitio, a little within a small stream entering the Rio Negro on the left bank. . . . [Spruce here describes how his crew proposed to murder him as related in his Journal. The letter then continues:—] During the course of five years' travel among Indians, I had been accustomed to repose the utmost confidence in them: to sleep unarmed in the midst of them in the most lonely places, and frequently to stroll into the forest alone when we stopped by the shore to cook, when,

FIG. 47.—*Mauritia subinervis*, Spruce. In low forest between San Carlos and Solano. (R. S.)

had they been so disposed, they might easily have embarked and left me to almost certain destruction. But on this last voyage I found it necessary to adopt an entirely different plan, and not desiring to be in such society a moment longer than I could help, I did not stop on the way for a few days at certain points, as I had previously intended doing.

. . . . . .

I propose therefore (*D. V.*) to ascend by the next steamer from Pará for Peru. One is now up at Nauta, and the next is not expected to start before the first of March; in the latter I hope to be a passenger.

My notion is to get to a place called Tarapoto,[1] among the mountains on the left bank of the Huallaga. The steamer goes as far as Yurimaguas, which is 70 miles, or seven days' journey, below Chasuta. From Chasuta overland to Tarapoto is five hours' journey on foot or mule-back. It is the most easy of access of any place really in the mountains; its population is considerable, and as a good many cows, sheep, and pigs are bred there, one may count on experiencing no lack of victuals. Its site is described as very picturesque—in a small plain among lofty mountains, from which trickle down several streams, abounding, it is said, in *shells*, which will be something new to me. If unfortunately the small steamers have ceased running, then it would take two months to reach Chasuta in an ordinary boat, and in my present weak state I do not feel myself competent for such a voyage, especially when the unceasing plague of mosquitoes by day and night is taken into account. . . .

[1] So called from the abundance of the Tarapoto palm (*Iriartea ventricosa*, Mart.; *Paxiuba barriguda*, Braz.).

For a long while after I came out I rarely allowed myself to rest in a hammock by day, but latterly, and especially since I have been so ill, I have been obliged to yield more to the weakness and languor which too frequently comes over me, and to repose from my labour at short intervals. My friends in the Barra wonder to see that I still go on working, and tell me that the most industrious European in less than five years generally accommodates himself to the *far niente* to which the climate and the example of all around him so temptingly invite. Five years' experience has also pretty well disgusted me with drunken Indians for workmen.

To Mr. George Bentham

BARRA DO RIO NEGRO, *Jan.* 12, 1855.

. . . . . .

The Barra is much changed since I left it in 1851. The employees connected with the newly-formed province decidedly outnumber the rest of the white inhabitants (male), yet there is not an acre more of ground under cultivation, and the products of the soil are far from sufficing for the consumption of the population. Hence living here is much dearer than it was. We sometimes sit down to a meal where every article at table is imported either from Europe or North America. Biscuit from Boston, U.S., butter from Cork, ham or codfish from Oporto, potatoes from Liverpool, etc. . . .

The changes of political boundaries and names of provinces so frequent here, and which have now been added to, are puzzling to students of botanical geography, and may lead to important mistakes. The earliest division of that portion of Amazon land which belongs to Brazil seems to have been this: the Capitania do Rio Negro included all the country north of the Amazon, to the very mouth of the latter; that of Pará the country south of the Amazon as far as the Rio Madeira; and the remainder of the territory south of the Amazon, *i.e.* from the Madeira to the Peruvian frontier, formed the Capitania do Solimoẽs. Afterwards the three capitanias were reduced to two, that of the Rio Negro comprehending all the land on both sides of the river to the westward of Parentins (a little below Villa Nova), and that of Pará the land to the eastward of the same point. After the separation of Brazil from Portugal, the two were combined to form the Provincia do Pará. Thus it remained at the time of my arrival in Brazil; but it has lately reverted to its former division into two parts, the eastern under the name of Provincia do Pará, and the western under that of Provincia do Amazonas.

I may add that what in Venezuela was formerly the Misiones del Alto Orinoco is now the Canton del Rio Negro, embracing all that portion of Spanish Guayana extending northward to the foot of the Cataracts of Atures, westward and southward to the frontiers of New Granada and Brazil, and eastward to Demerara.

Mark now some results of this instability of landmarks. Von Martius puts the habitat "Rio Negro" to the plants found by him on the Solimoẽs and other rivers belonging to the Capitania do Rio Negro (as it existed at the time of his voyage), very many of which probably do not exist at all on the Rio Negro, and were certainly not seen there by him, for he did not ascend that river.

When I was at San Fernando and Maypures, my notions of geography were continually shocked by hearing the people say "aqui em Rio Negro" ("here in the Rio Negro"). "Why do you say Rio Negro," I would ask, "when here we are on the Orinoco?" "Because," said they, "we are in the *Canton* do Rio Negro."

Finally, as I came down the Rio Negro, the people were already beginning to say "aqui en Amazonas" ("here in the Amazon").

The only limits which can be counted constant are those formed by the rivers and mountains. Even the term "North Brazil" may in a few years cease to have any significance. "Guayana," as it was anciently understood by the Spaniards and Portuguese, namely, all the tract between the ocean, the rivers Amazon, Negro, and Orinoco, is a quite natural division, and is still well known in common parlance under the same name.

. . . . . .

[The following Notes written during Spruce's latest residence at Barra may follow here :—]

Contrast between the Shores of the Amazon and those of the Rio Negro

In the former river the receding waters in many places leave broad margins of mud hard enough to walk on, and becoming sparsely clothed with annual grasses and Cyperaceæ as summer advances. In ascending it I have sometimes walked half a mile across these annual meadows, and at the farther side have come only on the common willow, *Salix Humboldtiana*, two or three Psidia with a willow-like aspect at a distance, and *Mimosa Asperata*.

On the Upper Rio Negro nothing similar is seen; the forest

often skirts the water with a permanent edge, not falling away in masses at the commencement of the dry season as on the Amazon, and especially on the Solimoẽs; and when the river is low the steeply sloping bank, whether of rock or earth, is for miles uninterruptedly clad with rootlets from the bases of the trees which form to it a dense fringe.

On the Amazon and Solimoẽs in the wet season, when the current in the river is most rapid, in the inundated forest on each side there is usually scarcely any current, and the farther the gapó is entered the stiller the water becomes. Each day as the river rises the inundation widens, but silently and insensibly. Should there be a slight fall in the ground, as there sometimes is towards an inland lake, then for a time there is a rapid flow from the river until the lake becomes filled to the level of the river. I have been entangled in one of these currents when we were glad to pull the montaria across it by catching hold of the twiners that hung from the trees, the oars being of scarcely any use.

When the river, having reached its height, begins to descend, there is generally a perceptible flow from the gapó. The floating plants (various Marsileaceæ, Naiades, Ceratophylla, minute Hydrocharideæ, and the new Euphorbiacea, *Phyllanthus fluitans*) which had covered the still waters of the gapó during the rising flood, and had just attained perfection as the river reached its height, now begin slowly to move out, and I have observed them floating away throughout the breadth of the river, though from the masses being broken up and the minute size of the individuals, a voyager, whose attention had not previously been called to their existence, would hardly notice them. These little plants afford shelter to several small univalve shells and to not a few winged and wingless insects, and in pushing the montaria through dense beds I have sometimes been startled by the springing up of a cloud of grasshoppers, nor is it infrequent to see the black snout of an alligator peering out and rapidly withdrawn as he becomes aware of the approach of his worst enemy, man. In some places the water is emptied out of the gapó with great rapidity, as in the extreme angle between the Solimoẽs and Rio Negro, where the sound of the waters rushing against the trees is as that of a cataract. In passing this place by moonlight on my return from Manaquirý, I received several blows against the tree-trunks and had my clothes torn in many places. Instances like this are, however, rare, and the water in general subsides from the forest as placidly as it had entered it.

It has been stated in Europe that the rapid rising of the waters of the Amazon causes trees to be torn down and sometimes large portions of earth to be carried away; but all the falls of land and trees I have seen (and I have been witness of several) occurred shortly *after the river began to ebb*, and they are owing to the water undermining the high banks, yet still supporting them whilst the flood lasts; afterwards, when the waters receding no longer sustain them from beneath, large portions of earth and forest fall simply by their own weight.

The Ilhas de Caapím or Grass-islands I now find are chiefly floated out of lakes by the rising of the waters of the Amazon, which thus fulfil the double purpose of calling certain species into existence and afterwards bearing them away to the ocean. The chief, often the sole, constituents of the Grass-islands are the Canna-rana (wild sugar-cane) and Piri-membéca (Brittle-grass), *Panicum spectabile* and *Paspalum pyramidale.* For the production of these plants white water is essential, as is proved by their absence from the whole Rio Negro and from the Rio Trombetas above the Furo, through which the waters of the Amazon are poured into the lower part of the Trombetas, where both these grasses are abundant.

Lakes are usually of black water, but those into which white water enters in the rainy season invariably produce these two grasses, sometimes in such abundance that they become periodically choked up, as I witnessed in two small lakes near Manaquirý, through which it is every year necessary to cut a passage.

In returning from Manaquirý, the Solimoẽs was so violently agitated by an easterly wind as to render it perilous crossing it in my small and deeply-laden montaria. We contrived, however, to reach a small Ilha de Caapím which was floating down at about three hundred yards from shore, put ourselves into the centre of it, so that the force of the waves was broken ere they reached us, and thus floated onward at our ease. Whilst my man composed himself to sleep, I amused myself with examining the composition of my novel berth. It consisted of but one species of grass, *Paspalum pyramidale*, and after several futile attempts I succeeded in drawing out an entire stem, which measured 45 feet in length and contained seventy-eight joints. It was simple, though others possessed two or three branches; all the joints save the three or four uppermost ones gave out rootlets, and several of the lowest internodes were in a semi-putrid state. Floating on the water, and kept in by the grass-stems, were an Azolla, two Salviniæ (one of them new to me and both in fruit), and a few barren plants of a small Hydrocharidea, and a small Pistia. It is to be noted that these Ilhas de Caapím are quite different from the floating rafts encountered by Humboldt on the Orinoco: similar rafts I have seen on the Amazon.

[The following Notes were also written during the same period as the last; and with a short account by myself of the present state of the rubber industry in the Amazon valley, form a fitting close to this portion of Spruce's Travels:—]

Notes on the India-rubber Trees of the Rio Negro

(*Journal*).

In the mouth of the Uaupés (on my return voyage) I found a rancho erected and a person employed in extracting india-rubber from the species I had discovered there (*Siphonia lutea*). All the way down the Rio Negro the smoke was seen ascending from recently opened seringales, principally in the islands. The extraordinary price reached by rubber in Pará in 1853 at length woke up the people from their lethargy, and when once set in motion, so wide was the impulse extended that throughout the Amazon and its principal tributaries the mass of the population put itself in motion to search out and fabricate rubber. In the province of Pará alone (which includes a very small portion of the Amazon) it was computed that 25,000 persons were employed in that branch of industry. Mechanics threw aside their tools, sugar-makers deserted their mills, and Indians their roças, so that sugar, rum, and even farinha were not produced in sufficient quantity for the consumption of the province, the two former articles having to be imported from Maranhaõ and Pernambuco, and the latter from the Upper Rio Negro and Uaupés.

The species of trees from which rubber is extracted on the Upper Rio Negro and Lower Casiquiari are two, *Siphonia lutea* and *S. brevifolia*, known respectively as the long-leaved and short-leaved Seringa. The former yields most milk, but neither is so productive as that of Pará (*S. brasiliensis*).[1] Both are straight, tall, and not very thick trees, with rather thin and smooth bark, and their average height may be about 100 feet.

Near the Barra some milk is taken from a species common on the river banks (*S. elastica* ?), but there is another species growing in the interior of the forest said to yield more milk. This I have not seen.

The species of Siphonia I have gathered on the Amazon and Rio Negro amount to seven or eight, and it is probable that two or three times as many yet remain to be discovered. On the Uaupés I met with two trees (2427, 2479, hb.) of a genus apparently not far removed from Siphonia, which yield pure rubber, and are also called by the Indians Xeringui; but the single (not ternate) leaves and the clustered trunks (often as many as ten from a root) give these trees an aspect very different from the Siphonia.[2]

When I ascended the Rio Negro in 1851, I showed the inhabitants the abundance of rubber trees they possessed in their forests, and tried to induce them to set about its extraction, but they shook their heads and said it would never answer. At length the demand for rubber, especially from the United States, began to exceed the supply; the price consequently rose rapidly, until early in 1854 it reached the extraordinary price of 38 milreis (£4 : 8 : 8) the arroba, a little over 5s. a pound.

The extraction of caoutchouc from the various

[1] [The name Hevea is now usually adopted for the trees formerly known as Siphonia.—ED.]

[2] [In Spruce's MSS. (Plantæ Amazonicæ) he makes these a new genus Muranda, and the species *M. siphonoides* and *M. minor*.—ED.]

Fig. 48.—India-rubber Trees in Flower near Pará.

species of Siphonia was, at the time of my arrival at Pará (July 1849), a branch of industry limited to the immediate environs of that city, being principally carried on in the island of Marajo and about the mouth of the Tocantins. The price it fetched in the Pará market (10 milreis the arroba, about 10d. a pound), and the great gains which those who trade in forest produce expect on their outlay, prevented the people of the interior from employing themselves in its extraction, to which must be added the general apathy of the Indians in the matter of undertaking any new kind of labour.

When the trees are flowering nearly all the milk goes to the nourishment of the flowers and none can be obtained from the trunk, while if a flower-panicle is wounded the milk starts out in large drops. It is customary to leave the trees untouched for a few months, until the fruit has attained its full size. About Pará the collection of rubber seems limited to the dry season, June to December. In the Upper Rio Negro the rubber trees flower from November to the end of January, and when I left San Carlos on November 23 little milk was to be obtained.

The usual mode of drying the milk by smoke applied to successive coatings on a mould is followed by most rubber-collectors. Some have filled a small square box with the milk and allowed it to coagulate, but as the milk does not harden till the end of ten days or more, and the mass then requires to be cut into slices and subjected to heavy pressure in order to free it from the water and air entangled in its substance, this mode is by no means popular.

It is found that the addition of alum hastens the coagulation of the milk, while ammonia has a contrary effect and is accordingly useful when the milk is required to be kept some time in a liquid state.

[At the time Spruce wrote the preceding notes the industrial use of rubber had just commenced that remarkable development which has continued to the present time. The increased demand from America in 1853, which Spruce referred to, was due to its more extensive use for waterproof clothing, goloshes, etc., but still more to the extension of its application to many of the arts, and to its great value in making water-tight and air-tight tubes, belts and washers for machinery. But the greatest increase in its use has been for the tyres of bicycles, first solid, then about 1888 pneumatic, which latter soon became universal both for cycles and motor-carriages, and has led to an enormous consumption of this remarkable natural product. A short *résumé* of the present state of the rubber-trade in Pará and the Amazon valley may be interesting to our readers.

At the time Spruce and myself were in Pará, what was called bottle-rubber was the common form in which it was made. This was done by means of a ball of clay 3 or 4 inches in diameter, which by means of a stick was dipped in the milky sap and dried in successive coats till about an inch thick, when the clay was extracted, leaving a hollow ball of rubber with a short neck. The smoking was done by means of a fire of the fruits of two kinds of palms, generally abundant in the rubber forests, and whose thick and acrid vapour

Fig. 49.—An Indian smoking the Rubber.

coagulated the milk more quickly and effectually than smoke from any other kind of fuel. These are still used, but the rubber itself is made into a more convenient shape and size and in a more cleanly manner.

For this purpose an oval wooden spade is used, something like a canoe paddle but with a rather longer handle, the surface of which is made quite smooth. This is dipped in the bowl of rubber-milk, and each layer held over the smoke to dry by means of the long handle. This is repeated till a large mass is formed nearly twice as large as a man's head, and of a subglobular shape somewhat like that of a Dutch cheese. The paddle is removed by slitting the half circumference next the handle, when it can be withdrawn and a nearly solid mass of clean rubber is left. From four to six of these balls is a load for a man.

During the last fifty years the supply of rubber from the Amazon valley has fairly kept pace with the demand, so that the price has rarely if ever again been so high as in the year Spruce referred to. But, owing to the scanty native population of these forests, the enormous quantity exported from Pará of about 30,000 tons annually is only obtained by drawing upon an immense extent of country. Not only is the whole of the Amazon itself from its mouth up to the very roots of the Andes everywhere full of seringuiros (as the men who extract the rubber are called), but all its chief tributaries are more or less devoted to the same industry. Steamers run regularly up to the town of Iquitos, now the centre of the whole rubber trade of the great Andean rivers—Ucayáli, Huallaga, Napo,

Pastasa, and many others. Other steamers ascend the Rio Negro as far as Sta. Isabel, collecting the rubber from its highest tributaries; while others ascend the Tocantins, the Tapajoz, and the Madeira as far as their respective cataracts; while the Purus, which has no such obstructions, is navigated for a distance of 2555 miles from Pará.

All these lines of steamers are mainly supported by the rubber trade, and there is reason to believe that, if required, many times the quantity now exported could be obtained without difficulty. Although a large number of trees and climbers in all parts of the world produce rubber, it is generally admitted that none yet discovered produce it of such good quality and so economically as the rubber trees of the Amazonian forests. Nor is there, in this region, any danger of the supply becoming exhausted, because it has been found that if the trees are cut down and then tapped at different points (as has been done under the expectation of getting a larger amount of rubber) a much smaller quantity can be extracted than can be obtained from the living tree in a single season. So long as the trees are not tapped during the flowering and fruiting season (when the flow from the bark is but scanty), it does not appear that the annual tapping in any way injures the trees, nor has it been noticed that any diminution occurs in successive years. As, therefore, the tree is a large and long-lived one, and in its native forests is freely reproduced by seed, we may consider that so long as the forests continue to exist the supply of this valuable product will be almost inexhaustible. One cannot but marvel at the extraordinary reserve of power everywhere manifested in nature. In the north the Sugar Maple secretes a sugary sap so abundantly that many gallons can be annually drawn away from a single tree without diminishing its production in future years or perceptibly shortening its life. In the tropics, other trees produce so different a substance as caoutchouc (or india-rubber) which can be indefinitely extracted in the same manner. It is impossible to believe that these, and a hundred other diverse kinds of sap, were not primarily developed to further the growth and vigour of the plant itself and to aid it in its struggle for existence with other plants. Yet whenever man draws off this precious fluid for his own purposes, nature seems always ready to make up the deficiency, so that the plant shall not suffer injury. It may perhaps be the case that this wonderful recuperative power has been developed for the purpose of guarding against the chance injuries inflicted by boring insects, wood-pecking birds, or scratching, biting, and goring mammals, whose combined attacks might otherwise destroy the vigour of the species and thus endanger its existence. Perhaps even we may trace the gummy or milky nature of so many saps, and their coagulation on exposure to the air, to the need for checking the loss that might occur if wounds, which may be made by hundreds on the smaller branches, twigs, and buds, were not rapidly self-healing. To this simple need of vegetative life we may owe that wonderful diversity in the products of the plant world which renders it an inexhaustible storehouse to supply the ever-growing needs of civilised man, whether for purely sensuous enjoyments as in fruits and spices and odoriferous substances, for the more æsthetic pleasures of varied flower or glossy veined wood, or to aid him in the development of the arts and sciences to which his ever progressive nature impels him.

Among these strange and varied products none perhaps have such remarkable and useful physical properties as that now familiar substance, india-rubber, the need for which is leading to the whole world being ransacked and entire populations being employed to obtain it.]

END OF VOL. I

Printed by R. & R. CLARK, LIMITED, *Edinburgh*.